Problems in Engineering Mechanics

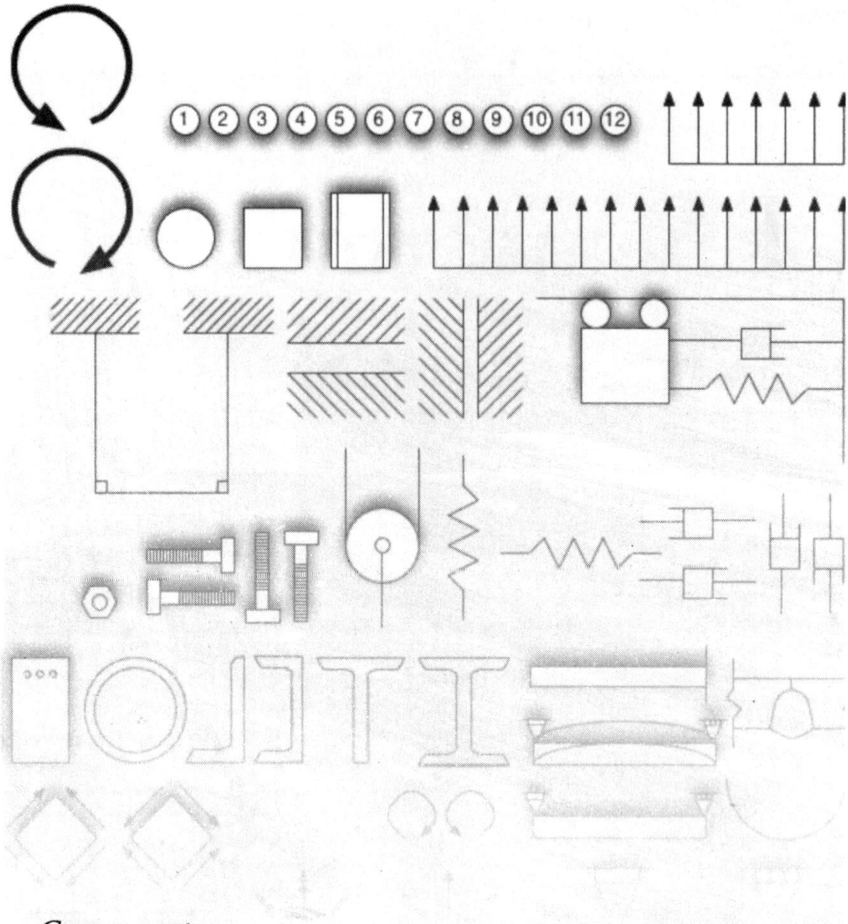

Comments

In this book the author has taken tremendous efforts to discuss comprehensive problems on the subject of mechanics. I am confident that this book will definitely be useful for first year engineering students and also the faculty in the field of mechanics.

Prof. S.M. Shiyekar ME, MIE, LMISTE, LMIASE, LMindACM, MISSE
Principal
Sinhgad Institute of Technology and Science (SITS)
Narhe, Pune

This book will be beneficial and fruitful for all the engineering students in the country. The way he [the author] has solved the problems is very good and can be understood by engineers as well. I wish him good luck for his future.

Prof. Dr. A.S. Padalkar ME (Mech), PhD
Principal
Sinhgad College of Engineering (SCOE)
Vadgaon (Bk), Pune

Problems in Engineering Mechanics

Manish Bhaskar

Computer Engineer
Founder, Director
Robotics Research
Institute of India

CBS

CBS PUBLISHERS & DISTRIBUTORS PVT. LTD.

New Delhi • Bengaluru • Chennai • Kochi • Mumbai • Pune

ISBN: 978-81-239-1907-2

First Edition: 2010
Reprint: 2015

Published by:
Satish Kumar Jain for CBS Publishers & Distributors Pvt. Ltd.,
4819/XI Prahlad Street, 24 Ansari Road, Daryaganj, New Delhi - 110002
delhi@cbspd.com, cbspubs@airtelmail.in • www.cbspd.com
Ph.: 23289259, 23266861, 23266867 • Fax: 011-23243014

Corporate Office: 204 FIE, Industrial Area, Patparganj, Delhi - 110 092
Ph: 49344934 • Fax: 011-49344935
E-mail: publishing@cbspd.com • publicity@cbspd.com

Branches:
- ***Bengaluru:*** 2975, 17th Cross, K.R. Road, Bansankari 2nd Stage, Bengaluru - 70
  Ph: +91-80-26771678/79 • Fax: +91-80-26771680
  E-mail: cbsbng@gmail.com, bangalore@cbspd.com
- ***Chennai:*** No. 7, Subbaraya Street, Shenoy Nagar, Chennai - 600030
  Ph: +91-44-26681266, 26680620 • Fax: +91-44-42032115
  E-mail: chennai@cbspd.com
- ***Kochi:*** 36/14, Kalluvilakam, Lissie Hospital Road, Kochi - 682018
  Ph: +91-484-4059061-65 • Fax: +91-484-4059065
  E-mail: cochin@cbspd.com
- ***Mumbai:*** 83-C, Dr. E. Moses Road, Worli, Mumbai - 400018
  Ph: +91-9833017933, 022-24902340/41 • E-mail: mumbai@cbspd.com
- ***Pune:*** Bhuruk Prestige, Sr. No. 52/12/2+1+3/2,
  Narhe, Haveli (Near Katraj-Dehu Road Bypass), Pune - 411041
  Ph: +91-20-64704058/59, 32342277 • E-mail: pune@cbspd.com

Representatives:

- Hyderabad: 0-9885175004
- Nagpur: 0-9021734563
- Vijayawada: 0-9000660880
- Kolkata: 0-9831437309, 0-9051152362
- Patna: 0-9334159340

Printed at:
India Binding House, Noida (UP)

to

my grandfather late Shri Babu Lal Sinha (freedom-fighter, Khabhainee, Arwal)

my mother Saroj Sinha and

my best friend Aarya

Acknowledgements

The publication of this book would not have been possible without the guidance, and technical and personal support of a number of people. I would like to thank all those individuals, researchers, professionals and technicians who have contributed to this book directly or indirectly. I went through many challenges and problems during the whole process that helped me in learning a few new things. Some of the colleagues who have directly contributed to my work and whose names I would like to mention are Sam, Jack, Vikash and Niru, who helped me in every situation when I felt lonely.

I have referred to many articles and magazines on the web as well as various journals that have contributed to this work.

I would like to thank the following

- My students who helped me in collecting the problems and checking the answers.
- My friends Rajeev, Niraj, Om, Ranjeet, Rahul and Robin for their encouragement and support.
- My elder brothers Kundan and Shashi who inspired and supported me to conquer the world.
- Mr YN Arjuna (CBSPD, New Delhi) and Mr RN Mandal (CBSPD, Pune) for reposing faith in me and publishing this book.

All abovementioned people have not only inspired me to undertake the work of writing this book but have also encouraged me to complete it.

Manish Bhaskar

Foreword

This gives me an immense pleasure in writing the Foreword to the book ***Problems in Engineering Mechanics*** written by Mr Manish Bhaskar, a final year student in computer engineering course studying at Sinhgad College of Engineering. Mr Manish accepted the herculean task in writing this book on mechanics at the young age of 22.

This is a unique case I have known during my entire career of teaching profession of 44 years. I have observed teaching staff having postgraduate qualifications many times are afraid of teaching this subject. In fact engineering mechanics can be made interesting and phobia in the minds of the students and, sometimes in the teachers, can be removed if they succeed in getting conceptual understanding of the subject matter. The author understands the different levels of the students studying this subject and barring for syllabus of different universities, he has nicely prepared the layout of this book. Mr Manish Bhaskar has made unstinted efforts to present the subject matter attractive to the entire student community.

The book is to be seen, referred to and to be recognized in altogether new perspective. I wish author Manish for its success and I am sure that he will make efforts of this nature in future also.

Dr MR Shiyerkar ME, PhD, FIE
Professor and Head
Department of Civil Engineering
Sinhgad College of Engineering
Pune 411 041

Foreword

I am extremely delighted and proud in writing the Foreword to the book *Problems in Engineering Mechanics* written by Mr Manish Bhaskar, the youngest author in India on engineering mechanics.

This in one of the biggest achievements for an engineering student to write the book which is meant for first year engineering students and the students who are preparing for competitive examinations also. The way he has solved the problems is very good and can be understood by the engineers as well.

This book will be beneficial and fruitful for all the engineering students in the country.

I wish him good luck for his future.

Dr AS Padalkar ME, PhD
Principal
Sinhgad Technical Education Society's
Sinhgad College of Engineering
off Sinhgad Road
Pune 411 041

Preface

Having been a student of computer engineering, it gives me an immense pleasure in presenting this book. This book is for all such students who like to solve different types of problems — and guess, you are one of them. This book is mainly intended for the use of engineering students. It may prove to be of great interest to the college teachers, and also the engineers who find mechanics a fascinating subject.

It was my goal to solve problems with different methods, find out the easiest one, and share that method with everyone. I am very much inspired by the books written by JL Meriam/LG Kraige, Beer/Johnston, and Ramsey. These three books put together may be called 'Bible of Engineering Mechanics'. Till now, I have taught two hundred and fifty students and I saw that majority of the students have a good knowledge of theory but fail in cracking the problems. Do you face the same problem? Be cool, this is natural. This book is all about how to utilise the theoretical knowledge in solving the problems in the easiest way.

Budding engineers from IITs, NITs and all other colleges will enjoy solving the problems because there is too much variety and brain twister problems which have been asked in competitive examinations too.

The language used in this book is simple and the text is written in a way that every student can understand it in a one-time reading.

You will be unable to solve some problems if you do not just try repeatedly because someone has said "you are never a loser until you quit trying", and you will get hold of the concept. No worries, answers are provided at the end of the chapter. I did not compromise on the quality of the problems so that they do contain many times more than one subject in the same problem.

The way I have expressed my thoughts may be funny for some students, rude for some, humorous for some and God knows what other things come to our mind. But the ultimate thing is how you perceive it and will you be able to enjoy the book.

On a serious note, I hope that the subject matter of the book will instil confidence among the readers and the book will help them like an ideal teacher and a friend.

A feedback in the form of suggestions, comments, phone calls or emails (bhaskar.sinhgad@gmail.com) from you for further improvement of the book will be highly appreciated.

Manish Bhaskar

Contents

Part I

Statics

1 Equilibrium

Summary and Concept

Force triangle

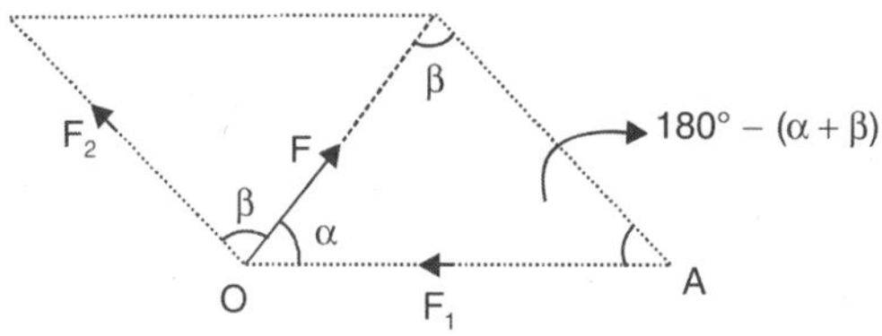

Fig. 1

By sine rule $$\frac{F_1}{sin\beta} = \frac{F_2}{sin\alpha} = \frac{F}{sin[180-(\alpha+\beta)]}$$

Resultant of two forces

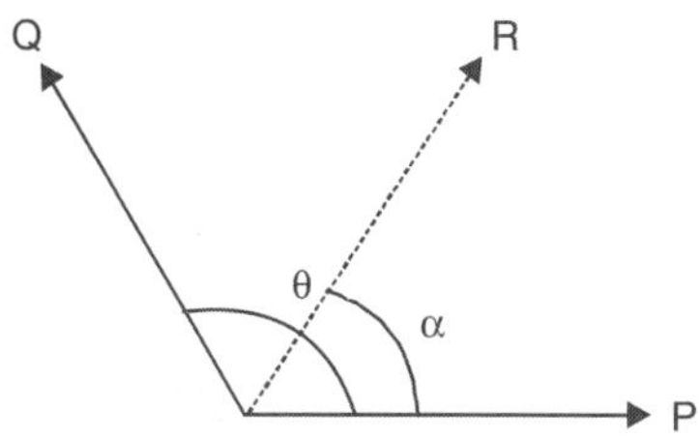

Fig. 2

$$R = \sqrt{P^2 + Q^2 + 2PQ\cos\theta}$$

$$\tan\alpha = \frac{Q\sin\theta}{P + Q\cos\theta}$$

Varignon's theorem

The moment of resultant of a force system about any point is equal to the algebraic sum of moments of all other forces about the same point.

$$R \cdot X = \sum M$$

Where X = Perpendicular distance of resultant 'R' about a point where moment is taken.

Equations of condition of equilibrium

$$\sum F_x = 0, \ \sum F_y = 0$$

$$\Sigma M_0 = 0$$

'o' is the any arbitrary point having maximum number of unknown forces (generally)

Co-ordinates of centroid for distributed mass system

For one-dimensional body ⇒

$$\bar{x} = \frac{\Sigma L_i X_i}{\Sigma L_i}, \quad \bar{y} = \frac{\Sigma L_i y_i}{\Sigma L_i}$$

for two dimensional body ⇒

$$\bar{x} = \frac{\Sigma A_i X_i}{\Sigma A_i}$$

$$\bar{y} = \frac{\Sigma A_i y_i}{\Sigma A_i}$$

A represents area of the body.

For three-dimensional body ⇒

$$\bar{x} = \frac{\Sigma V_i X_i}{\Sigma V_i}$$

$$\bar{y} = \frac{\Sigma V_i Y_i}{\Sigma V_i}$$

V represents volume of the body.

Co-ordinates of centroid for continuous mass system

$$\bar{x} = \frac{\int x \cdot dL}{L}$$

$$y = \frac{\int y \cdot dL}{L}$$

Similarly for 2D and 3D body also.

Pappus-Guldinus's theorem

(i) The area of a surface of revolution is equal to the length of the generating curve times the distance traveled by the centroid of the curve while the surface is being generated.

$$A = 2\pi \bar{y} L$$

Where $2\pi\bar{y}$ is the distance traveled by the centroid of the line L.

(ii) The volume of a body of revolution is equal to the generating area times the distance traveled by the cenroid of the area while the body is being generated

$$V = 2\pi \bar{y} A$$

Where $2\pi\bar{y}$ is the distance traveled by the centroid of area A.

Level 1

Example 1

Forces acting on the cross section of a steel channel are shown in Figure. Replace this system by an equivalent single force at point 'G'. Also determine the distance x of the point 'G' from point 'C'.

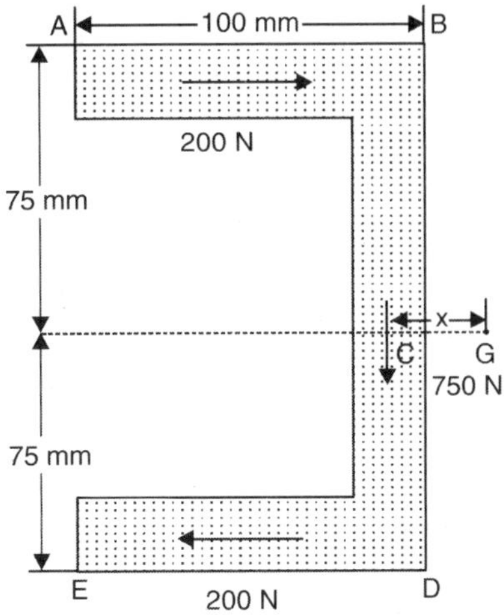

Fig. E1

Example 2

The moments of a given plane system of forces about three points (0, 1), (2, 0) and (2,2) are 108 and 63 units respectively. Find the magnitude direction and position of resultant force.

Example 3

Three forces *P*, *Q* and *R* are acting on a equilateral triangular plate *ABC* as shown in figure. The resultant of 200 kN is known to pass vertically up from point B. Find the magnitudes of the three forces such that the moment of the three forces about B is zero. Points *E*, *F*, *G* are the centers of the sides of the triangular plate.

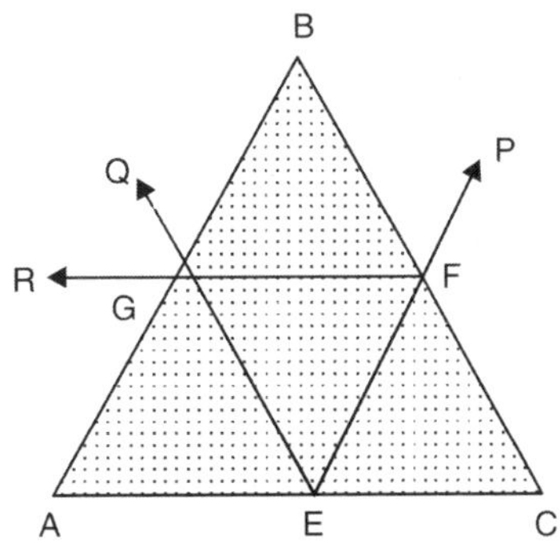

Fig. E3

Example 4

Replace the couple and force shown in figure by an equivalent single force applied to the lever. Determine the distance from the center of the shaft to the point of application of this equivalent force. (Radius of the shaft is 60 mm)

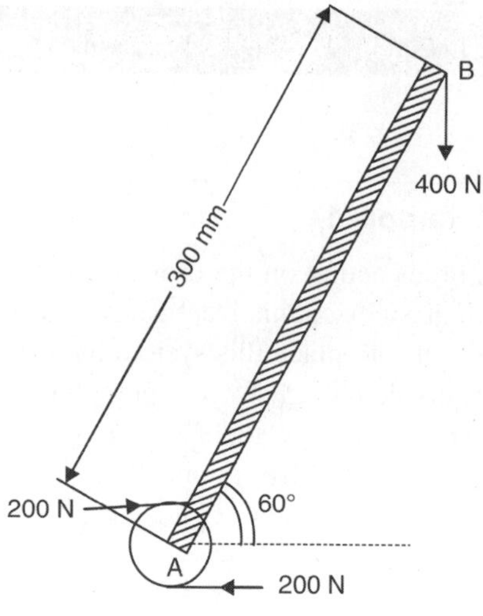

Fig. E4

Example 5

A multiple-drilling machine is used to drill simultaneously six holes in the steel plate. Each drill exerts a 4 N-m clockwise couple on the plate. Determine an equivalent couple formed by the smallest possible forces acting (*a*) At *A* and *C* (*b*) On the plate.

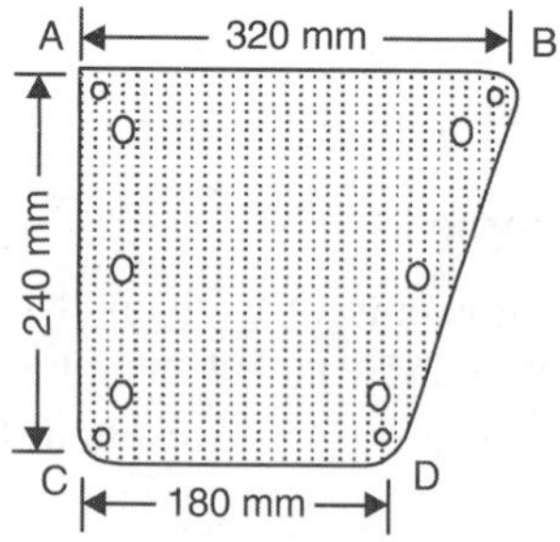

Fig. E5

Example 6

The ends of a light string *ABCD* are fixed. Weights are fastened to the string at *B* and *C* and the parts *AB*, *BC*, *CD* are inclined to the horizontal at angles of 60°, 30° and 55°. The weight at B is 22 N, Determine that at *C*.

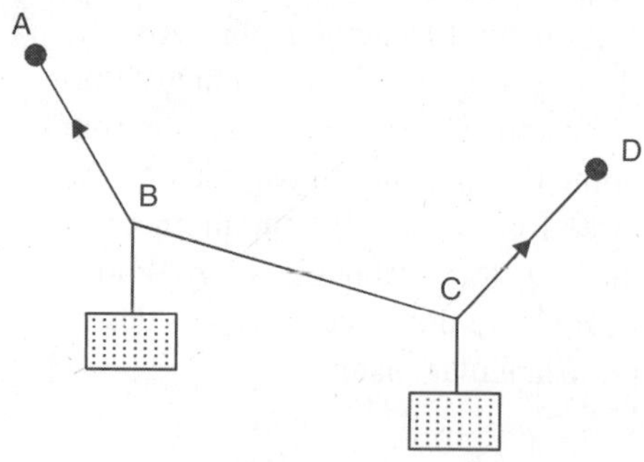

Fig. E6

Example 7

Locate centre of gravity of model as shown in figure from base *AB*.

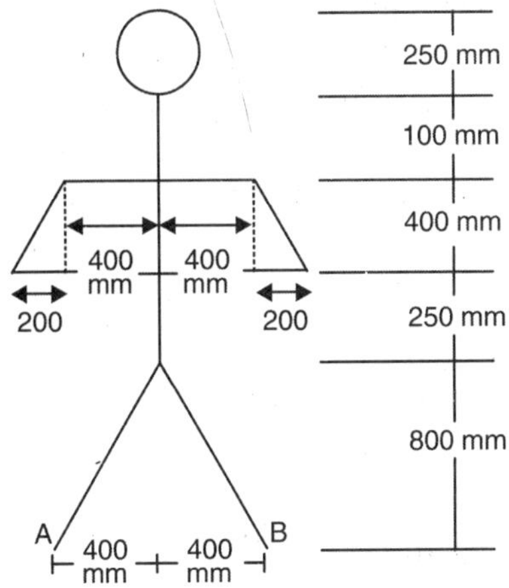

Fig. E7

Example 8

The homogeneous wire *ABC* is bent into a semicircular *arc* of radius '*r*' and straight section as shown in figure and is attached to a hinge at *A*. Determine the value 'θ' for which the wire is in equilibrium for the indicated position.

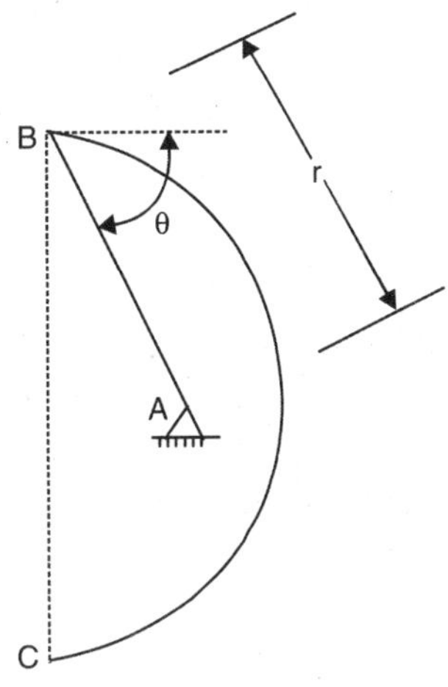

Fig. E8

Example 9

The 100 kg wheel rests on a rough surface and bears against the roller *A* when the couple *M* is applied. If *M* = 60 Nm and the wheel does not slip, compute the reaction on the roller *A*.

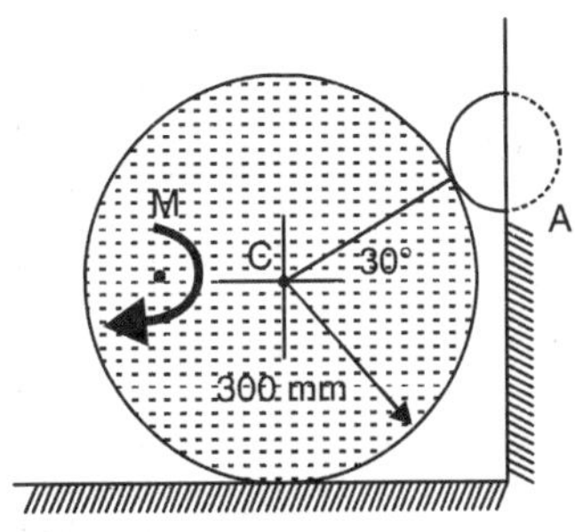

Fig. E9

Example 10

Calculate the tension T in the cable which supports the 1000 N load with the pulley arrangement shown. Each pulley is free to rotate about its bearing and the weights of all parts are small compared with the load. Find the magnitude of the total force on the bearing of pulley C.

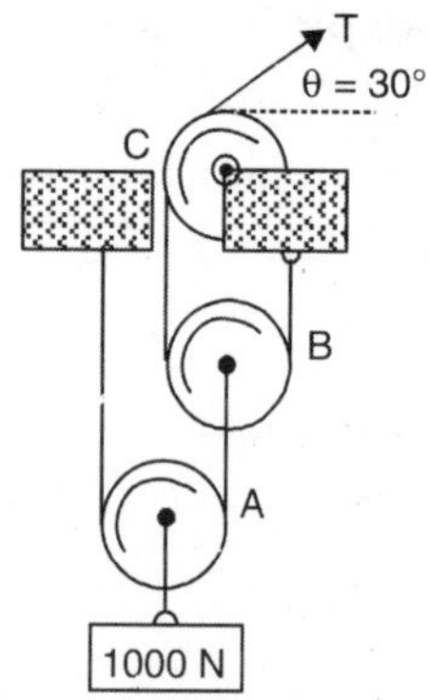

Fig. E10

Example 11

Two spheres of radius R_1 and R_2 respectively rest inside a smooth hollow cylinder of external radius R_0 and internal radius R where $R < R_1 + R_2$. If the sphere of radius R_2 is lower and sphere of radius R_1 is upper as shown in the figure. Determine the minimum weight of the cylinder that it will not overturn. The weight of upper sphere is W.

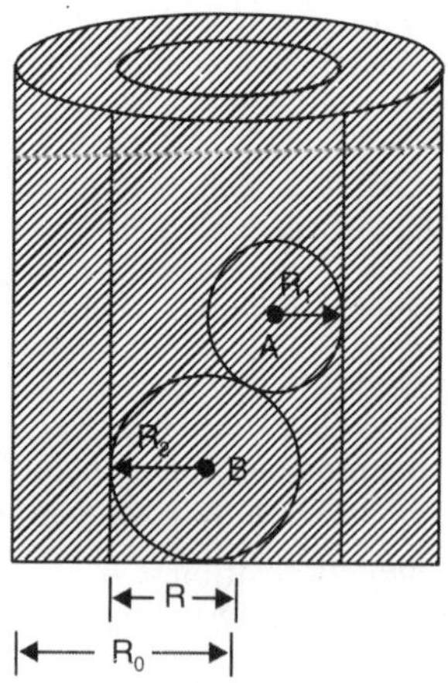

Fig. E11

Example 12

The device shown in the figure is useful for lifting drywall panels into position prior to fastening to the stud wall. Estimate the magnitude P of the force required to lift the 25-kg panel.

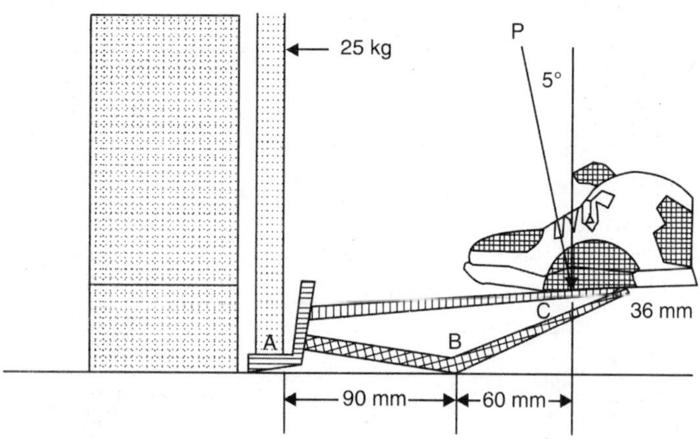

Fig. E12

Example 13

Determine the force P required to begin rolling the uniform cylinder of mass m over the obstruction of height *h*.

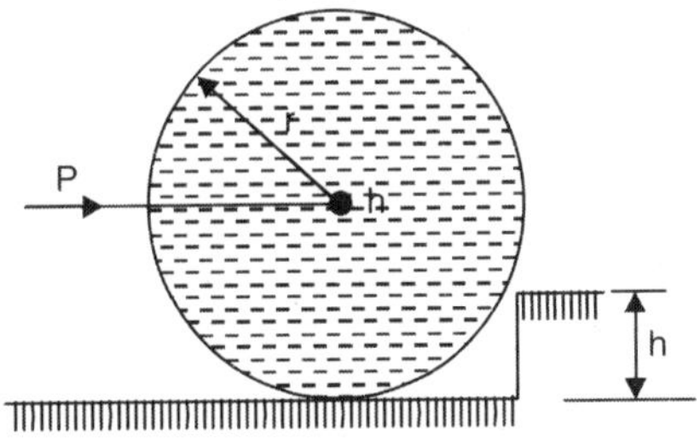

Fig. E13

Example 14

The uniform 100 kg beam is supported initially by its end rollers on the horizontal surface at *A* and *B*. By means of the cable at *C* it is desired to elevate end *B* to a position 3 m above end *A*. Determine the required tension P, the reaction at *A* and the angle θ made by the beam with the horizontal in the elevated position.

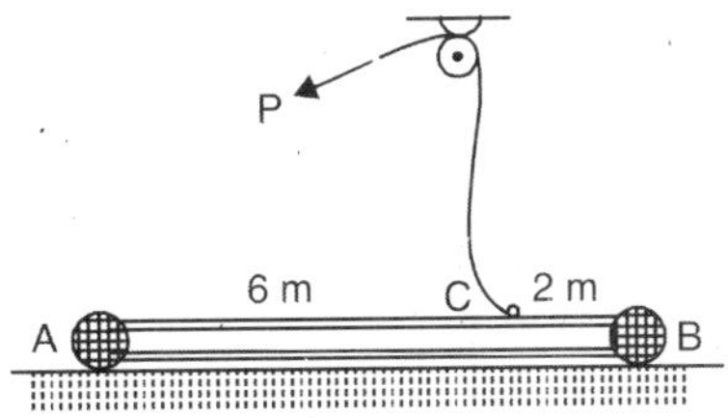

Fig. E14

Example 15

Two cylinders A and B of diameters 8 cm and 12 cm respectively are held in equilibrium by separate strings as shown in figure. Cylinder B rests against vertical wall. If weights of cylinder A and B are 20 N and 40 N respectively, determine tension in strings and reactions at all point of contacts.

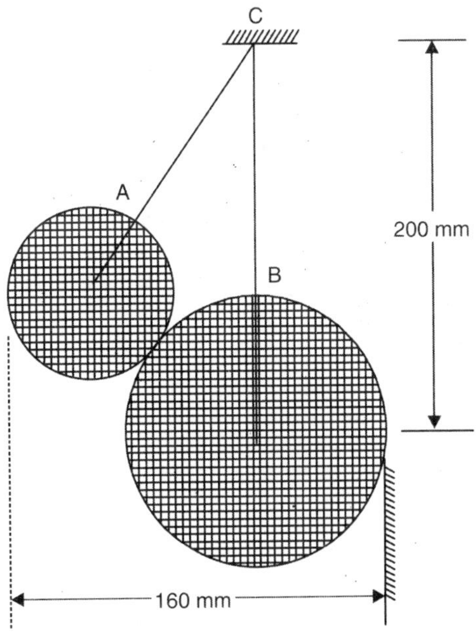

Fig. E15

Example 16

The figure shown is formed of thin homogeneous wire. Determine the length l of portion DE of the wire for which the centre of gravity of the figure is located at D.

If (*a*) $\theta = 30°$ (*b*) $\theta = 90°$

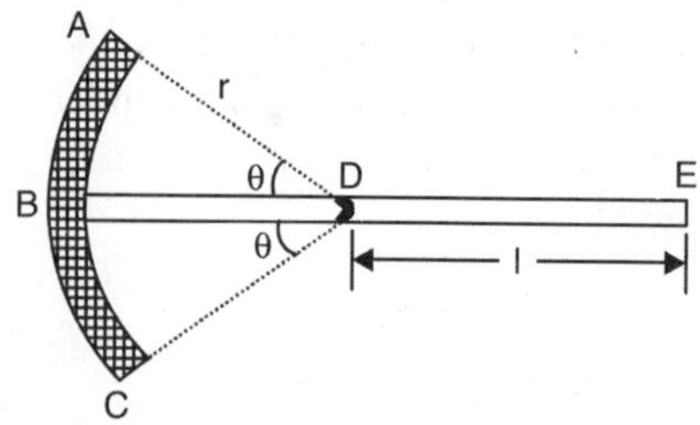

Fig. E16

Example 17

A right circular cone of base '*r*' is attached to a hemisphere of radius '*r*'. Determine the height '*h*' of the cone so that the centroid is located in the plane between the cone and hemisphere.

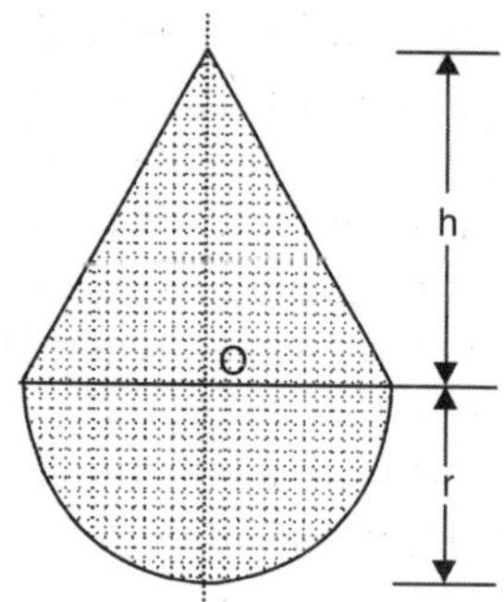

Fig. E17

Example 18

Five forces are acting on a regular hexagon. A clockwise moment of 2 kNm is required to be applied at point A so that the hexagon will not rotate about A. Determine the side of the hexagon.

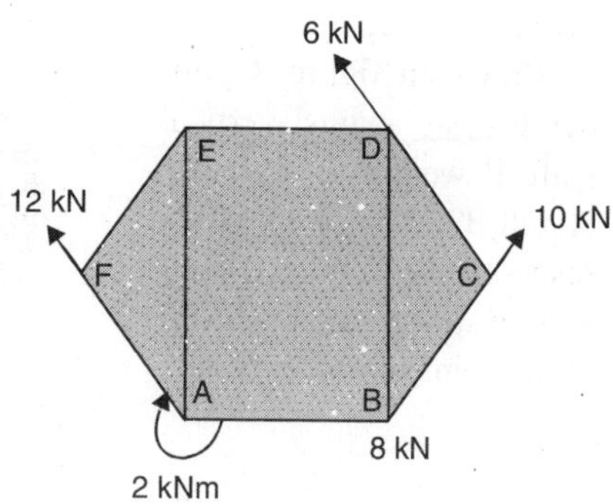

Fig. E18

Example 19

Locate the centroid of the area shown in figure. The radius of the circle is 30 cm.

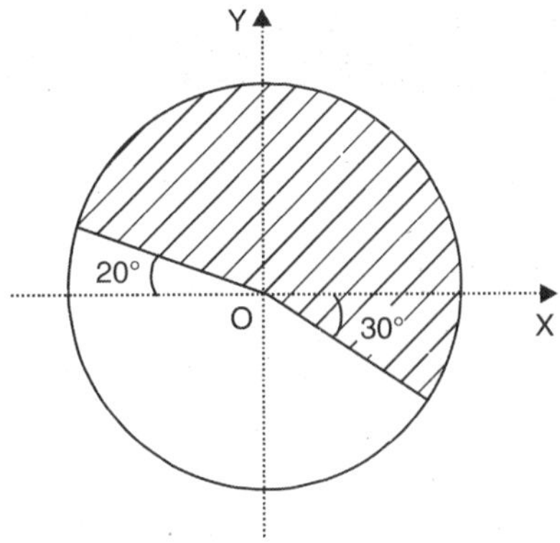

Fig. E19

Example 20

A small ring of mass m can slide on a smooth plane. The ring is attached by a light thread to the highest point of the wire and in equilibrium; the thread is taut and makes an angle θ with the vertical. Find the tension of the thread and reaction of the wire on the ring.

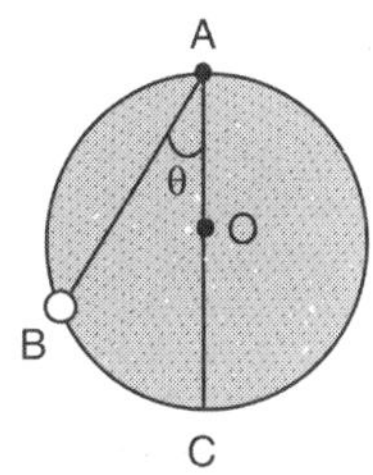

Fig. E20

Example 21

Determine the magnitude *T* of the tension of the supporting cable and the magnitude of the force on the pin at *A* for the jib crane shown. The beam *AB* is standard 0.5 m beam with a mass of 95 kg per meter of length.

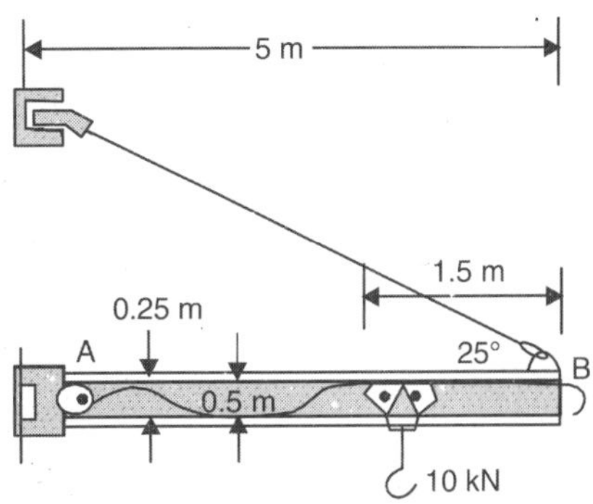

Fig. E21

Level 2

Example 1

Three uniform boxes, each having a weight W and length a are stacked as shown.

Determine the maximum distance l the top box can extend out from the bottom so that stack does not topple over.

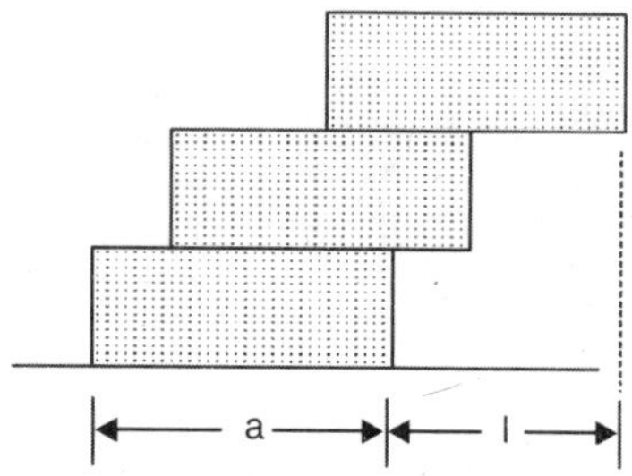

Fig. E1

Example 2

Locate the centroid of the area bounded between line $y = 2x$ and parabola $y^2 = 4x$.

Example 3

A chain loop of length 1.25 m is placed around a 250 × 250 mm piece timber. The mass of timber is 175 kg determine the tension in the chain for the arrangement shown in figure.

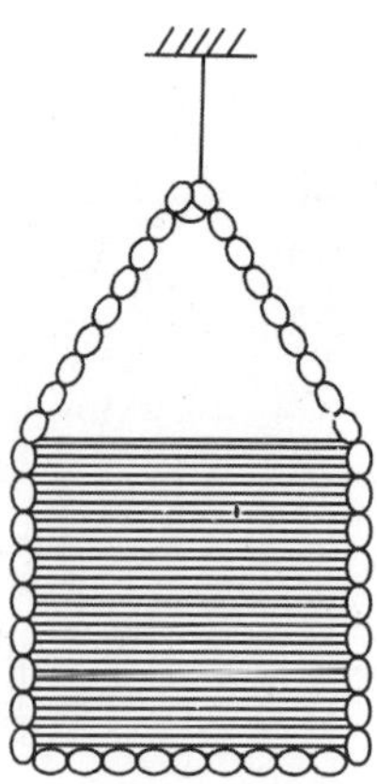

Fig. E3

Example 4

Inside a fixed hollow cylinder of Radius R whose generators are horizontal, there are placed symmetrically two equal cylinders each of radius r. A third cylinder of radius r of same mass is placed symmetrically on them. Determine the condition for equilibrium in terms of R and r.

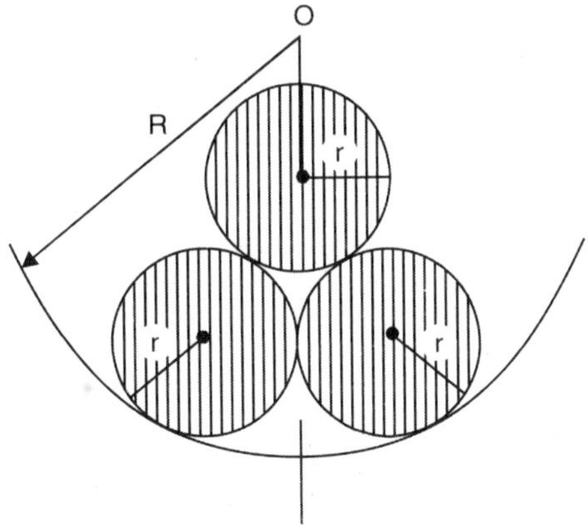

Fig. E4

Example 5

The mass per unit length of a ladder increase uniformly form the top to the bottom of the ladder and is twice as great at the bottom. The ladder stands on oriented plane. Determine distance x_c of the center of mass of the ladder from the end B.

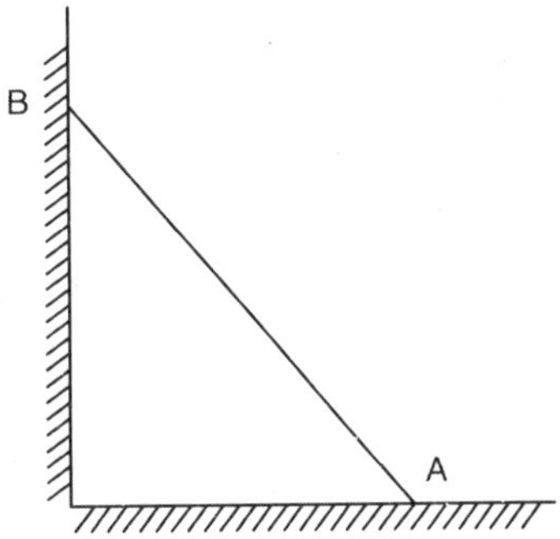

Fig. E5

Example 6

A uniform rod AB of length $3R$ and weight W rests inside a hemispherical bowl of radius R as shown in figure neglecting friction, find the angle θ corresponding to equilibrium.

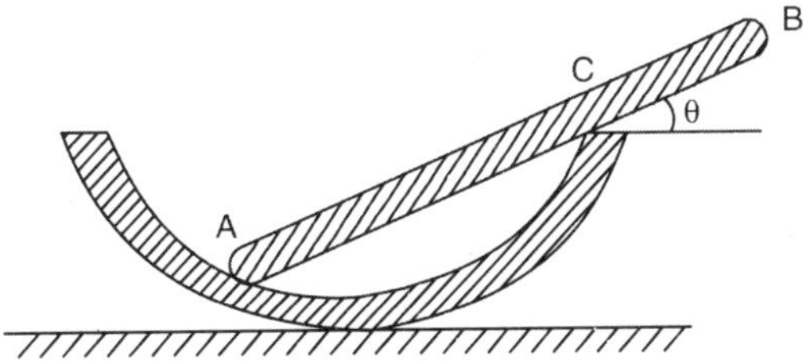

Fig. E6

Example 7

Find the co-ordinates of the centre of gravity of a quadrant of the area of the curve $x^{2/3} + y^{2/3} = a^{2/3}$ bounded by the axes of co-ordinates.

Example 8

Rod *AB* is bent into shape of an arc of a circle and is loaded between two pegs *D* and *E*. It supports a load *P* at *B* Neglecting friction and the weight of rod, determine the distance *C* corresponding to equilibrium when a = 20 mm, R = 100 mm.

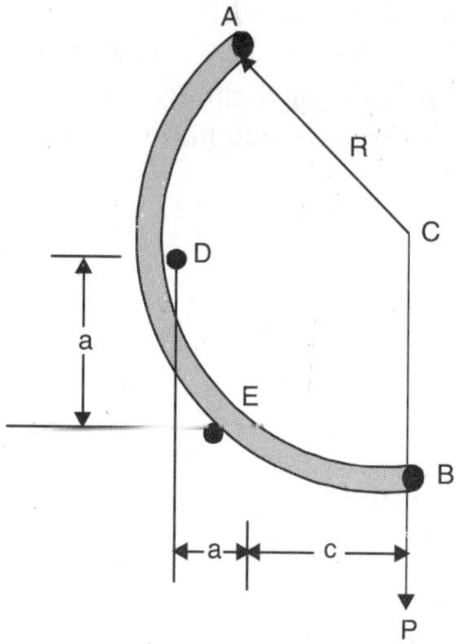

Fig. E8

Example 9

The hinged gate ABC closes an opening of width b (perpendicular to the paper) in a water channel. The water has free access to the underside as well as the right side of the gate. When the water level rises above a certain value of h, the gate will open. Determine the critical value of h, Neglect the mass of the gate.

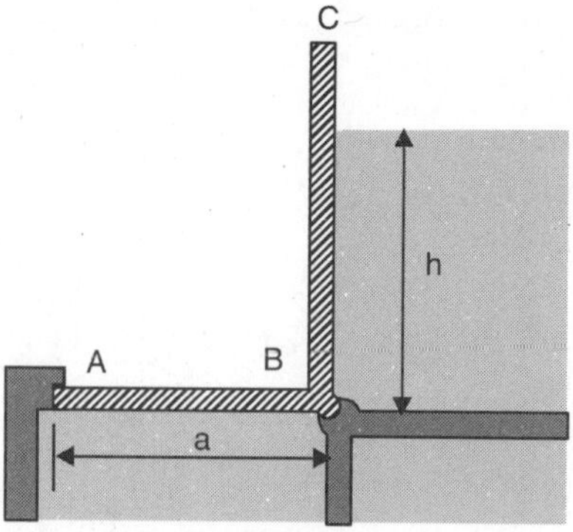

Fig. E9

Example 10

A thin lamina of uniform thickness is as shown in figure. If it is suspended from the point *A* by a wire what angle will line *AB* make with vertical?

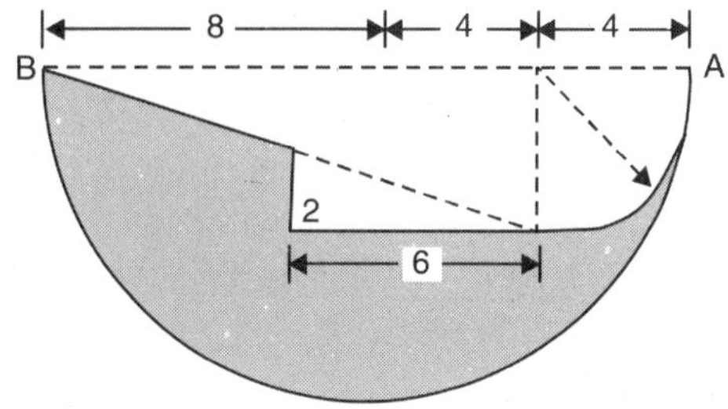

Fig. E10

Example 11

The 3-m plank shown in section has a density of 800 kg/m^3 and is hinged about a horizontal and through its upper edge O. Calculate the angle θ assumed by the plank with the horizontal for the level of fresh water shown.

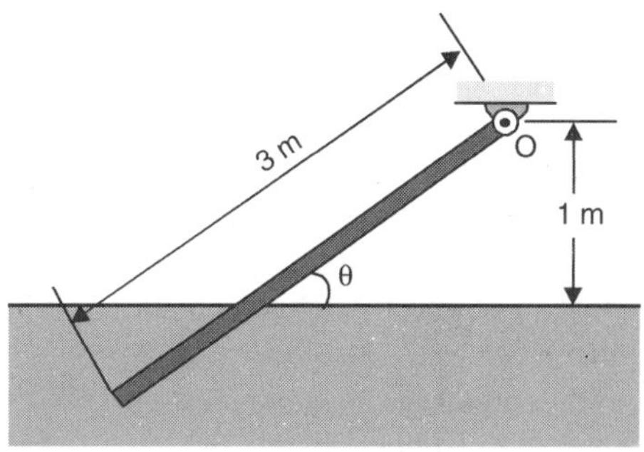

Fig. E11

Example 12

Four bars of uniform cross-section are connected to each other as shown in figure. Locate the center of gravity of the composite figure. Also find the centroid of the area falling under these bars.

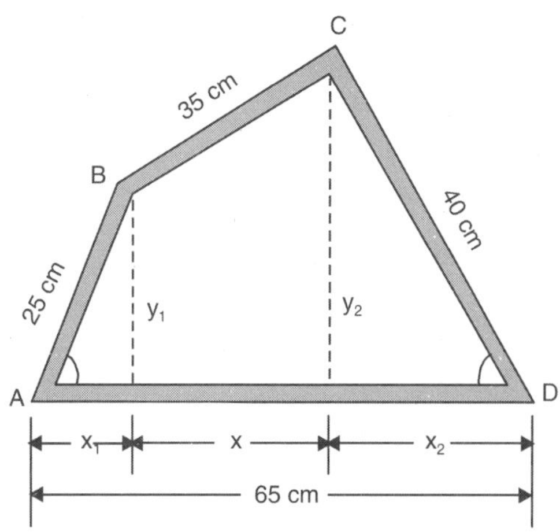

Fig. E12

Example 13

A slender rod of mass m_1 is welded to the horizontal edge of a uniform semi cylindrical shell of mass m_2. Determine an expression for the angle θ with the horizontal made by the diameter of the shell.

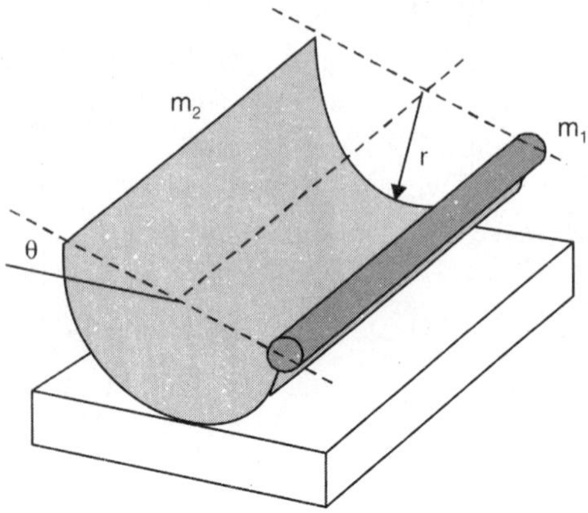

Fig. E13

Example 14

A body of weight *W* is suspended by two equal threads *AP, BQ* the points of support *A, B* are on the same level at a distance a apart and the threads are fastened to two points *P, Q* on the body so that *PQ* is horizontal and *PQ* = *b*. Determine the couple value *M* about a vertical axis so that body is deflected through on angle θ with *PQ* at a depth h below *AB*.

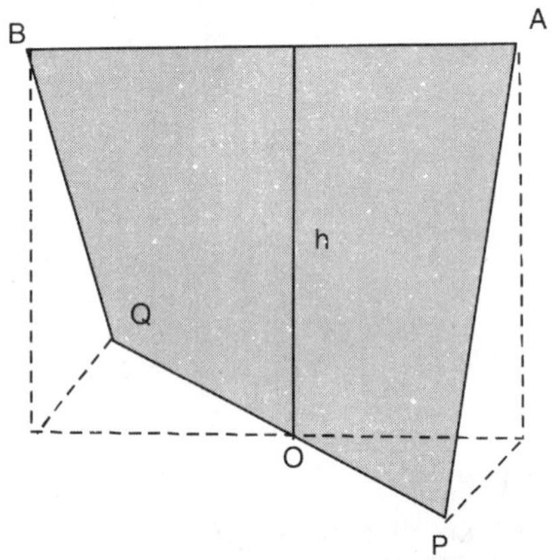

Fig. E14

Example 15

Determine completely the resultant for *R* exerted on the cylindrical dam surface by the water. The density of fresh water is 1.000 Mg/m^3, and the dam has a length *b*, normal to the paper of 30 m.

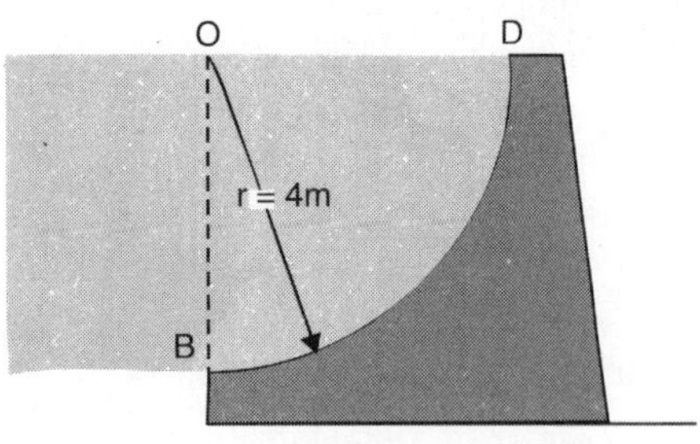

Fig. E15

Example 16

A uniform rod has a ring at one end which slides along a smooth vertical wire the rod rests touching a smooth cylinder of radius r whose axis is horizontal and at a distance c from the wire. If θ is the angle the rod makes with the horizontal find its length in terms of c, r and θ.

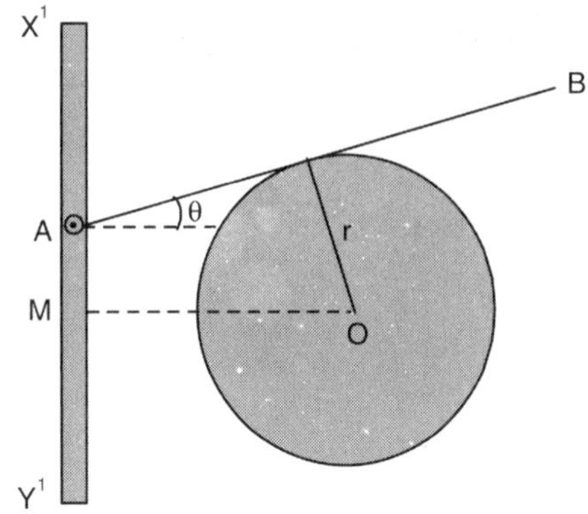

Fig. E16

Example 17

A prismatic bar AB of length l and weight w remains in a vertical plane and is supported at B by a vertical smooth wall and at A by a smooth cylindrical surface OC which is perpendicular to the plane of the figure. Find the equation of the curve OC referred to the coordinate axes x and y if the bar is in equilibrium for any position as defined by the angle θ.

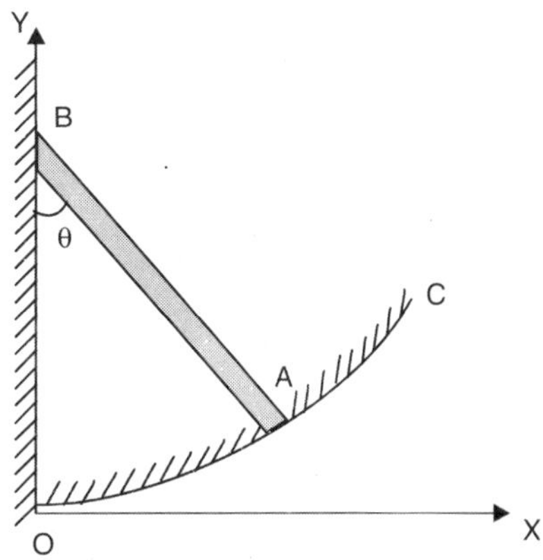

Fig. E17

Example 18

Determine the resultant force R exerted on the semicircular end of the water tank shown in the figure if the tank is filled to capacity. Express the result in terms of the radius r and the water density p.

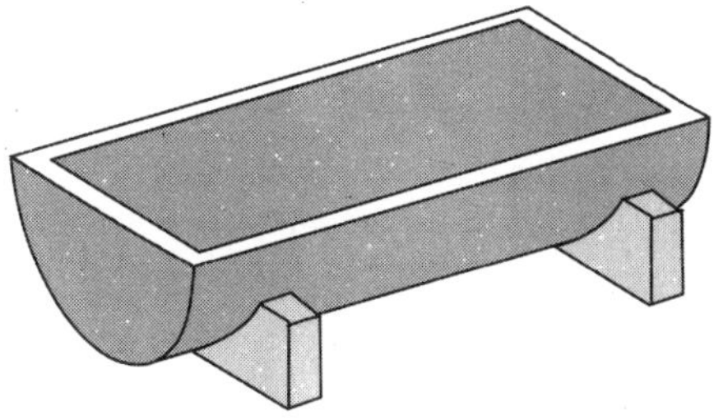

Fig. E18

Example 19

Locate the centroid of the area bounded by curve $x = y^3$, the y-axis and line $y = 1$.

Level 3

Example 1

Find the centre of gravity of a loop of the lemniscate $r^2 = a^2 \cos 2\theta$

Example 2

Determine the location of the centroid of the half right circular cone as shown in figure.

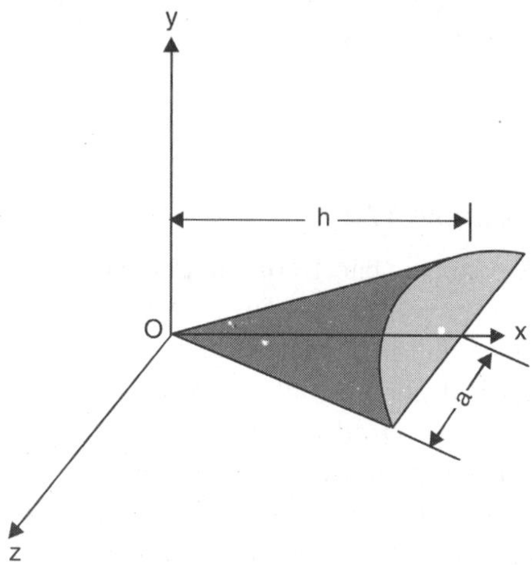

Fig. E2

Example 3

Three weighted rods of length l each are hinged at points A and B lying on the same horizontal and joint through hinges at point C and D. The length $AB = 2l$. A load of mass m is suspended at the hinge C. Determine the minimum force applied to the hinge D for which middle rod remains horizontal. The mass of hinges can be considered.

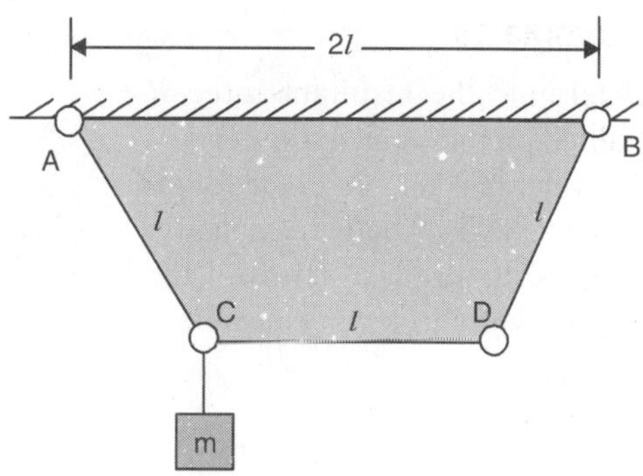

Fig. E3

Example 4

A uniform triangular Lamina ABC of weight w is suspended from a fixed point by strings of length l_1, l_2, l_3 attached to its corners A, B, C. Determine the tensions in the each string.

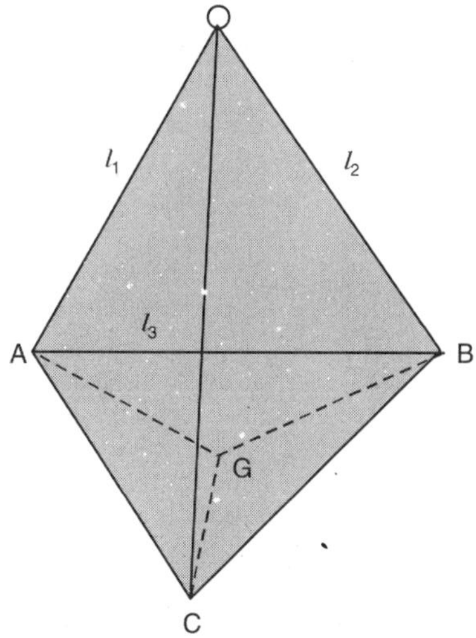

Fig. E4

Example 5

A uniform ring of mass m and radius r carries an eccentric mass m_0 at a radius b and is in an equili-brium position on the incline, which makes an angle α with the horizontal. If the contacting surfaces are rough enough to prevent slipping. Write the expression for the angle θ, which defines the equilibrium position.

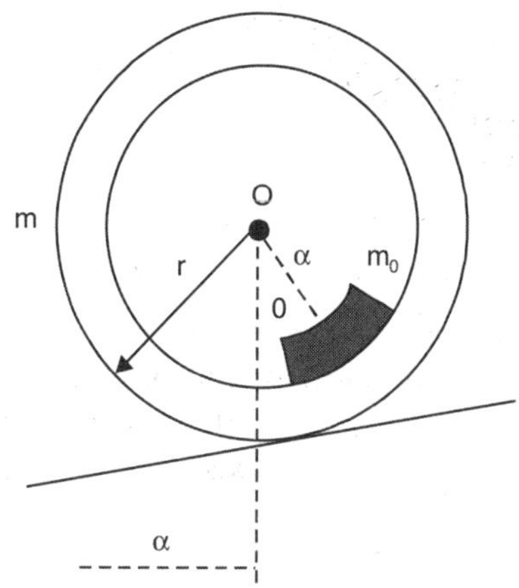

Fig. E5

Example 6

Two rings of mass m_1 and m_2 can slide on a smooth circular wire in a vertical plane. They are connected by a light string, which subtends an angle 2β at the center of the circle when the rings are in equilibrium on the upper half of the wire. Determine the angle made by the string with the horizontal.

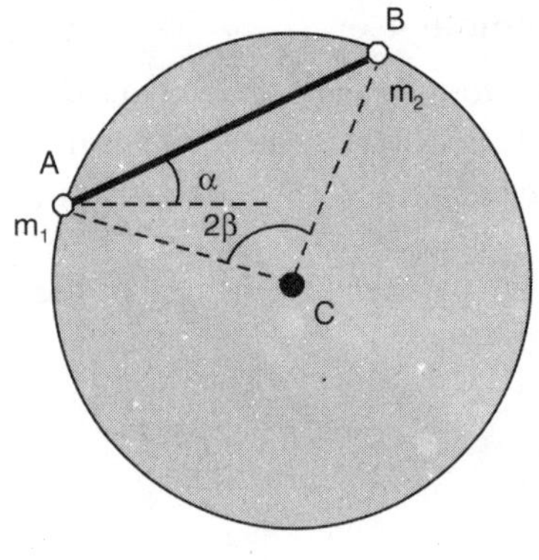

Fig. E6

Example 7

A uniform triangular lamina *ABC* is suspended from the corner A as shown in figure. Determine the angle made by side *BC* with horizontal in equilibrium.

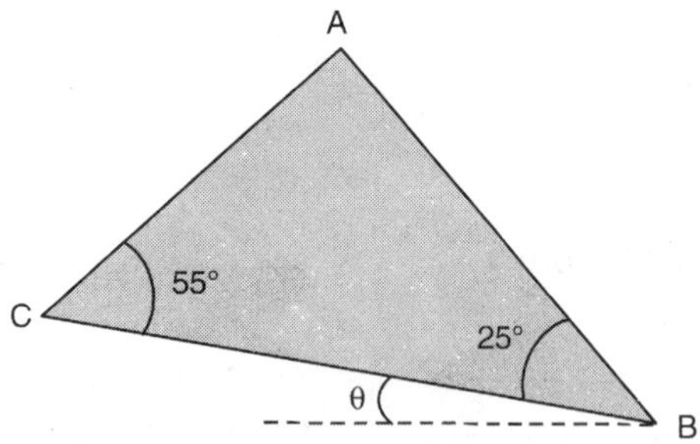

Fig. E7

Example 8

Determine the area of the surface of revolution as shown, obtained by rotating a Quarter-circular arc about a vertical axis.

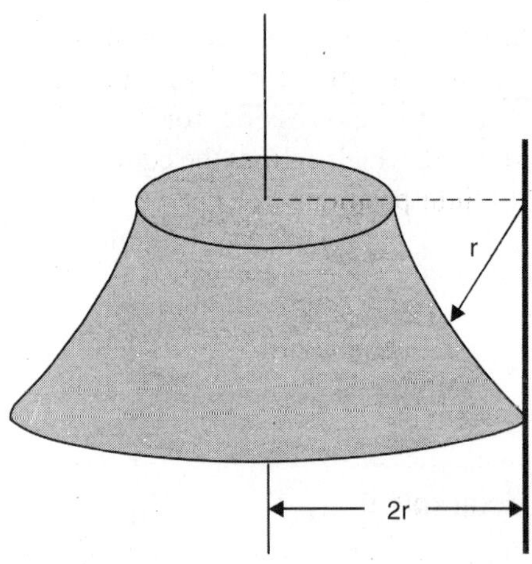

Fig. E8

Example 9

A homogeneous rod of length $2l$ leans against a vertical wall at one end and against a smooth stationary surface at another end. What function $Y(X)$ must be used to describe the cross section of this surface for the rod. Rod remains in equilibrium in any position even in the absence of friction?

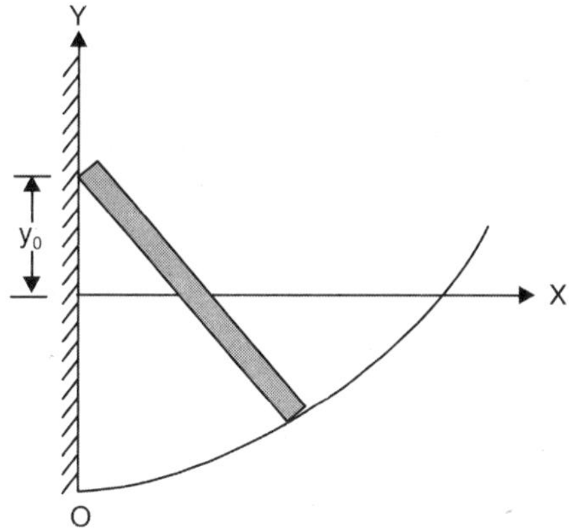

Fig. E9

Example 10

The Hut shown in fig subjected to a horizontal wind and the pressure p against the circular roof is approximated by $p_0 \cos \theta$. The pressure is positive on the windward side of the hut and is negative on the leeward side. Determine the total horizontal shear Q on the foundation per unit length of roof measured normal to the papers.

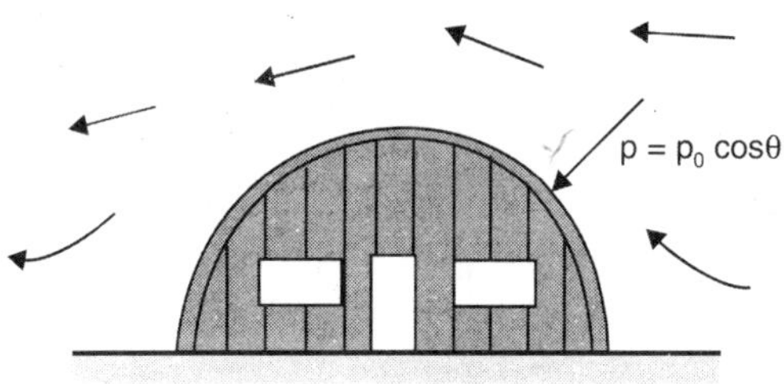

Fig. E10

2 Friction

Summary and Concept

Friction

Friction is a type of resistive force that opposes the relative motion between the surfaces in contact.

Static friction

It opposes the tendency of relative motion but never reverse the tendency.

If two surfaces 1 and 2 are in contact then

$$V_{12} = 0$$

It is self-adjusting force

$$f_s = \mu_s N$$

Where f_s, μ_s, N represents static frictional force, coefficient of static friction and normal reaction acting by the surface respectively.

Kinetic friction

It is the friction experienced by a body when it moves.

(*a*) Sliding friction

(*b*) Rolling friction

For all drag forces

$$\vec{F}_{12} \uparrow\downarrow \vec{V}_{12}$$

For moderate relative speed between the surfaces.

$$\vec{F}_{12} = -k\vec{V}_{12}$$

Where k represent positive constant.

It is not a self-adjusting force in the case of drag friction.

$$f_k = \mu_k N < \mu_s N$$

Where f_k and μ_k represents kinetic frictional force and coefficient of kinetic friction respectively.

Belt friction

For a rough pulley having tensions T_1 and T_2 at both ends in the belt passing over it, the relation between T_1 and T_2 can be generated as

If $T_1 > T_2$

$$\frac{T_1}{T_2} = e^{\mu_s \beta}$$

Where $\mu = \mu_S$

When belt is not slipping

$\mu = \mu_k$

When belt is actually slipping

β = Lap angel (must be in radians)

Breaking Torque if $T_1 > T_2$

$M = (T_1 - T_2) R$

Where R = Radius of pulley

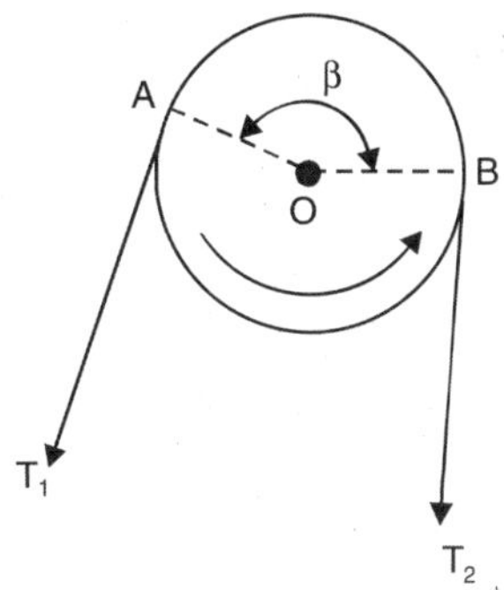

Fig. 1

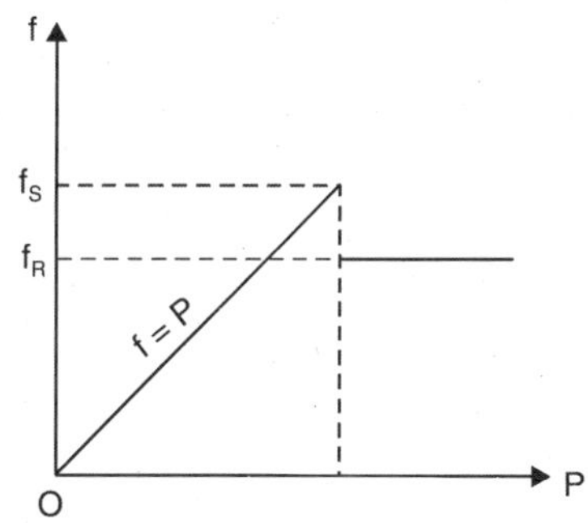

Fig. 2

Wedge and block

(*i*) Draw F.B.D of each wedge and block

(*ii*) Draw E.D (Equation diagram) for each F.B.D

(*iii*) Apply condition of equilibrium for each equation diagram

(*iv*) For frictional forces use $f = \mu N$

(*v*) Find number of unknowns and write corresponding equations

Level 1

Example 1

To adjust the vertical position of column supporting a weight of 2000 N, two 5° wedge are used as shown in figure. Determine the force '*p*' necessary to start the wedges if the angle of friction at all the surface is 25°. (Neglect friction at rollers)

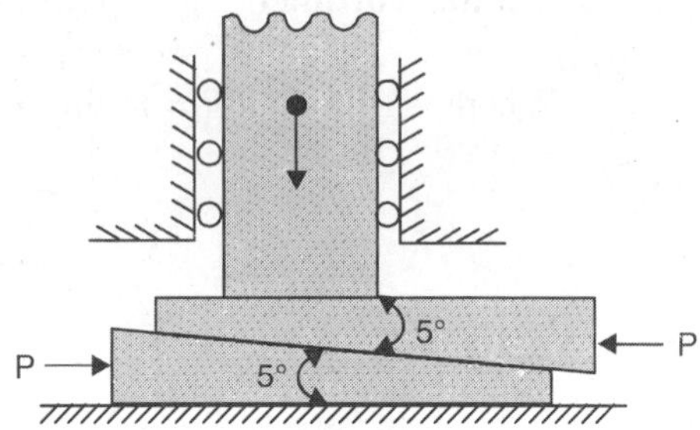

Fig. E1

Example 2

A wedge A of negligible weight is to be driven between two 51 kg blocks *B* and *C*. Knowing that coefficient of static friction is 0.35 between blocks and horizontal surface and zero between the wedge and each of blocks, determine the smallest force *P* required to start moving the wedge if the blocks are equally free to move and if block *C* is securely bolted to horizontal surface.

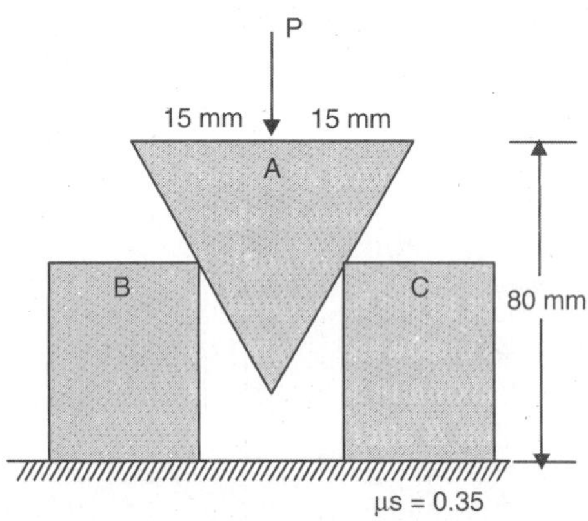

Fig. E2

Example 3

A pulley of diameter 80 cm is rotating due to torque of 100 Nm clockwise as shown in figure. To apply brake a force *P* is applied to brake lever *ABC* at *A*. The lever is hinged at *B*. Find *P* required if μ between belt and pulley is 0.3. *AB* = 60 cm and *BC* = 30 cm

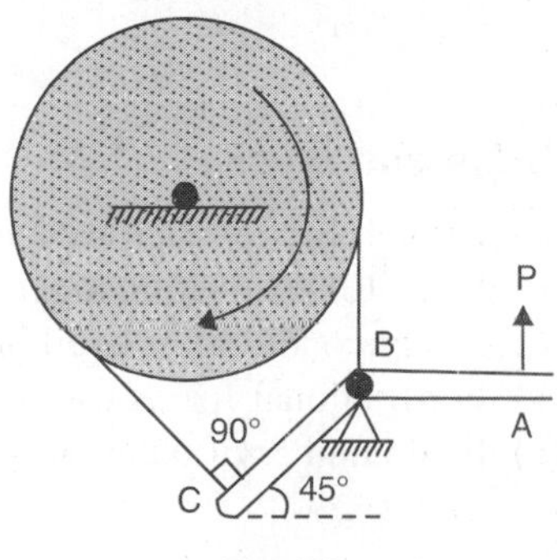

Fig. E3

Example 4

Determine the coefficient of friction value (equal to all contact surfaces) μ so that the window remains in the equilibrium. The value of P is given as $P = W/3$, where W is the weight of window.

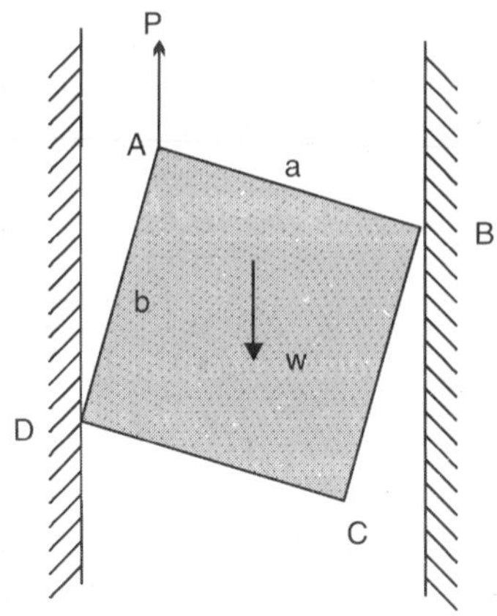

Fig. E4

Example 5

A disk of mass 'm' is restively on the rough surface, introduces a rolling resistance. A force 'P' is applied in horizontal direction passing through the center of the disk as shown in the figure. Find the value of coefficient of friction so that rolling can occur.

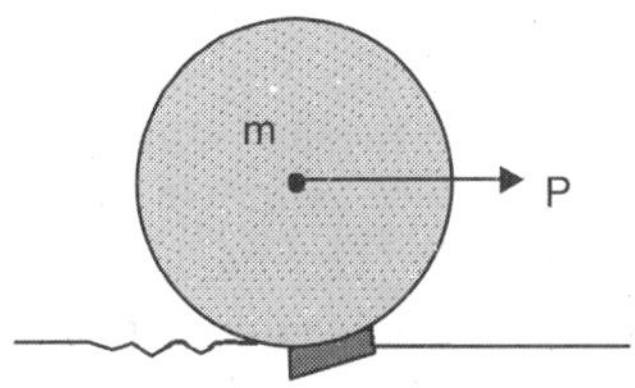

Fig. E5

Example 6

A rope is looped around two pulleys as shown. Determine force P required to hold the weight of 500 N.

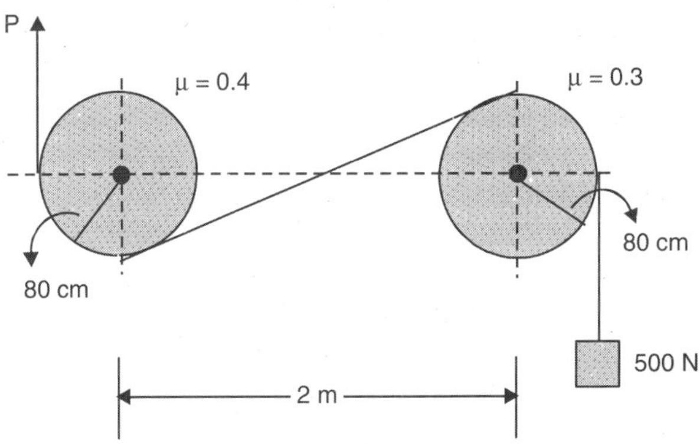

Fig. E6

Example 7

Find the least value of P such that the system will just start moving to the right. Take $\mu = 0.3$. Assume smooth pulley.

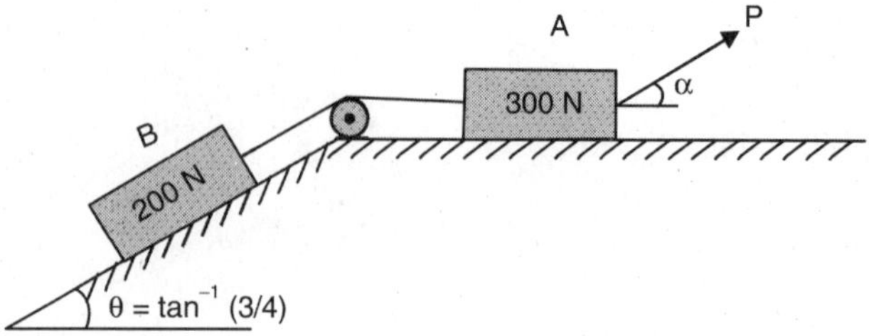

Fig. E7

Example 8

What should be the maximum value of M that can be applied on the cylinder of weight W and radius r between the corner of wall and the surface as shown so that it may not spin? μ is the coefficient of friction between any contact.

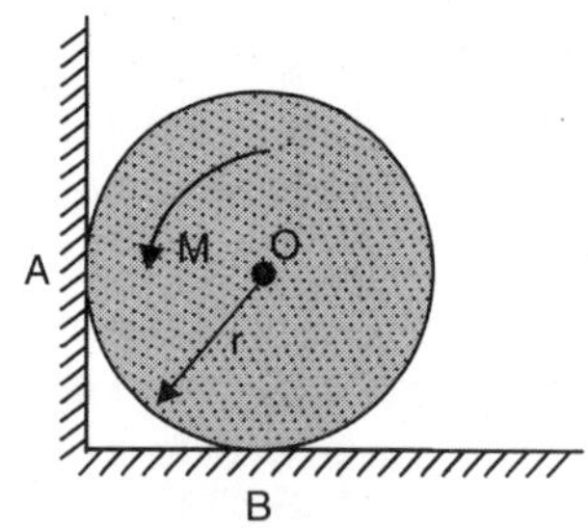

Fig. E8

Example 9

Block A of weight 40 kN is placed on another block B of weight 10 kN as shown in figure. Find the force P applied to the lower block B so that it just raise the block A. Angle of friction for all contact is $\phi = 16°$.

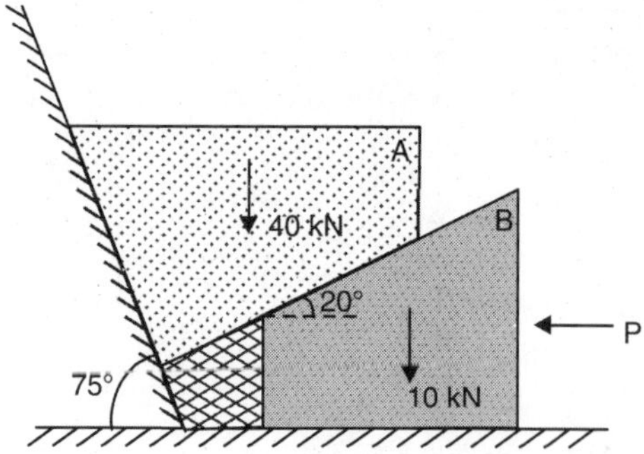

Fig. E9

Example 10

A rope is wrapped around a horizontal bar for 2.5 turns as shown in figure. By exerting a force of 800 N at the free end of the rope, one can just support a load of 60 kN on the other end. Determine the number of times the rope should be wrapped around the bar to support a load of 200 kN with the same force of 800 N applied at the free end.

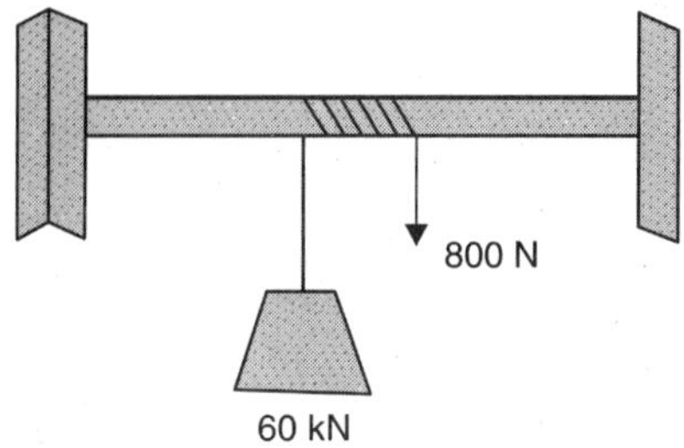

Fig. E10

Example 11

A body of weight W is placed on an inclined plane whose inclination α is greater than λ (angle of friction). Determine the least value of force P which will prevent the body from sliding down the plane.

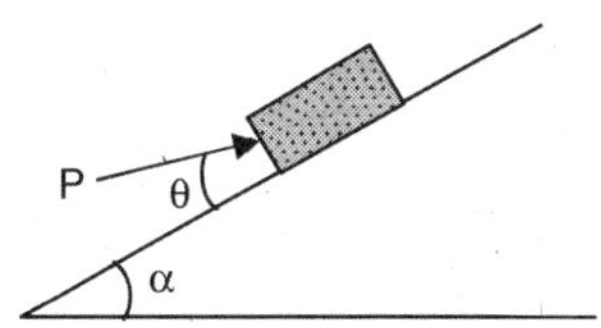

Fig. E11

Example 12

Determine how far 'x' the man of weight 890 N can walk up the 112 N plank without causing it to slip. Coefficient of friction is 0.5 at A and surface at B is smooth. The length of plank is 4.58 m and center of gravity is 0.91 m from plank.

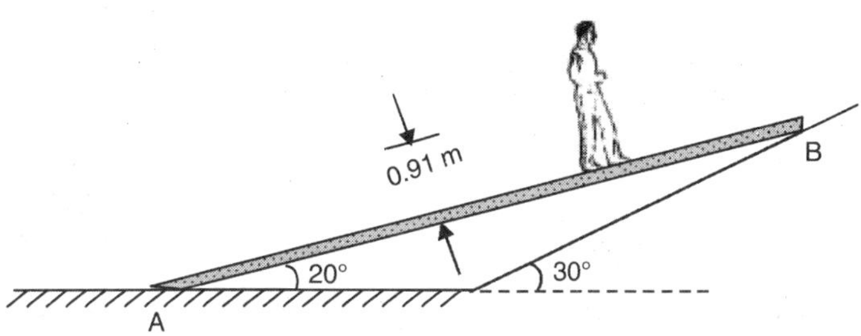

Fig. E12

Example 13

Determine the smallest force F required to raise end B of rod, $\mu_S = 0.4$ between the wedge and the rod and 0.2 between the wedge and the floor. Mass of rod AB is 6 kg.

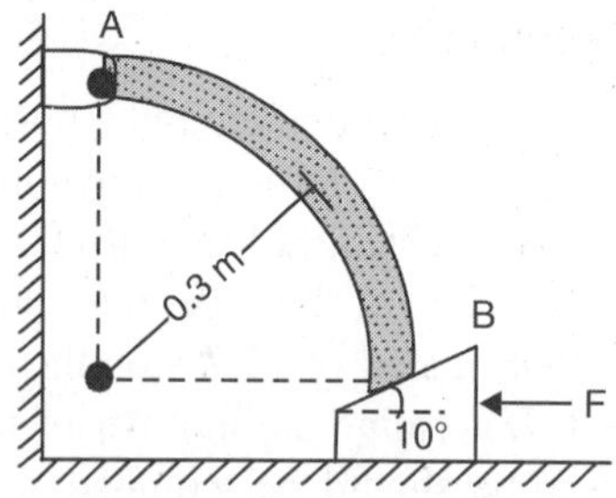

Fig. E13

Example 14

A uniform block of weight W rests on horizontal floor. It is pushed by a force as shown in figure. Determine the condition in P and W, so that

(a) Block does not slide.
(b) Block does not tip.

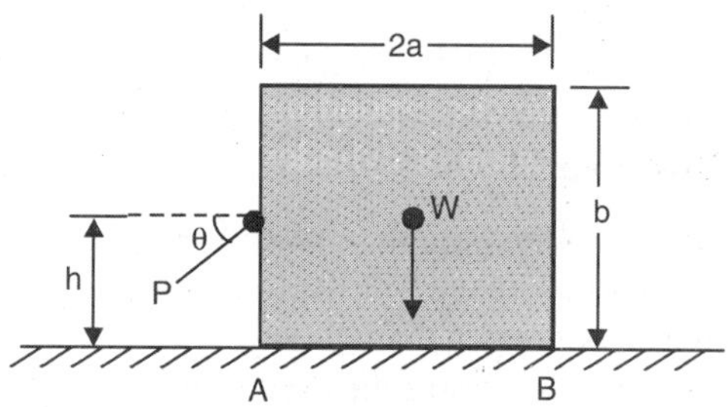

Fig. E14

Example 15

Two boxes of weights w_1 and w_2 are stacked on the floor as shown in figure. μ_1 is the coefficient of friction between the boxes and μ_2 is between the box and the floor. A force F is applied to the top of the boxes and gradually increased. What is the maximum force before the equilibrium destroyed? It is given that $W_1 = W_2 = 100$ N, $\mu_1 = \mu_2 = 0.3$

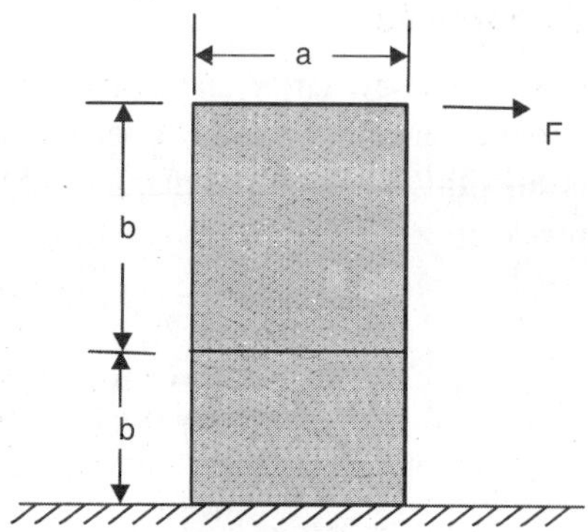

Fig. E15

Example 16

A flat belt connects pulley A which drives a machine tool to pulley B, which is attached to the shaft of an electric motor. The coefficients of friction are

$\mu_s = 0.25$ and $\mu_k = 0.20$ between both pulley and the belt. Maximum allowable tension in the belt is 600 N. Determine the largest torque, which can be exerted by the belt on pulley *A*.

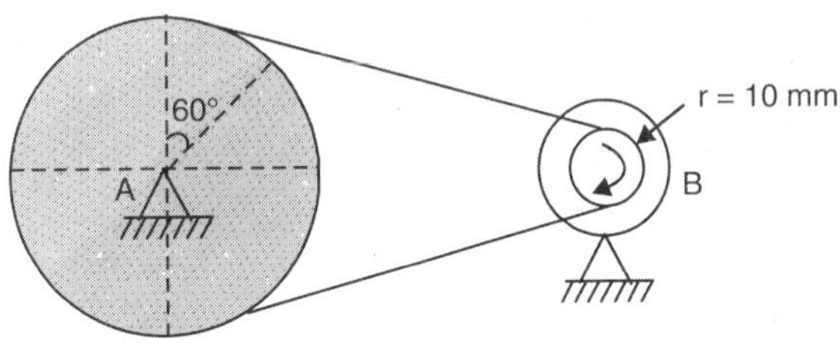

Fig. E16

Example 17

The pipe shown is gripped between two levers that are pinned together at *A*. If coefficient of static friction between the levers and the pipe is $\mu = 0.3$, determine the maximum angle θ at which the pipe can be gripped without slipping. Neglect weight of the pipe.

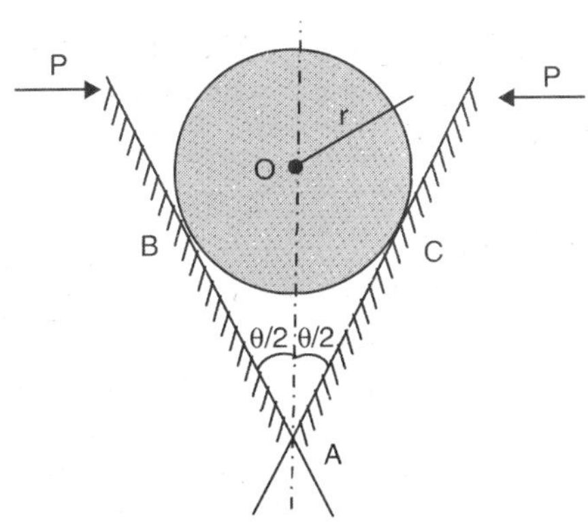

Fig. E17

Example 18

A 60-kg cabinet is mounted on casters, which can be locked to prevent their rotation. The coefficient of friction is 0.30. If $h = 0.8$ m, determine the magnitude of the force *P* required to move the cabinet to the right

(*a*) If all casters are locked.

(*b*) If the casters at *B* are locked and the casters at *A* are free to rotate.

(*c*) If the casters at *A* are locked and casters at *B* are free to rotate.

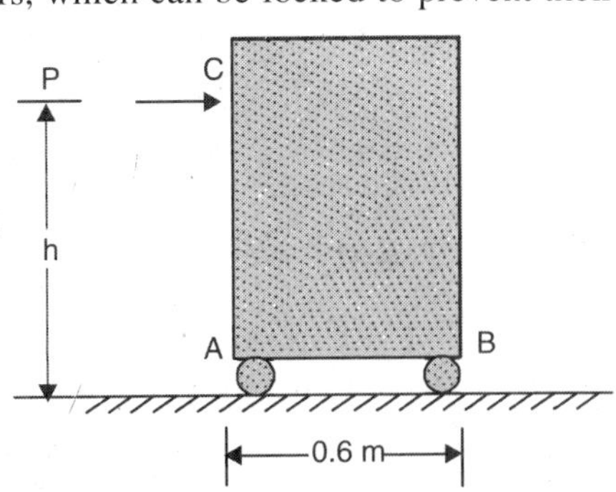

Fig. E18

Level 2

Example 1

The horizontal position of the 500 kg rectangular block of concrete is adjusted by the 5° wedge under the action of the force P. If the coefficient of static friction for both pair of wedges surfaces is 0.30 and if the coefficient of static friction between the block and the horizontal surface is 0.60, determine the least force P required to move the block.

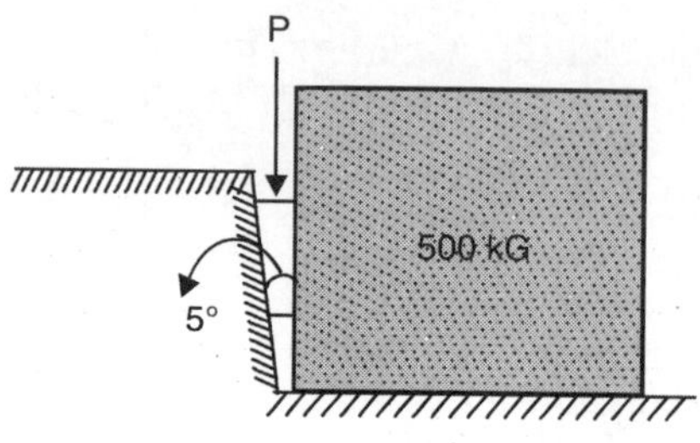

Fig. E1

Example 2

An isosceles triangular plate having weight w is on the verge of slipping. If $\mu = 0.25$ determine the angle θ made by the base of the triangle with the horizontal. Given that the base is half the side.

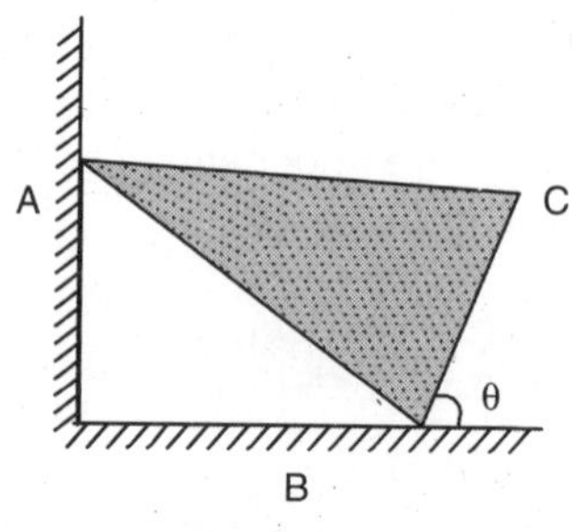

Fig. E2

Example 3

A square threaded pack as shown in figure under the action of the axial load W and a moment M is applied about the axis of the screw. The screw has a lead L (advancement per revolution) and a mean radius r. Determine the value of M so that it will be just sufficient to turn the screw. Angle of friction is ϕ at the screw contact.

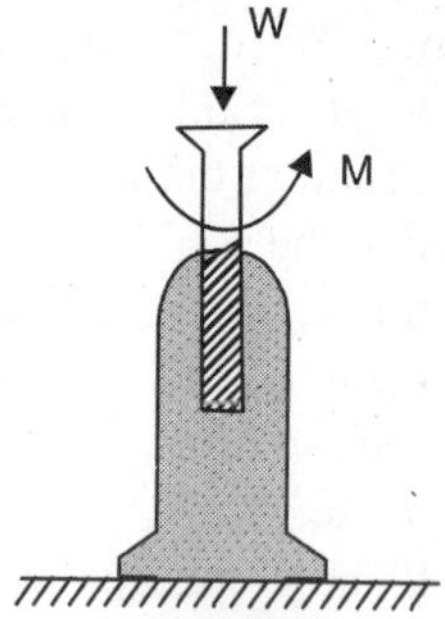

Fig. E3

Example 4

Two blocks *A* and *B* are to be held in position by means if an inextensible rope passing over a fixed drum as shown. If $\mu = 0.2$ between all the surfaces of contact, determine the minimum mass of the block *A* so that *B* is prevented from going downwards (mass of *B* is 500 kg).

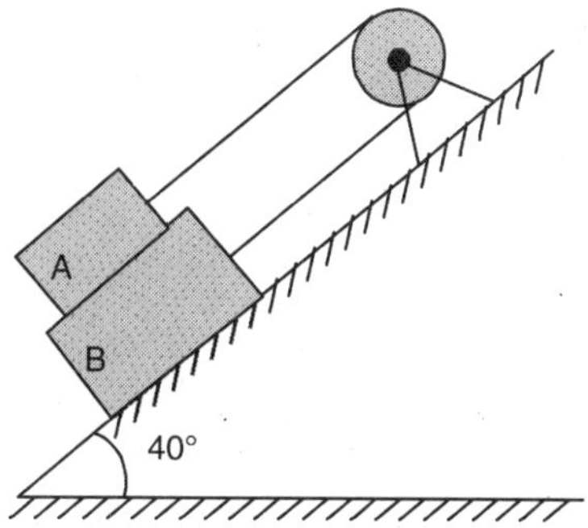

Fig. E4

Example 5

A semicircular plate having radius '*r*' is on the verge of sliding. If $\mu = 0.3$ between all the surfaces of contact determine the angle θ.

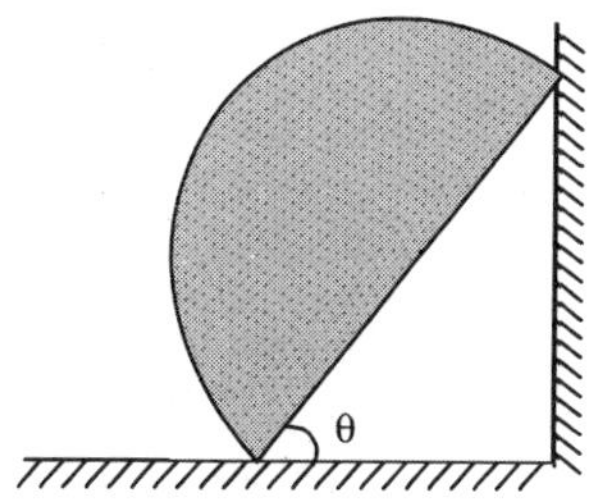

Fig. E5

Example 6

A uniform rod of length *L* and weight *w* rests as shown in figure. If $\mu_s = 0.35$ and $\mu_k = 0.5$, determine minimum tension in the cord and corresponding maximum angle θ for which equilibrium is maintained.

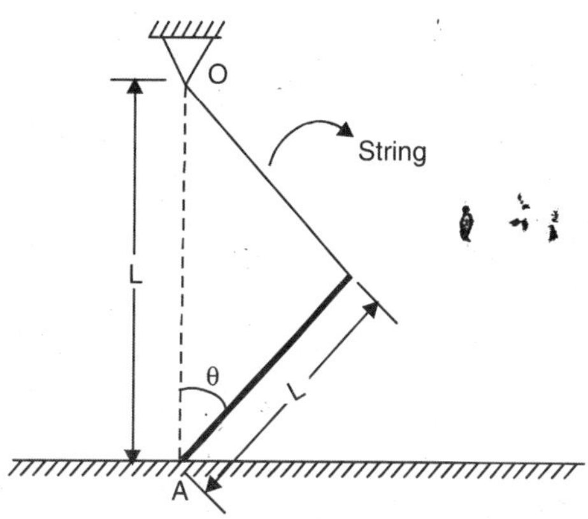

Fig. E6

Example 7

Find the maximum distance x from the horizontal centerline of the drawer at which the force P may be applied and still allow the drawer to the opened without binding the corners. Neglect friction on the bottom of the drawer and take the coefficient of static friction at the corners to be μ_s.

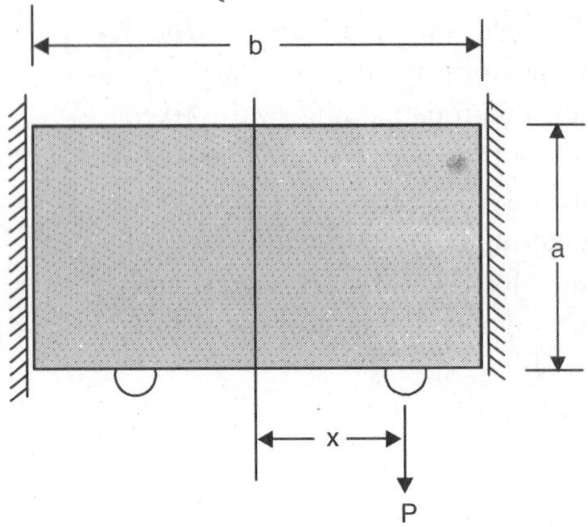

Fig. E7

Example 8

A uniform ladder of length L and weight W leans at an angle θ in a vertical plane against support of height h. The coefficient of static friction is μ_s. Determine the condition in terms of μ_s, L and h is so that ladder will not slip at any angle.

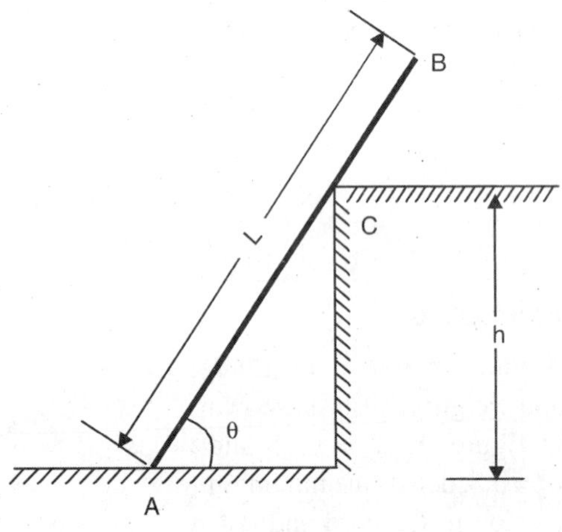

Fig. E8

Example 9

The uniform 400 kg drum is mounted on a line of rollers at A and a line of rollers at B. An 80-kg man moves slowly a distance of 700 mm from the vertical centerline before the drum begins to rotate. All rollers are perfectly free to rotate, except one of them at B which must overcome appreciable friction in its bearing. Calculate the friction force F exerted by that one roller tangent to the drum and find the magnitude R of the force exerted by all rollers at A on the drum for this condition.

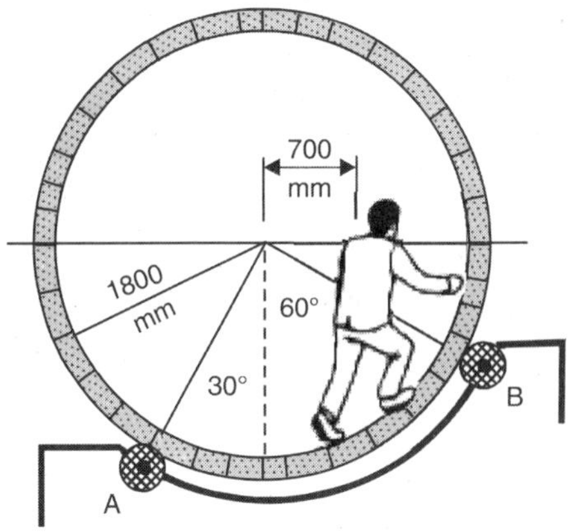

Fig. E9

Example 10

A cord having length 10 m and weight 0.5 N/m is suspended as shown. Find the longest length x without causing motion. Take $\mu_s = 0.5$ (neglect size of pulley).

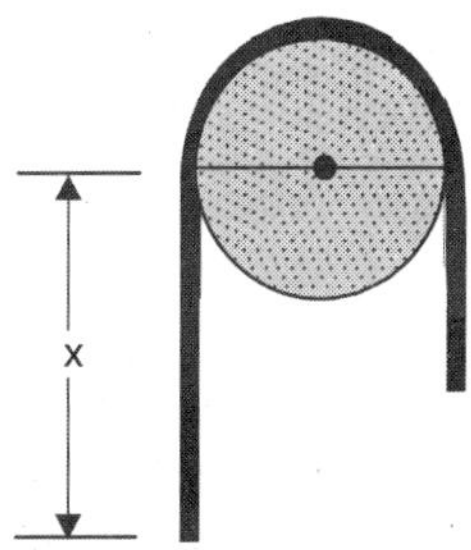

Fig. E10

Example 11

As an arrangement shown in the figure, a belt is used to control the speed of a flywheel. Determine the magnitude of couple being applied to the wheel if $\mu_k = 0.25$ and flywheel is rotating clockwise at a constant speed.

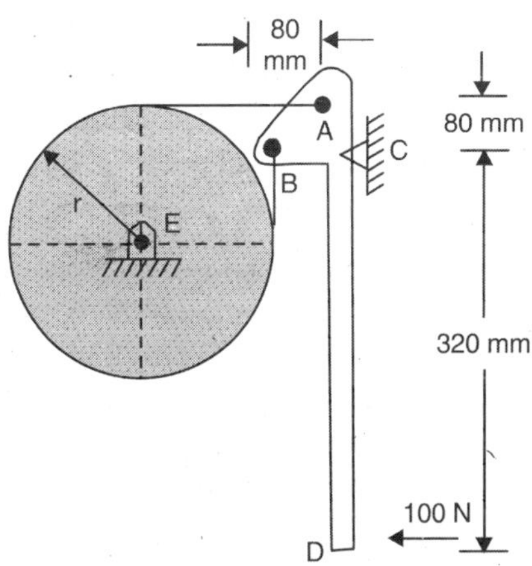

Fig. E11

Example 12

Two 8° wedges of negligible weight are used to move 1000 kg block. Determine the smallest force *P* which should be applied on *B* to move the block $\mu_s = 0.25$ at all surfaces of contact.

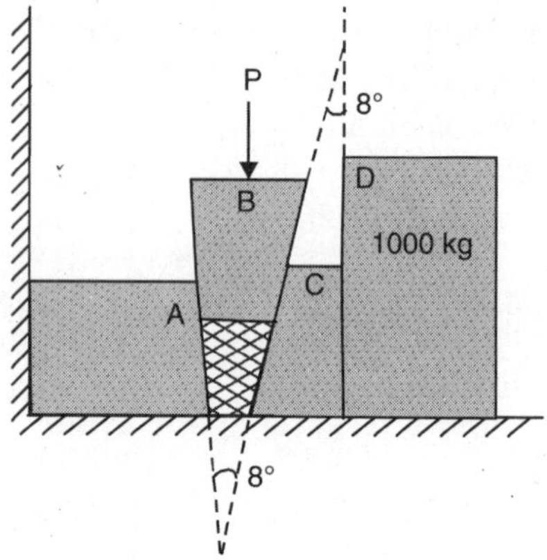

Fig. E12

Example 13

As an arrangement shown in figure, a cable is placed around 3 pipes. Determine

(*a*) The smallest mass for which equilibrium is maintained.

(*b*) The largest weight *W* which can be raised if pipe *B* is slowly rotated counterclock-wise while *A* and *C* remain fixed. It is given that $\mu_s = 0.25$ and $\mu_k = 0.2$.

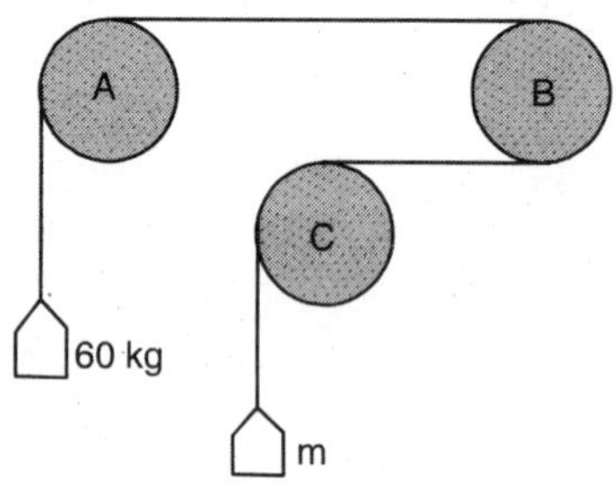

Fig. E13

Example 14

A cylinder of radius *R*, weight *W* is wedged between a vertical wall and a light hinged bar as shown in figure. What is the force *P* which when applied at the end of the bar at right angles to it, is just sufficient to cause the cylinder to move upwards the coefficient of friction between the cylinder and wall is 0.2 and that between the bar and cylinder is 0.4.

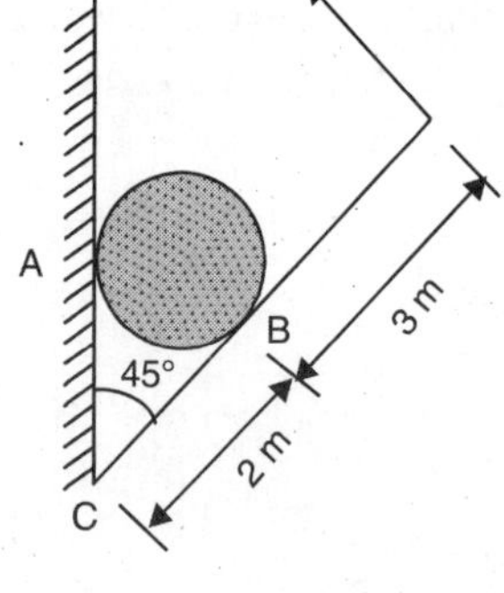

Fig. E14

Example 15

A cylinder rests on a horizontal plane. A plank lies across the cylinder with one end resting on the plane making an angle α with the horizontal. If the plane is just about to slide on the cylinder, determine the angle of friction between the plank and cylinder.

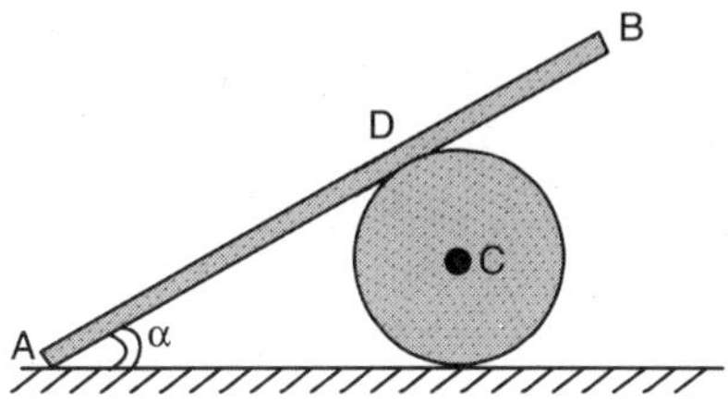

Fig. E15

Level 3

Example 1

A uniform heavy rod AB lying on a rough table has a force applied at the end A which is gradually increased until the rod is just about to move in such a way that the initial motion will be rotation about a point C in AB. Determine the distance AC and the least value of P at A that will just move the rod.

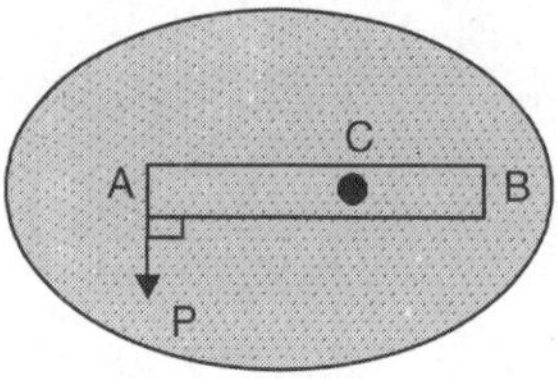

Fig. E1

Example 2

A half section of a pipe of weight 400 N is to be shifted to the right along the floor without overturning. If $\mu = 0.35$ find (*a*) largest allowable value of angle α and (*b*) corresponding tension T.

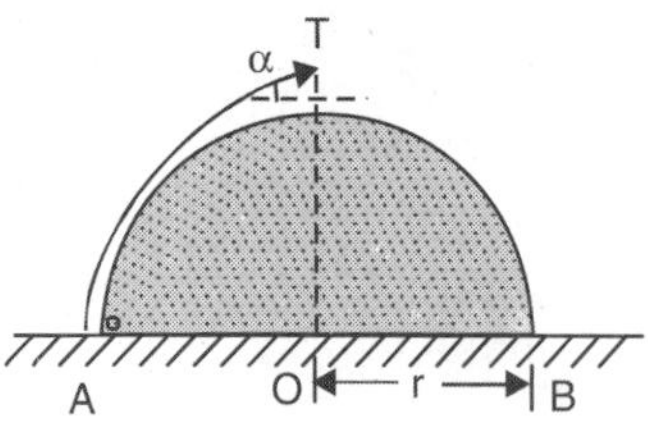

Fig. E2

Example 3

A uniform triangular lamina rests on a rough horizontal plane, supported upon short pegs of equal length at its corners. Coefficient of friction between peg and lamina is μ and weight of lamina is W. Determine the least couple that will cause the lamina to turn in its plane.

Example 4

A heavy circular hoop is hung over a rough peg. A weight equal to that of the hoop is attached to it at a given point. Find the coefficient of friction between the peg and the hoop so that system may hang in equilibrium whatever point of the hoop is placed in contact with the peg.

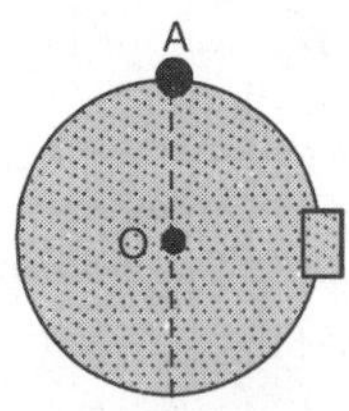

Fig. E4

Example 5

A uniform rod with its ends resting an two equally rough inclined plane having inclination α and β respectively in a vertical plane at right angle to the line of intersection of the inclined planes. Determine the angle formed by the rod AB with the vertical.

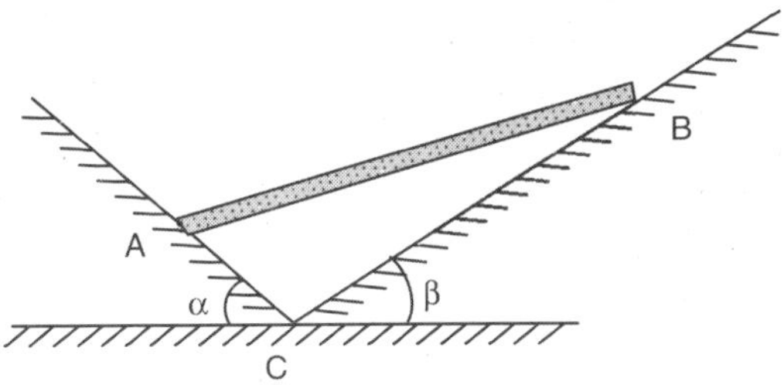

Fig. E5

Example 6

Assume vertical wall as smooth. Determine the smallest positive angle θ for which the ladder of length L and weight W can remain at rest (angle of friction at A is 26.56°).

(*i*) By using equilibrium equations

(*ii*) Without using equilibrium equations

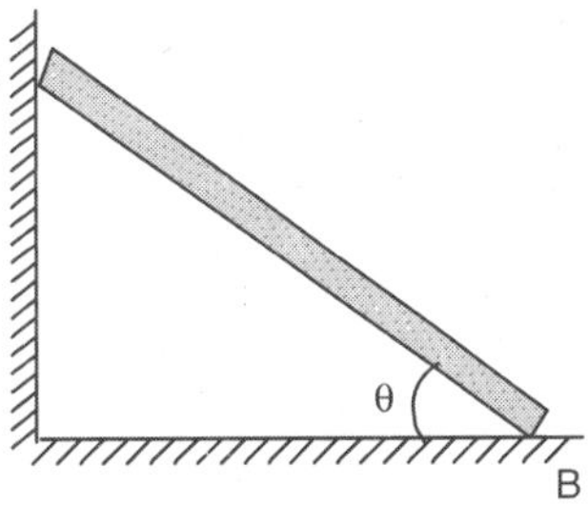

Fig. E6

Example 7

A uniform circular cylindrical log of radius a and weight W lies with its axis horizontal between two rough parallel horizontal rails at the same level. Determine the couple value required to the log in a plane perpendicular to the rails and axis so that

(*i*) Log will turn over one of the rails

(*ii*) Log will turn about the axis.

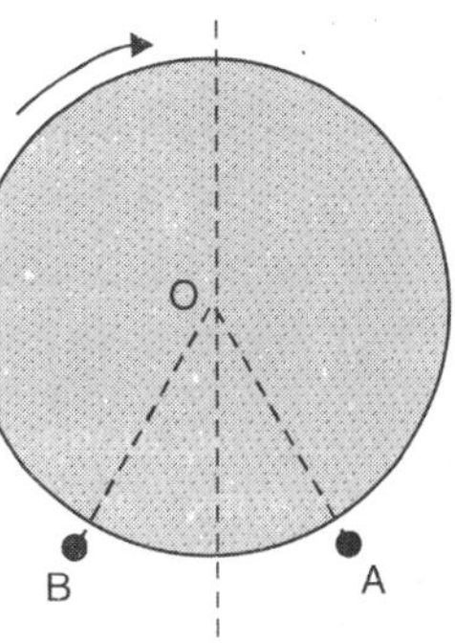

Fig. E7

Example 8

A hoop of mass M hangs from a rough peg in a vertical plane. An insect of mass μ starts from the lowest point of the hoop and crawls slowly upwards. Determine the condition so that insect can reach the peg. λ is the angle of friction between the hoop and the peg.

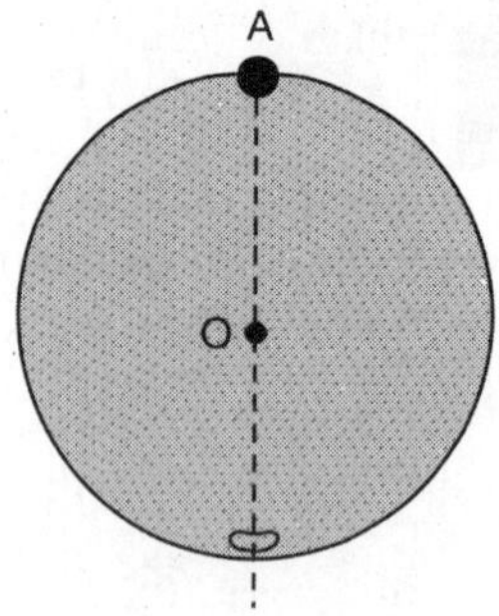

Fig. E8

Example 9

The uniform rod with center of mass at G is supported by the A and B, which are fixed in the wheel. If the co-efficient of friction between the rod and pegs is μ, determine the angle θ through which the wheel may be slowly turned about is horizontal axis, starting from the position shown, before the rod begins to slip. Neglect the diameter of the rod compared with the other dimensions points.

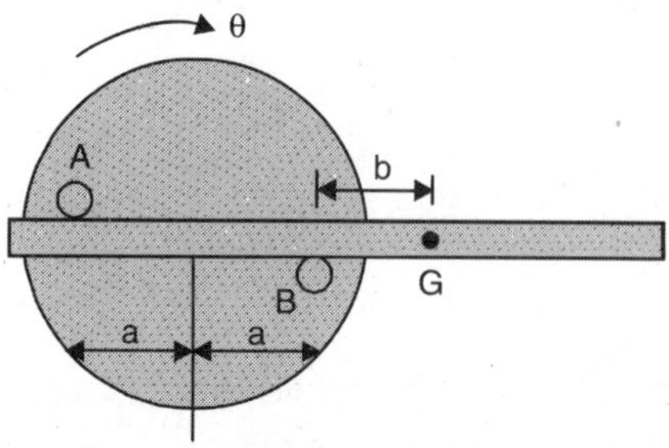

Fig. E9

3 Analysis of Structures

Summary and Concept

Equilibrium Conditions

$$\begin{aligned} \sum F_x &= 0 \\ \sum F_y &= 0 \\ \sum M_0 &= 0 \end{aligned}$$

Cable

(*i*) If number of unknown forces are '*n*' then consider $(n-3)$ parts of cable for finding the unknown quantities.

(*ii*) Consider part or parts of the cable, starting from any support up to the point where sag in known.

(*iii*) The tension is maximum in that portion of the cable, which has largest angle of inclination with the horizontal. Such part or portion carring maximum tension must be adjacent to one of the supports.

(*iv*) To find out sag of any point, consider part of the cable starting from support and upto the point where sag is required but with known reaction components use $\sum M = 0$.

(*v*) Use Lami's theorem whenever required.

Frame

(*i*) For simple frame (without pulley) use the conditions of equilibrium after F.B.D and then break it through pinned point and again apply the condition of equilibrium.

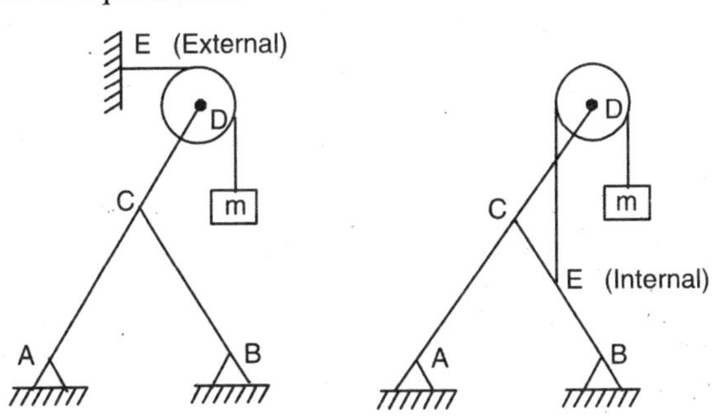

Fig. 1

(*ii*) If a string attached with a load at one end is passing over a smooth pulley and connected with any member of the frame, the tension is considered as internal, so while drawing F.B.D give one end tension where load is acted. The tension is considered as external when the end of the string is connected with the surroundings so while drawing F.B.D give both end tension.

(*iii*) If a string is connected between parts of the frame then tension develop in the string and if a rod is connected between parts of frame then compression develop in the rod.

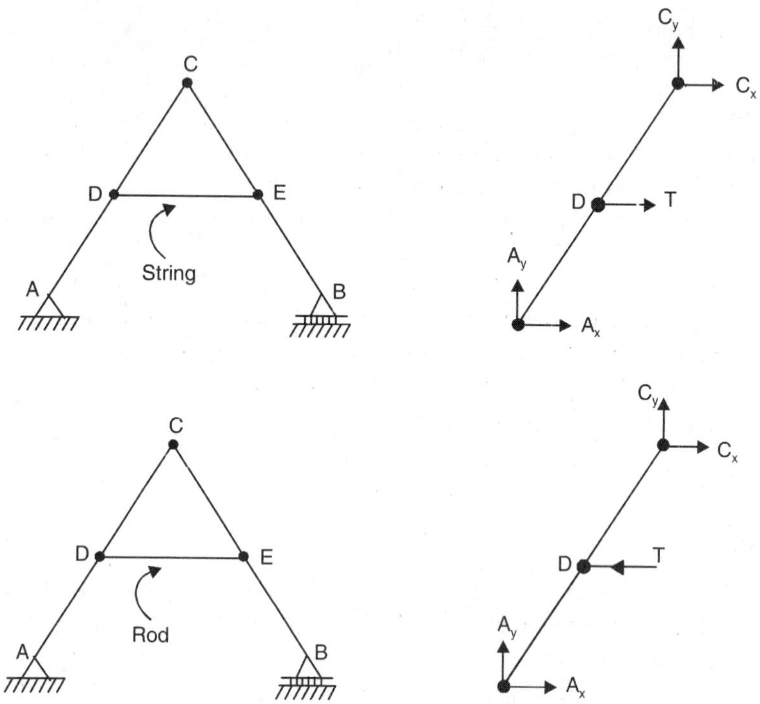

Fig. 2

Beam

(*i*) There are several types of beam in nature. Few important ones are

(*a*) Fixed beam

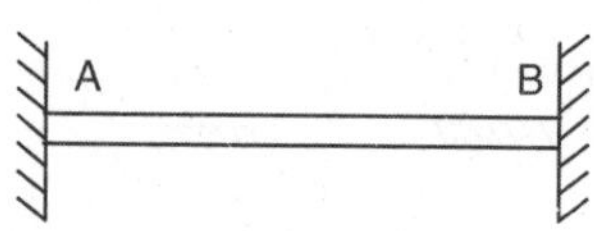

Fig. 3

(*b*) Continuous beam

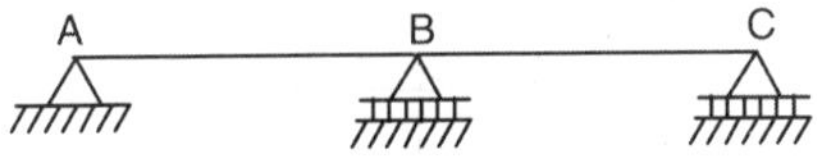

Fig. 4

(*c*) Cantilever beam

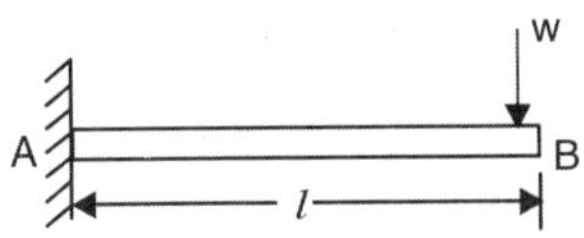

Fig. 5

(*ii*) Types of loads of the beam

(*a*) Point load

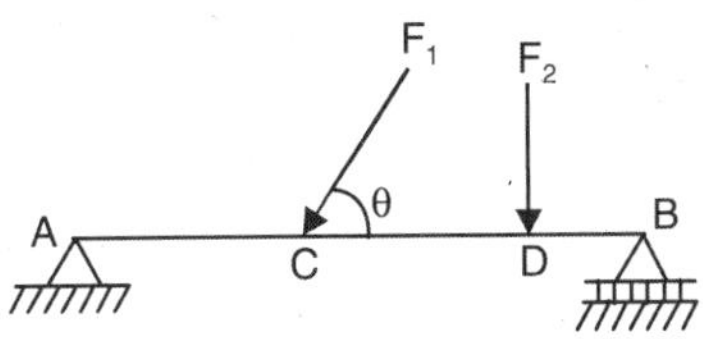

Fig. 6

(*b*) Uniformly distributed load

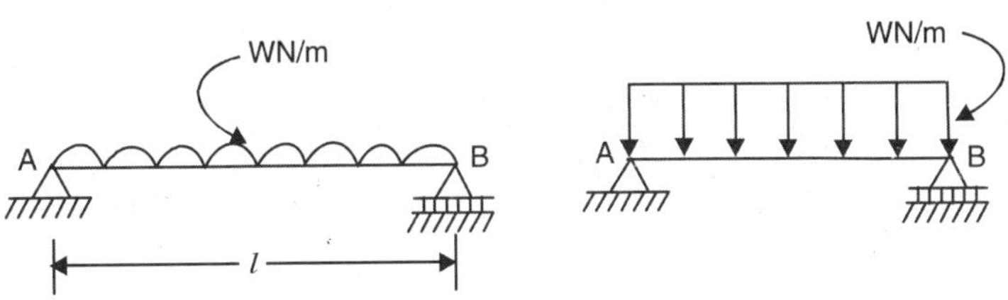

Fig. 7

(*c*) Uniformly varying load

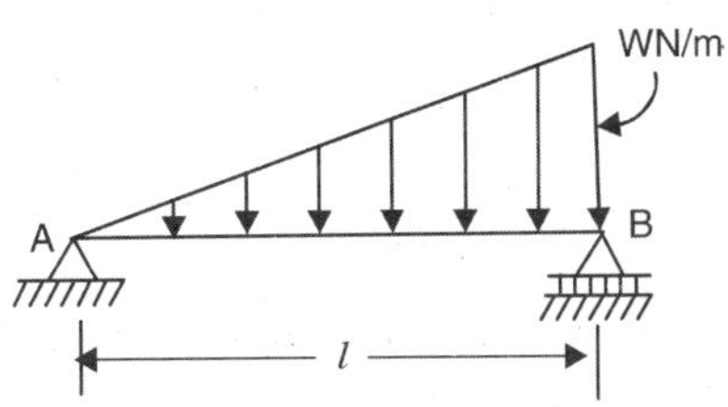

Fig. 8

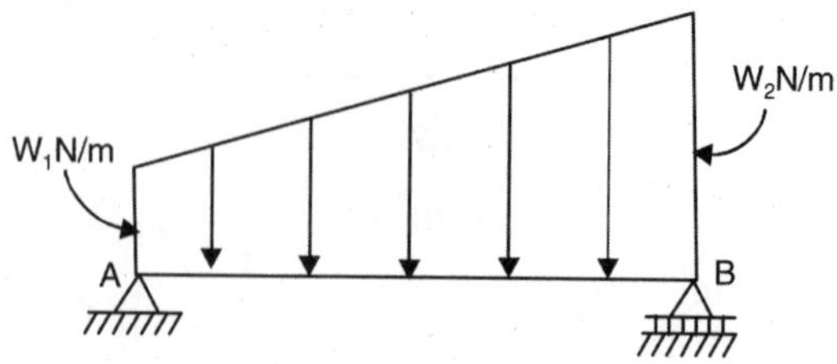

Fig. 9

(*d*) Varrying load

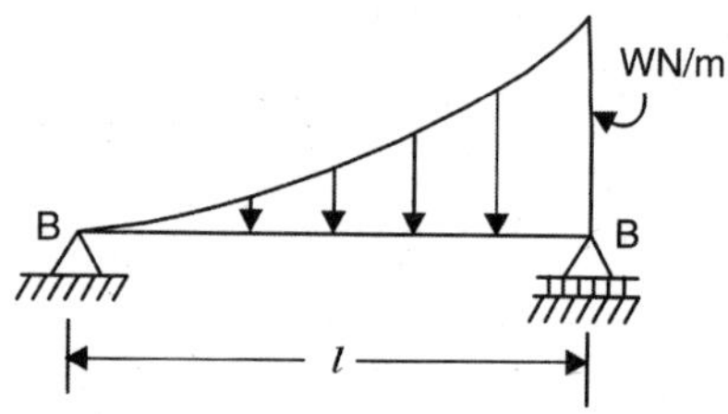

Fig. 10

(*e*) Compound beam

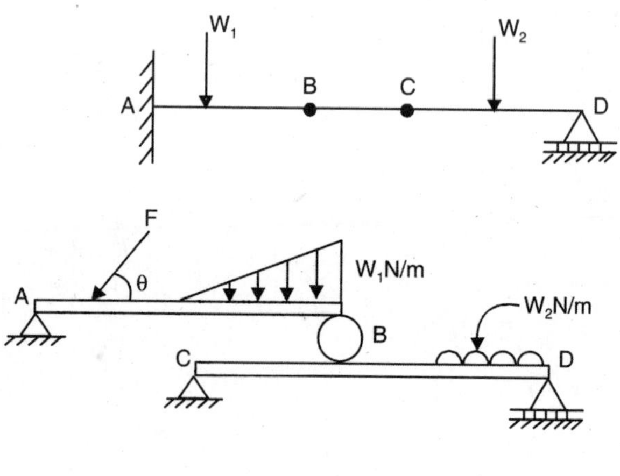

Fig. 11

(*iii*) For every problem apply conditions of equilibrium and if necessary apply conditions of equilibrium on the parts of beam using pinned point concept.

Truss

(*i*) Rigid structure is composed of number of straight members pin jointed to each other.

(*ii*) Apply conditions of equilibrium.

(*iii*) Break into number of small trusses and again apply the same concept (condition of equilibrium).

(*iv*) Force away from the point give tension and towards the point give compression.

Stable truss

A truss which does not collapse under the loading is called the stable truss.

If j = number of joints

n = Number of members

For stable truss

$$n = 2j - 3$$

For constable truss

$$n \neq 2j - 3$$

Method of sections

The concept if used for finding the unknown forces in members that belongs either left or right part of an imaginary line.

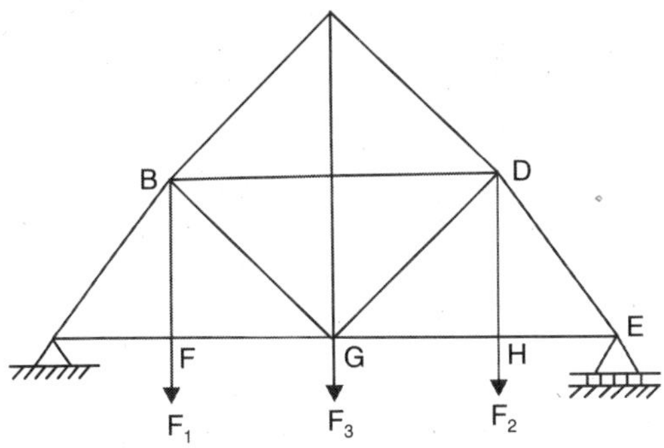

Fig. 12

For finding the forces in members *CD* and *DG* we can break the truss in small part through the imaginary line $a - b$ and then we can concentrate only right part of the truss.

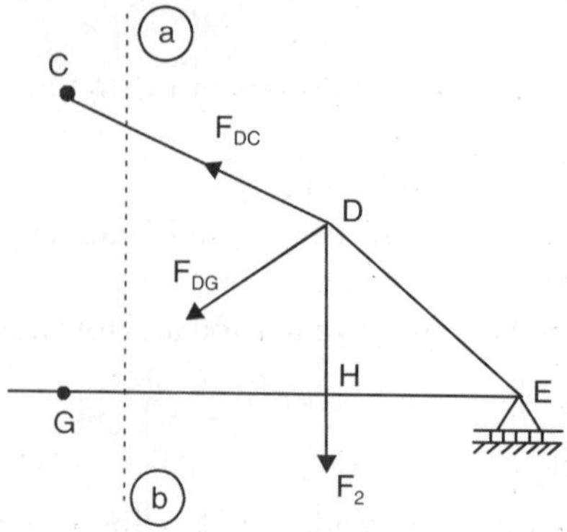

Fig. 13

Relations between load and shear

$$\frac{dV}{dx} = -W$$

Where dV represents change in shear force with respect to small increment in x, i.e. dx. The numerical value of the slope at an point is equal to the load per unit length at that point.

We can write the above equation in the integration form between two point 1 and 2 as

$$V_2 - V_1 = -\int_{x_1}^{x_2} W\,dx$$

Or $V_2 - V_1$ = – (Area under load curve between 1 and 2).

Relations between shear and bending moment

$$\frac{dM}{dx} = V$$

Where dM represents change in bending moment for the small increment in x, i.e. dx.

We can write the above equation in the integration form between two points 1 and 2 as

$$M_2 - M_1 = \int_{x_1}^{x_2} V\,dx$$

Or $M_2 - M_1$ = Area under shear curve between 1 and 2.

Level 1

Example 1

X frame is loaded and supported as shown in figure. Find the horizontal and vertical components of the reaction at *A* and *C*.

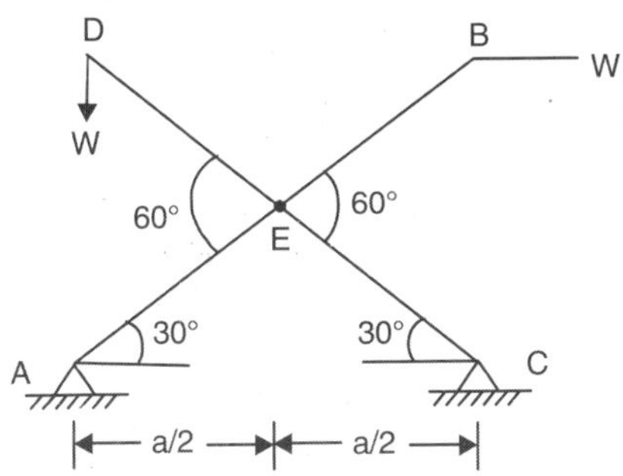

Fig. E1

Example 2

Frame shown in figure carries horizontal load of 2 kN. It is hinged at *E* and *F*. Neglecting friction calculate.

(*i*) Reactions at *E* and *F*.

(*ii*) Forces on the pins at *A* and *D*

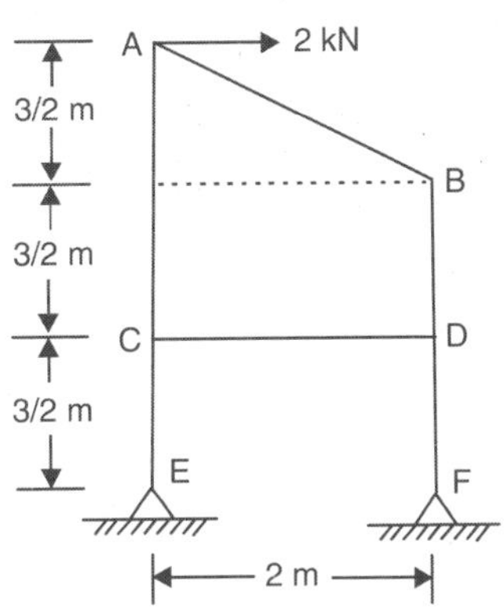

Fig. E2

Example 3

Calculate support reactions of *A* and *D* in the frame shown in the figure.

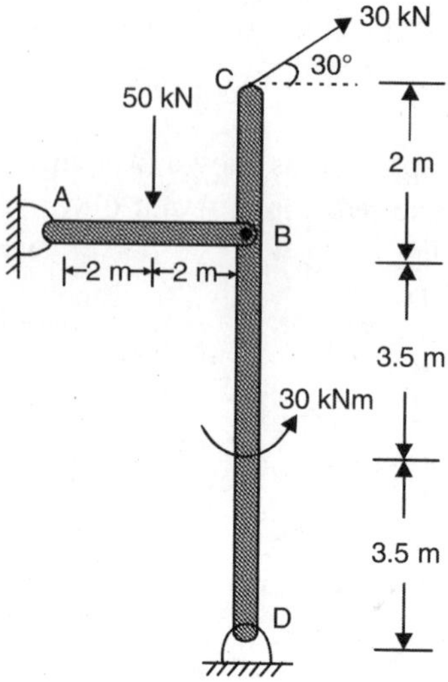

Fig. E3

Example 4

Find the reactions of the beam shown in figure.

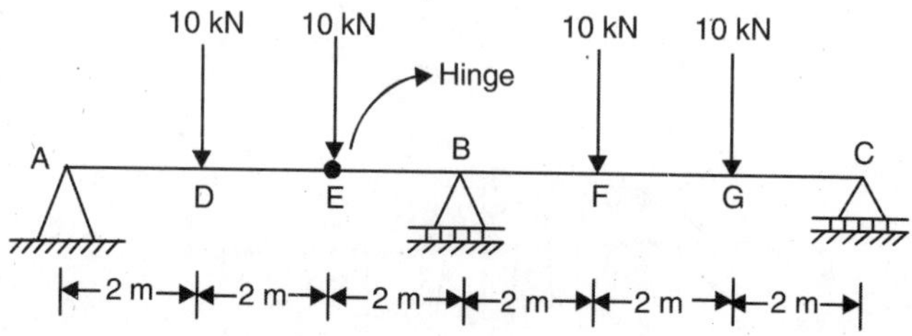

Fig. E4

Example 5

Determine the force in each member of truss shown. Also state whether each member is in tension or compression.

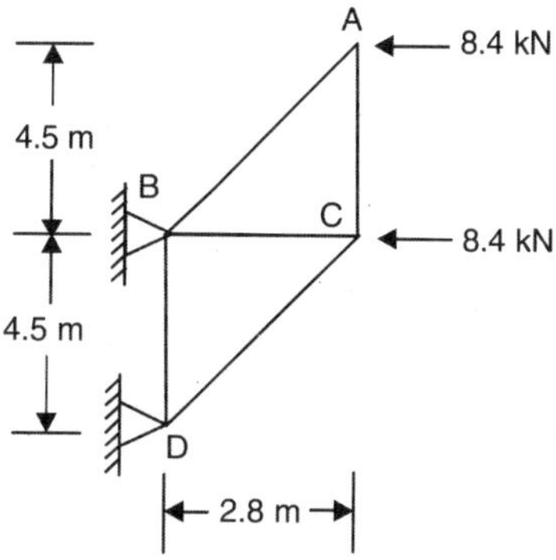

Fig. E5

Example 6

The structure in figure is loaded by 320 Nm counter clockwise couple applied to the member *AB*. Member *AB* and *BCD* are connected by pin at *B*. Neglecting the weights of the members, find support reaction at *A* and *D* and force at pin *B*.

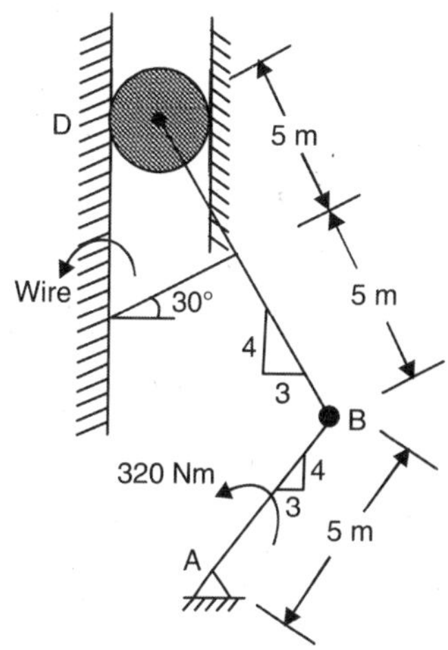

Fig. E6

Example 7

A slender rod *AB*, of weight *w* is attached to blocks *A* and *B* which move freely in the guides shown. The blocks as shown in figure are connected by an elastic cord that passes over a pulley at *C*. Express the tension in the cord in terms of *W* and θ. Determine the value of θ for which the tension in the cord is equal to 3*W*.

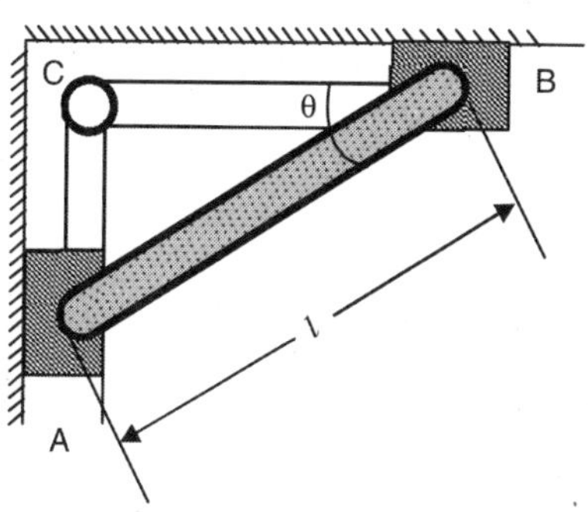

Fig. E7

Example 8

Analyze the truss completely as shown in figure. Obtain reactive components at A and B.

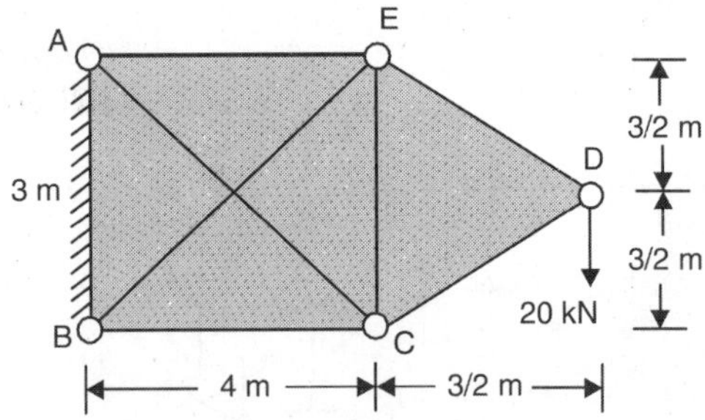

Fig. E8

Example 9

A truss is loaded and supported as shown in figure. If force in the member CE is 30 kN tensile then determine W. Hence obtain the force in the member AC.

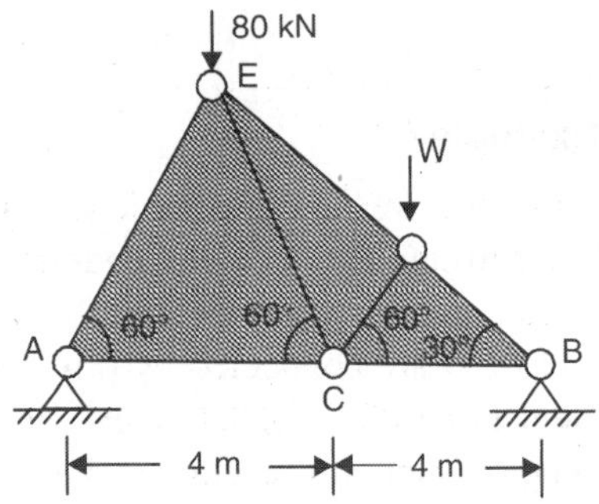

Fig. E9

Example 10

For the frame shown in figure, determine the components of all forces acting on members.

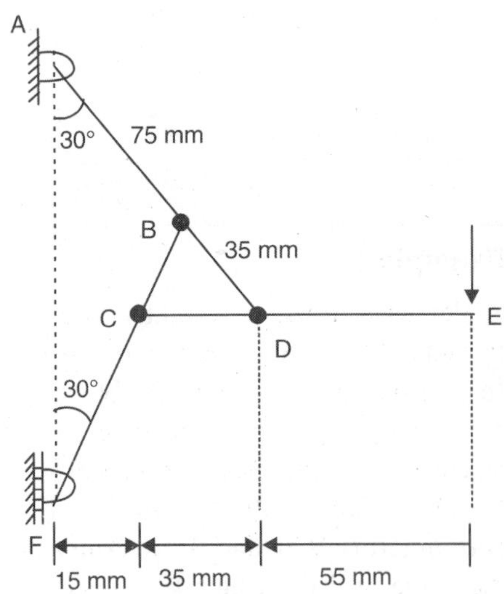

Fig. E10

Example 11

For the paper punch shown find the punching force Q corresponding to a hand grip P.

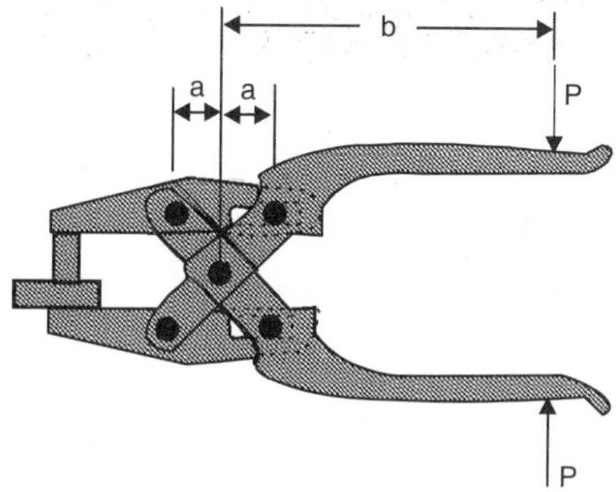

Fig. E11

Example 12

Determine the force in member AC of the loaded truss. The two quarter-circular members act as two force members.

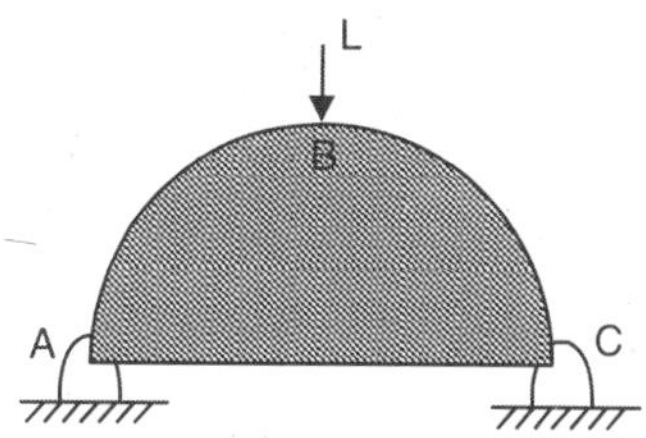

Fig. E12

Example 13

A wire of uniform weight 4 N/m is bent into a semi circular arc of radius 350 mm and three lines pieces each of length 350 mm. It is pin connected at A as shown in figure. Find the value of the horizontal force 'P' applied at D to keep the wire structure in equilibrium. Take AD vertical.

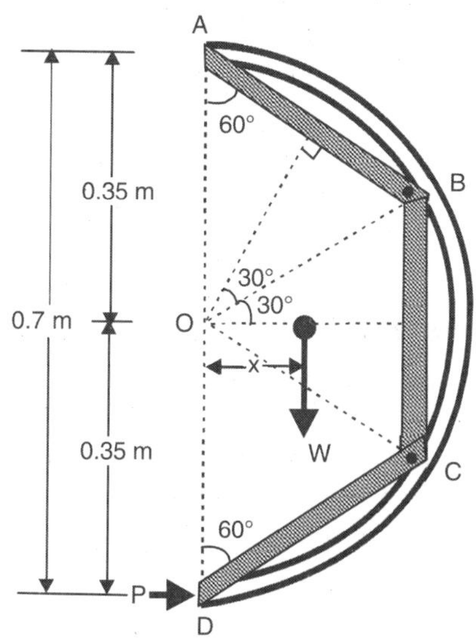

Fig. E13

Level 2

Example 1

Calculate the reactions of the beam loaded as shown in figure.

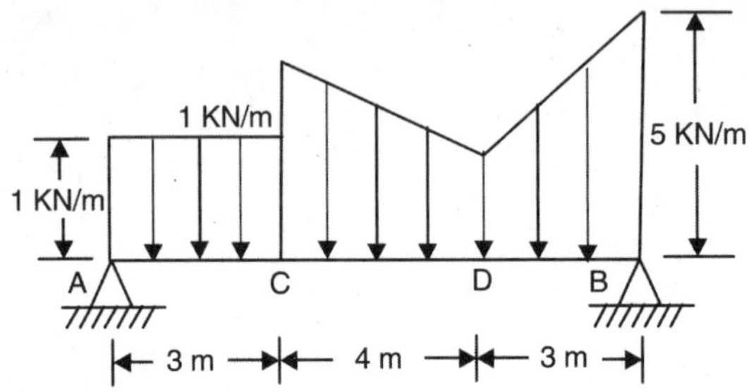

Fig. E1

Example 2

A beam supports a load distributed parabolic over its length. Determine the resultant of this distributed load and line of action. And hence find reaction at any one support.

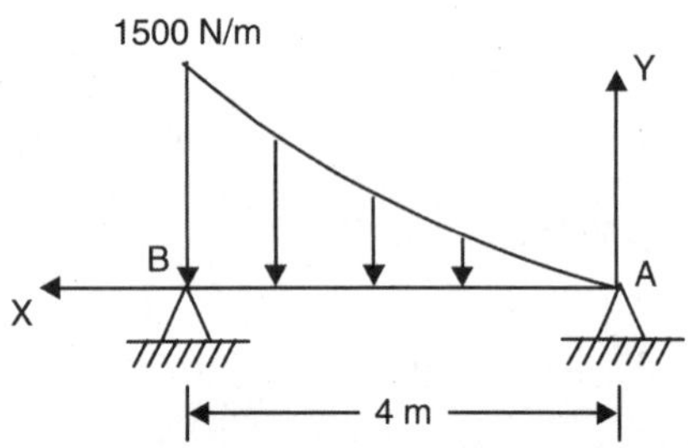

Fig. E2

Example 3

An oil pipeline is supported at 6 m intervals by vertical hangers attached to the cable shown in figure. Due to the weight of the pipe and its contents, the tension in each hanger is 4 kN. Determine

(*a*) Maximum tension in the cable and

(*b*) The height h.

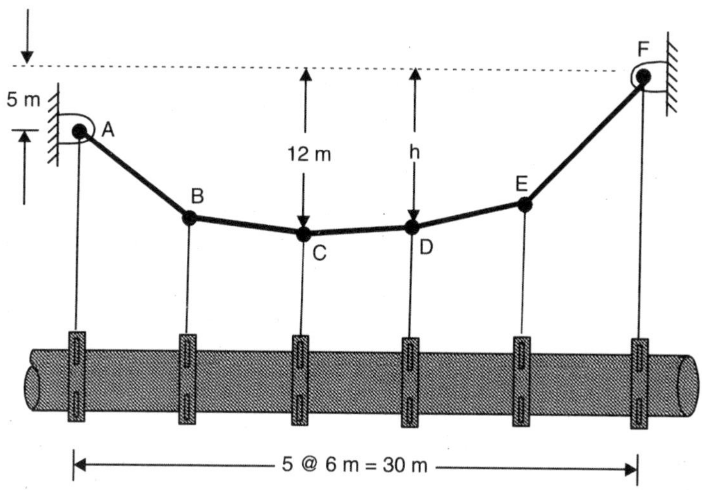

Fig. E3

Example 4

Determine the force in each of the members located to the right of DE for the inverted house roof truss as shown in figure.

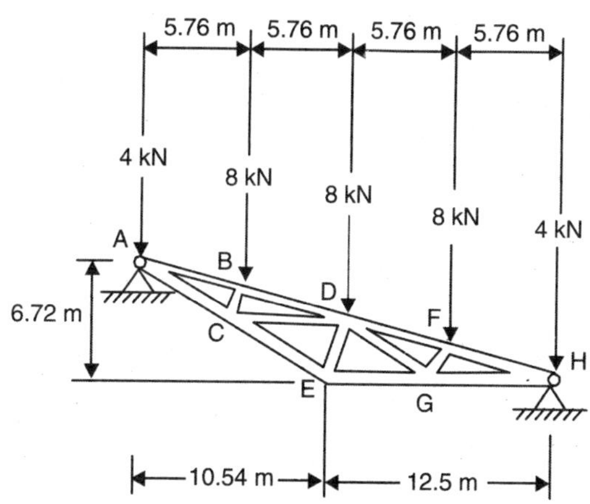

Fig. E4

Example 5

A folding camp-stood rests upon a smooth horizontal floor and is loaded as shown in figure. Determine the magnitude of the shear force on the pin at E and the position of the load P on the bar CD to make this force a maximum.

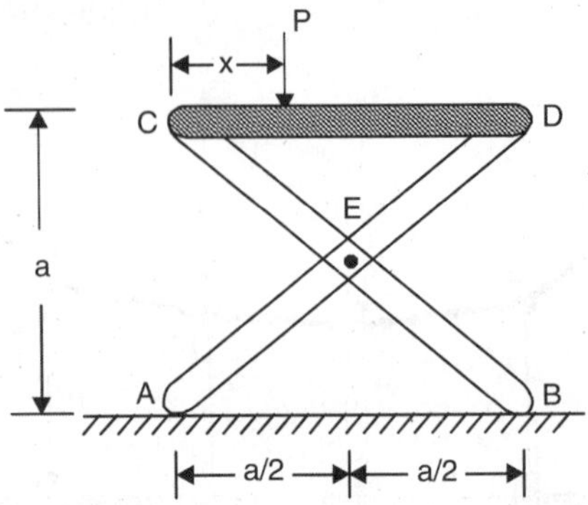

Fig. E5

Example 6

A dock crane running on a rail track shoes gauge is 3 m. Its lifting capacity is 400 kN at the hook hanging from its tip 10 m from the waterside rail. The dead weight of the crane is 200 kN and its acts a distance of 2 m from the same rail as shown in figure. What counter weight Q will be required and where should it be located so that the crane will be just stable when carrying the load and also when not doing so?

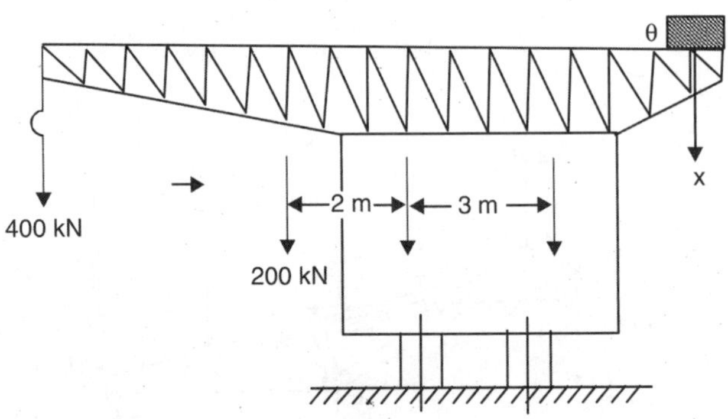

Fig. E6

Example 7

Determine the force in members FH, GI of the roof truss shown in figure.

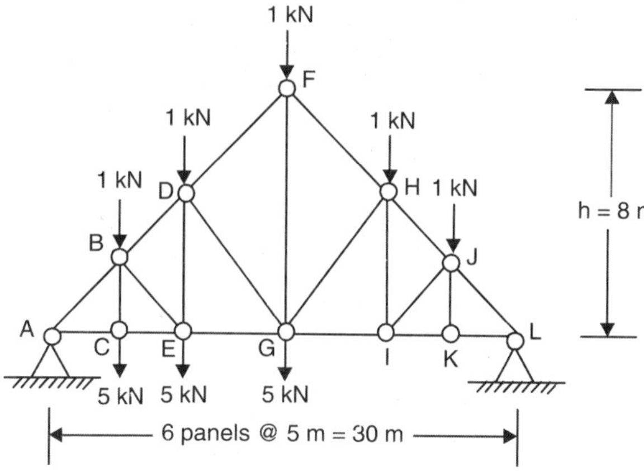

Fig. E7

Example 8

Calculate the horizontal and vertical components of reactions at *A* and *B* for the frame loaded and supported as shown in figure.

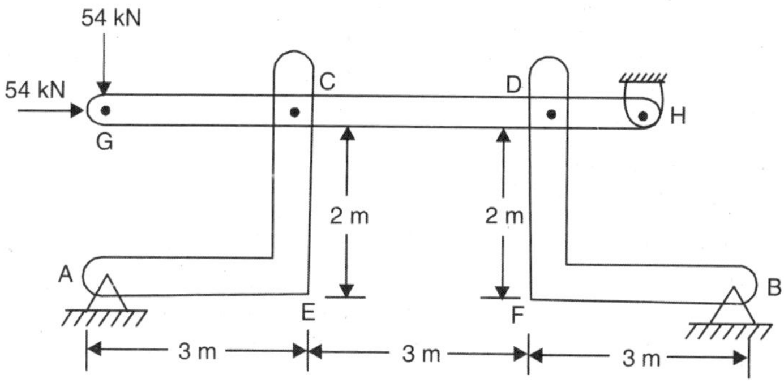

Fig. E8

Example 9

A framed structure shown in figure supports a load of 1 kN. Find the force in the bar BC and the shear force on the pin at *D*. The radius of the pulley is 0.25 m.

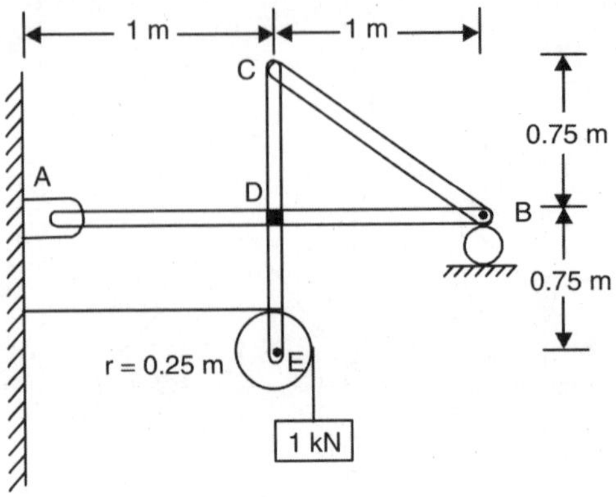

Fig. E9

Example 10

A carpenter builds the square frame *ABCD* and then braces it with member *EF* as shown in order to prevent racking (distortion into a rhombic shape) under the applied force *P*. Determine the tension *T* in the brace as a function of *x*. Take all connections to be pin joints. Member *DC* is firmly fastened to the floor.

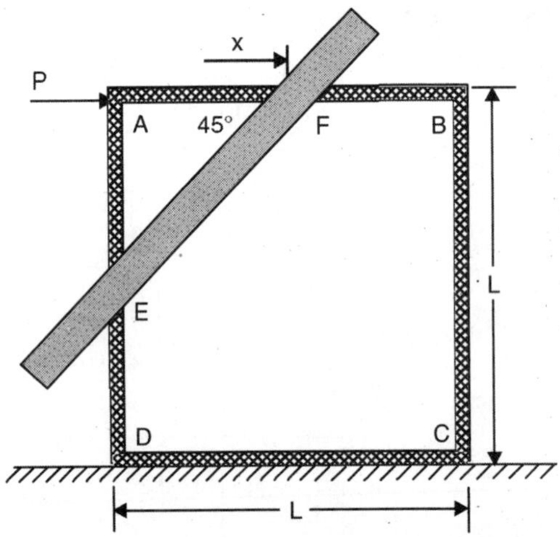

Fig. E10

Example 11

Draw the shear and bending moment diagram for the beam and loading shown.

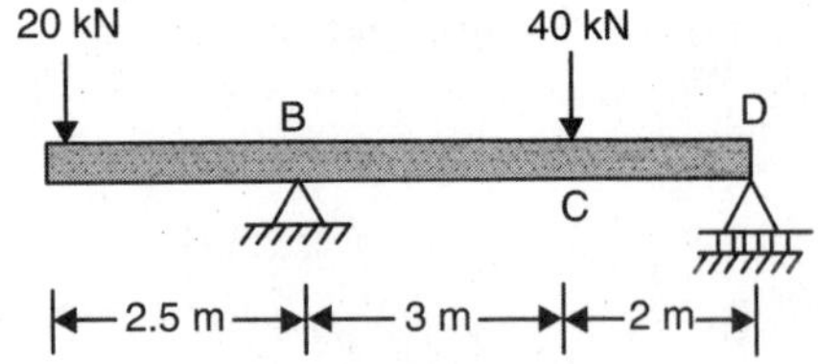

Fig. E11

Example 12

Find reactions at A, B and C for given compound beam.

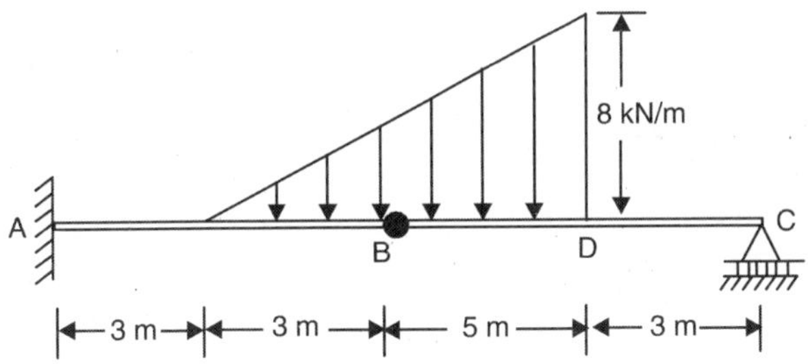

Fig. E12

Example 13

Draw the shear and bending-moment diagram for the beam *AB*. The distributed load 40 N/m extends over 12 m of the beam from *A* to *C* and the 400 N load is applied at *E*.

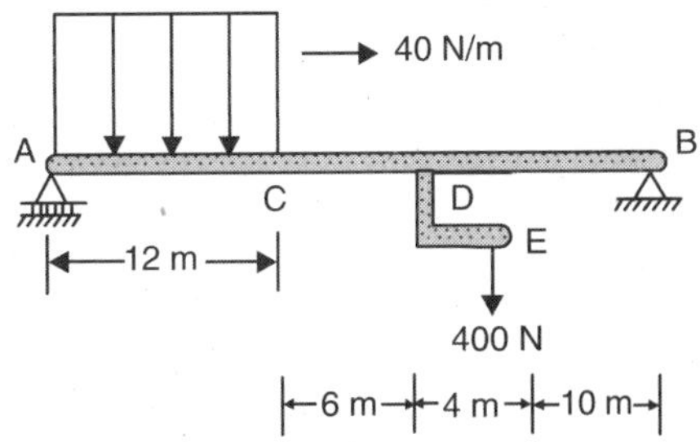

Fig. E13

Example 14

The uniform quarter circular member of mass *m* lies in the vertical plane and is hinged at *A* and supported against the vertical wall by its roller at *B*. For any section *S*, write expression for the shear force F_S, compression F_C, and bending moment *M* due to the weight of the member.

Example 15

A small bolt cutter operated by hand for cutting small bolts and rods is shown in the sketch for a hand grip $P = 150$ N, determine the force *Q* developed by each jaw on the rod to be cut. All dimensions in mm.

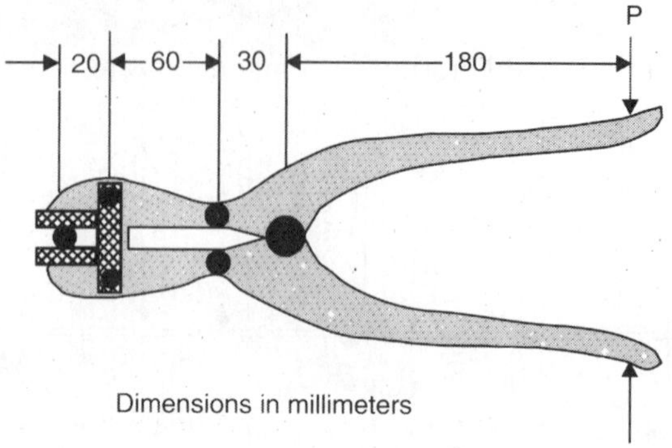

Fig. E15

Example 16

A 250-N force is applied to the foot operated air pump. The return spring *S* exerts a 3-N m moment on member *OBA* for this position. Determine the corresponding compression force in the cylinder *BD*. All dimensions in mm.

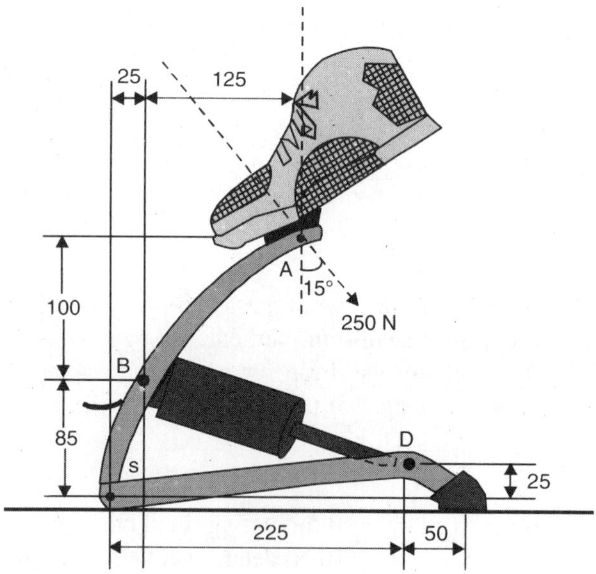

Fig. E16

Level 3

Example 1

Compound lever snips, shown in the figure are designed to replace regular tinners snips when large cutting forces are required. For the gripping force of 150 N what is the cutting force P at a distance of 30 mm along the blade from the pin at *A*?

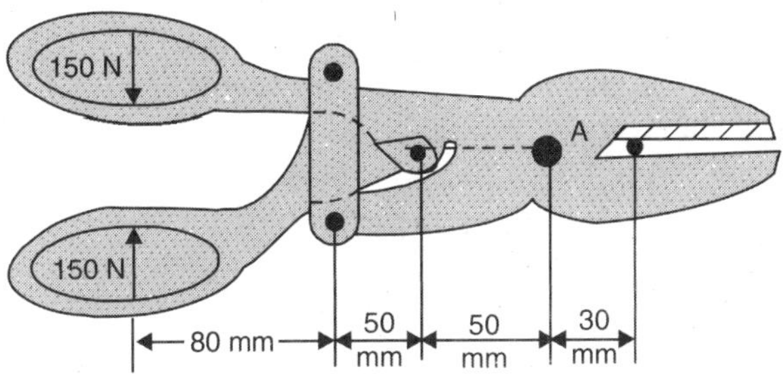

Fig. E1

Example 2

Determine the reaction component at *A*, *B*, *D*, *E* and *F*. Also find the value of load *W*.

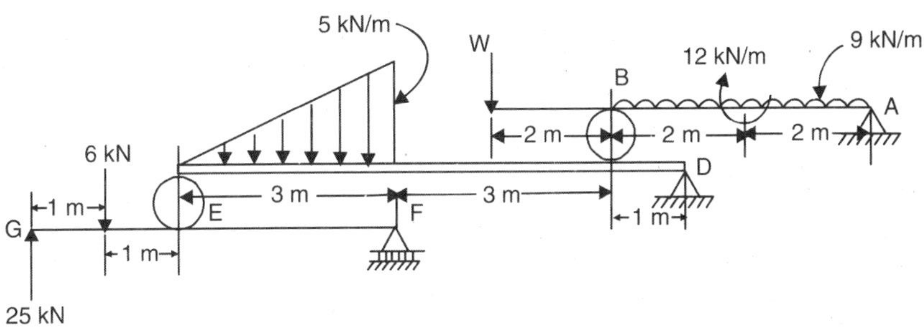

Fig. E2

Example 3

A beam 10 m long is loaded as shown in figure. Find reactions at supports.

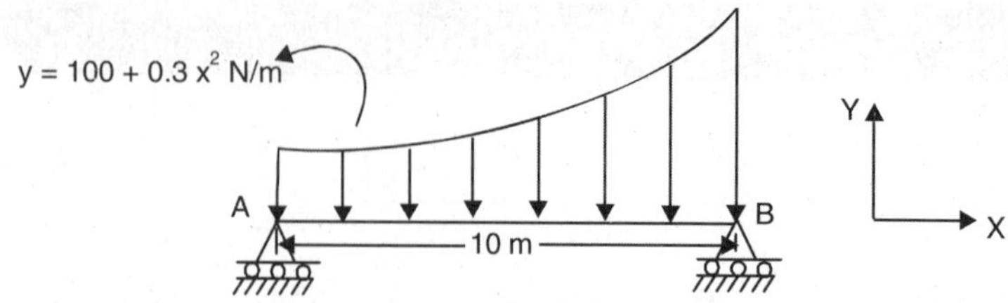

Fig. E3

Example 4

Two beam *AB* and *DE* are hinged at *A* and *D* respectively and supported as shown in figure. Find the magnitude and direction of reactions offered by supports *D* and *E* for the loading shown.

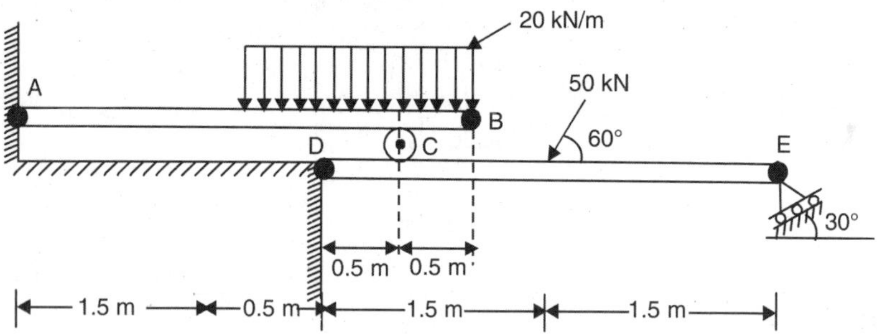

Fig. E4

Example 5

The cable ABCDE supports three vertical loads at the points *B*, *C* and *D* as shown in figure. If point *C* is 5 m below the left support *A*, determine

(*i*) The elevations of points *B* and *D* with reference to *A*

(*ii*) Maximum tension in the cable

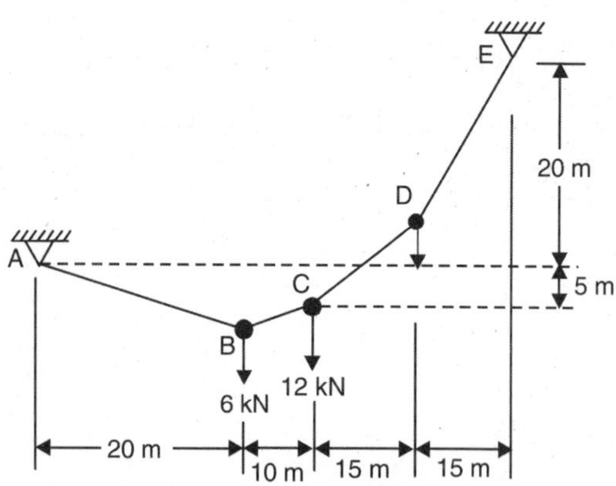

Fig. E5

Example 6

Sketch the shear and bending moment diagrams for the cantilever beam shown.

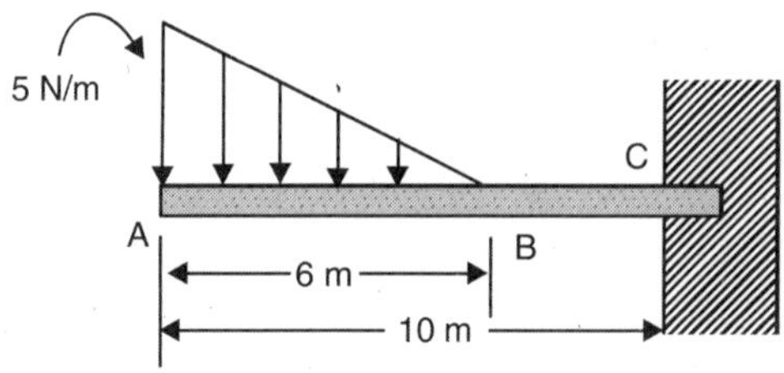

Fig. E6

Example 7

In a suspension bridge of 400 m span and 40 m dip the whole weight supported by the two chains is 200 kg per horizontal meter. Find the horizontal tension in the chains and the tension at the point of support.

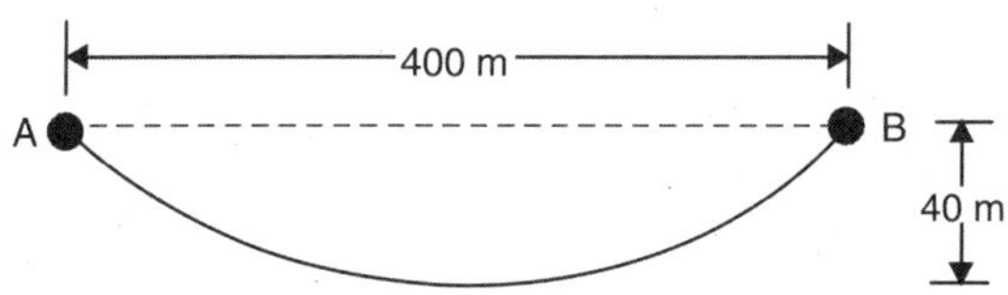

Fig. E7

Example 8

A light cable is attached to a support at *A*, passes over a small pulley at *B* and supports a load *P*. The mass per unit length of the cable is 0.75 kg/m; sag of the cable is 0.5 m. Determine slope of cable at *B* and total length of the cable from *A* to *B*.

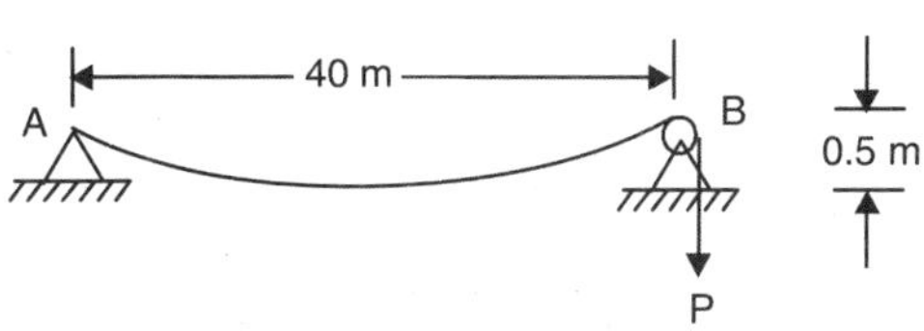

Fig. E8

Example 9

Determine the axial forces produced in link 1, 2, 3 and 4 of the beam shown in figure.

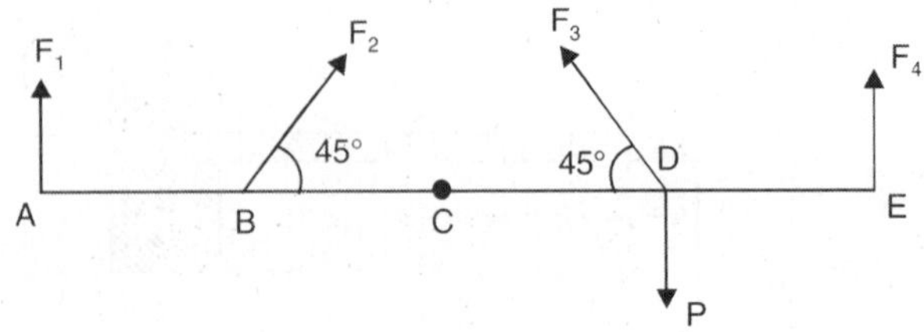

Fig. E9

Example 10

A 400-N load is attached to a small pulley which may roll on the cable *ACB*. The pulley and load held in the position shown by a second cable *DE* which is parallel to the portion *CB* of the main cable. Determine

(*i*) Reactions at *A* and *B*

(*ii*) The tension in cable *ACB*.

(*iii*) The tension in cable *DE*.

Neglect the radius of the pulleys and weight of the cables.

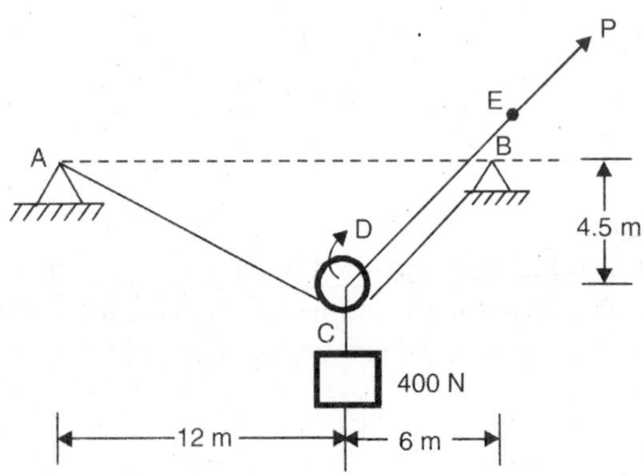

Fig. E10

Example 11

The cantilever beam is subjected to the load intensity (force per unit length) which varies as $\omega = \omega_0 \sin (\pi x/l)$. Determine the shear force F and bending moment M as functions of the ratio x/l.

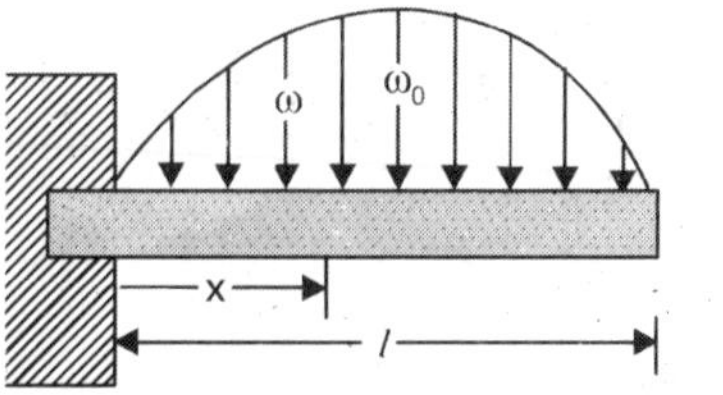

Fig. E11

Example 12

ABC is a beam supported at *B* and *C*. A variable point *P* is at a distance x from the one end *A*. Draw the graph of the shearing force *F* and the bending moment *M* at *P* as P moves along the beam from *A* to *C*. *W* denote the weight of beam per unit length.

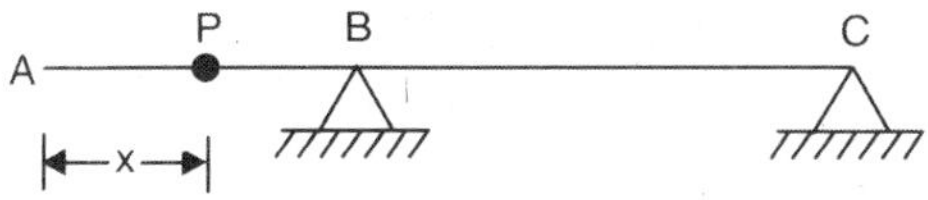

Fig. E12

Example 13

A light rod *ABCD* in which *AB* = 5 m, *BC* = 10 m and *CD* = 5 m is supported at B and *C* and carries uniformly distributed loads 10 N/m along *AB* and *CD* and a load of 20 N concentrated at the middle point E of *BC*. Draw diagrams to represent the shearing force and bending moment at all points of the rod.

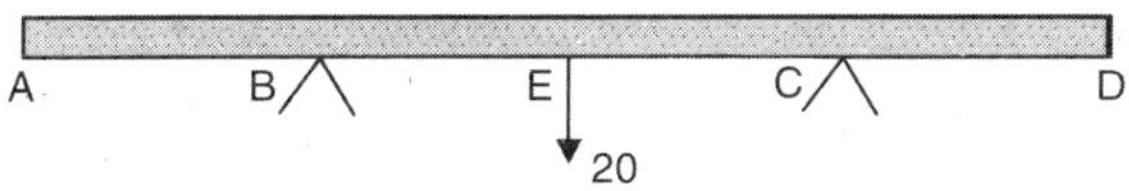

Fig. E13

4 Virtual Work

Summary and Concept

Work done by a Force

$$dW = \vec{F} \cdot d\vec{S}$$

$$|dW| = F ds \cos\theta$$

Where, θ is the angle between the force vector and displacement vector.

Work done by couple

$$dW = M \cdot d\theta$$

Virtual work

If the particle is in equilibrium, the total virtual work of the forces acting on the particle is zero for any virtual displacement of the particle.

$$\Sigma dW = 0$$

Principle of potential energy

The potential energy for a conservative force $F(x)$ acting in a direction x can be written as

$$\frac{dU}{dx} = -F(x)$$

The principle of potential energy for the equilibrium of a body or a system follows from the fact that the external force F should be zero.

$$\frac{dU}{dx} = 0$$

The potential energy must have stationary value at the position of equilibrium.

For each case $\frac{dU}{d\theta} = 0$ for equilibrium

Fig. 1

Stable state $U_{minimum}$

$$\frac{d^2U}{d\theta^2} > 0$$

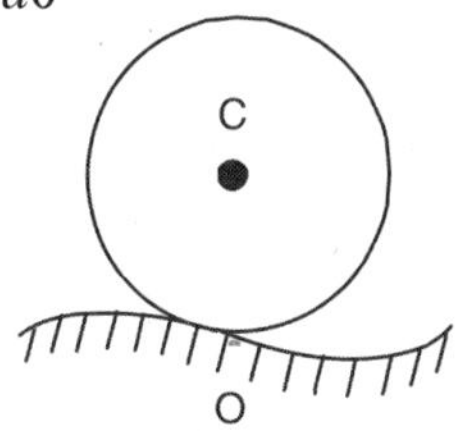

Unstable state $U_{maximum}$

$$\frac{d^2U}{d\theta^2} < 0$$

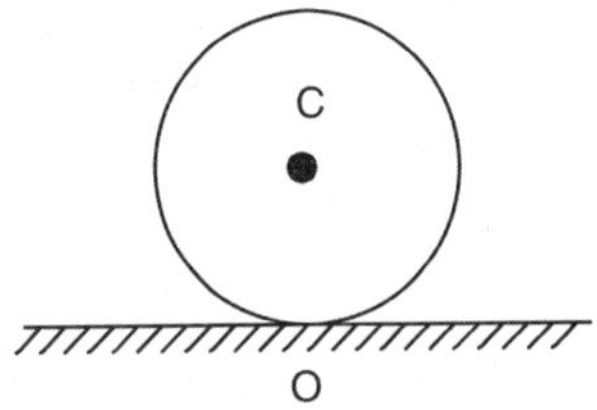

Fig. 2

Unstable state

$$\frac{d^2U}{d\theta^2} = 0$$

Neutral state

$$\frac{d^2U}{d\theta^2} = 0$$

Level 1

Example 1

Determine the magnitude of couple M required to maintain the equilibrium of the mechanism shown in figure.

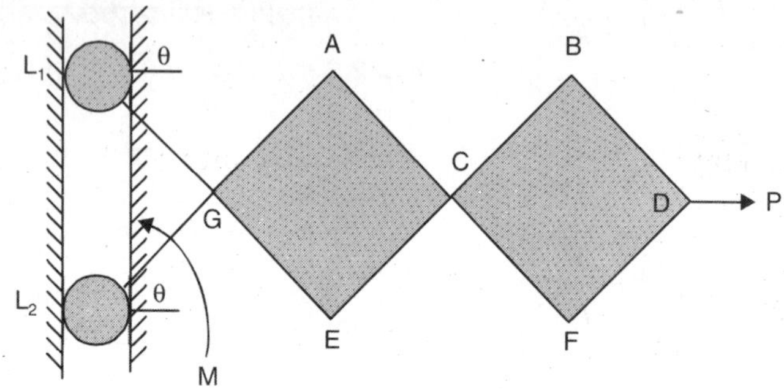

Fig. E1

Example 2

Determine reactions R_A and R_C for beam $ABCD$ loaded and supported as shown in figure.

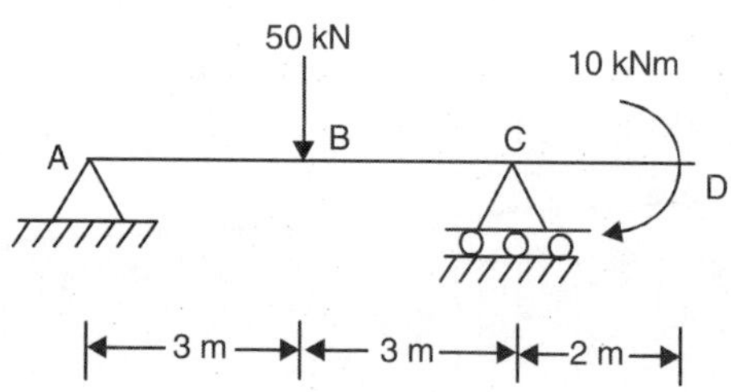

Fig. E2

Example 3

Four bars of equal length 'l' are hinged together at ends to form a rhombus $ABCD$ as shown in figure. Find relation of p and q for equilibrium.

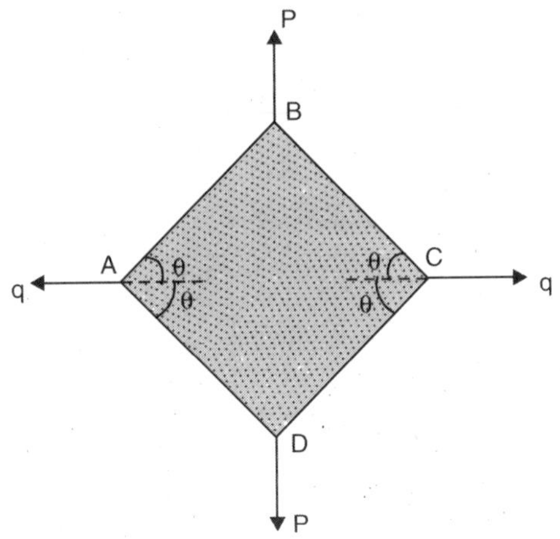

Fig. E3

Example 4

What is the force acting by the roller support so that wedge is in equilibrium. Assume mass of wedge is negligible.

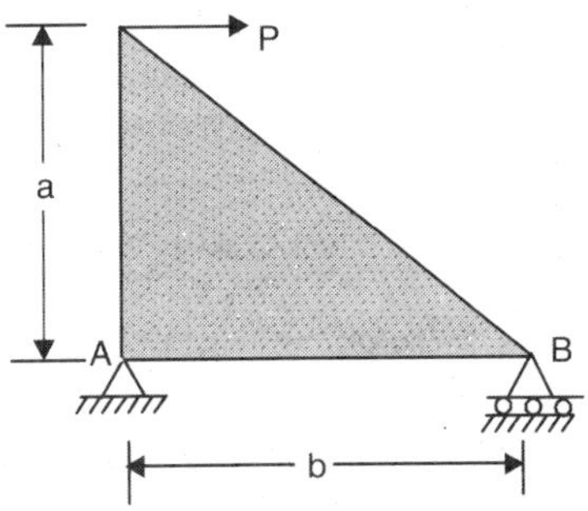

Fig. E4

Example 5

Determine m_1/m_2 in the figure shown if coefficient of friction between block and surface is μ.

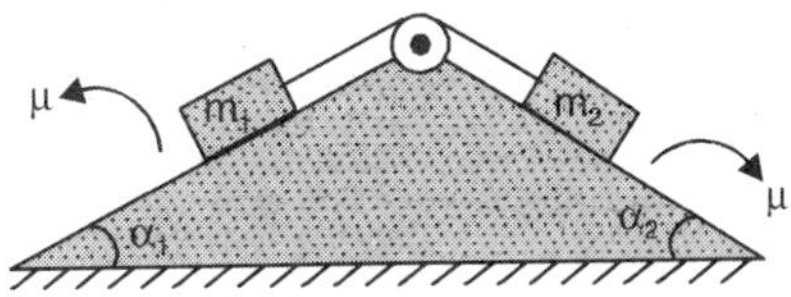

Fig. E5

Example 6

Determine the force P on the sliding collar which will prevent OA from rotating under the action of the couple M. Link OA in the horizontal position.

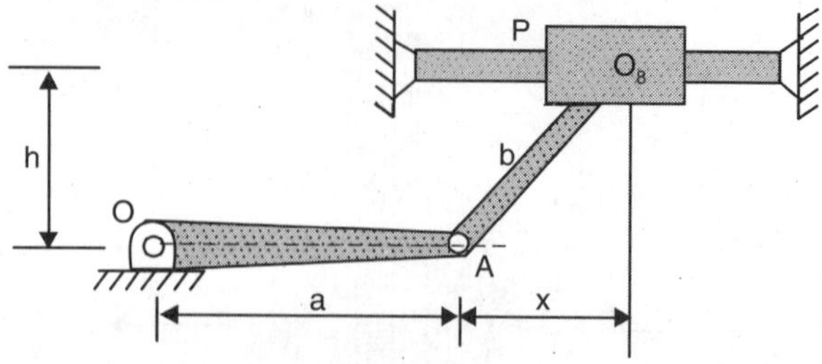

Fig. E6

Example 7

A ladder *AB* of weight *W* is held in equilibrium by a horizontal force *P* as shown in figure. Assuming the ladder to be idealized as a homogeneous rigid bar, relate *P* to *W*. assume the surfaces to be smooth.

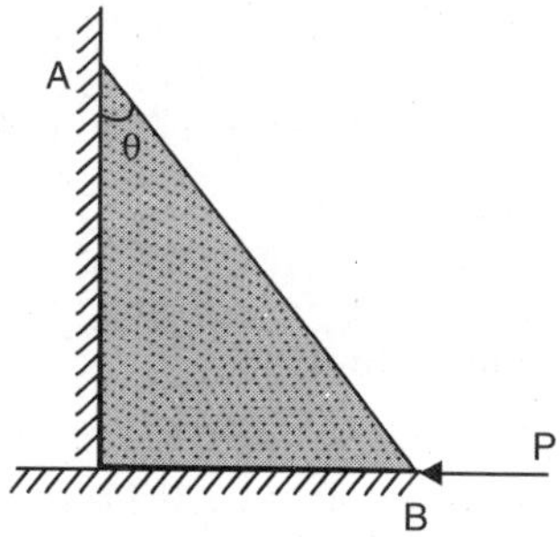

Fig. E7

Example 8

A linkage structure shown in figure consisting of two equal massless bars and a sliding block *B* is subjected to a set of two forces F_1 and F_2. For equilibrium, determine angle θ in terms of F_1 and F_2.

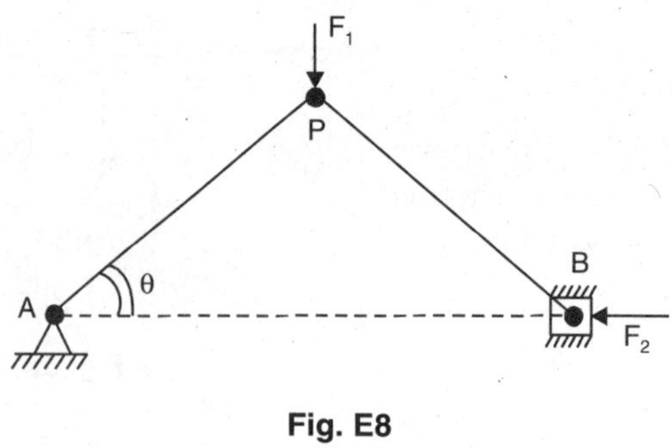

Fig. E8

Example 9

A plank of mass m is brought to an equilibrium position by the application of the couple M as shown in figure. Determine the maximum value of M that can hold the plank in equilibrium.

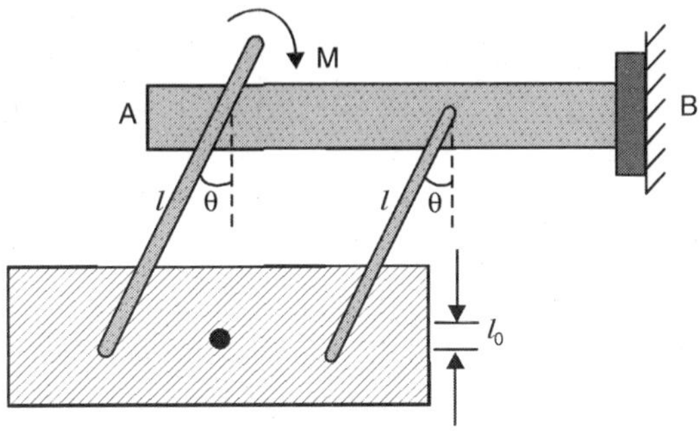

Fig. E9

Example 10

Find the magnitude of couple M to maintain equilibrium for given arrangement.

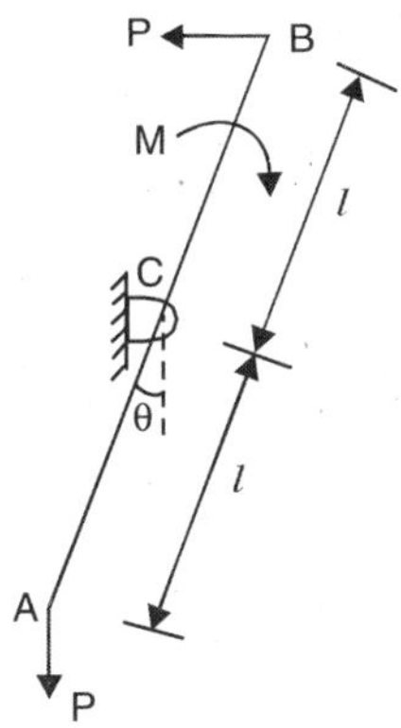

Fig. E10

Example 11

Determine the value of θ corresponding to the equilibrium position of rod AC when R = 10 m, p = 50 N and M = 2000 Nm.

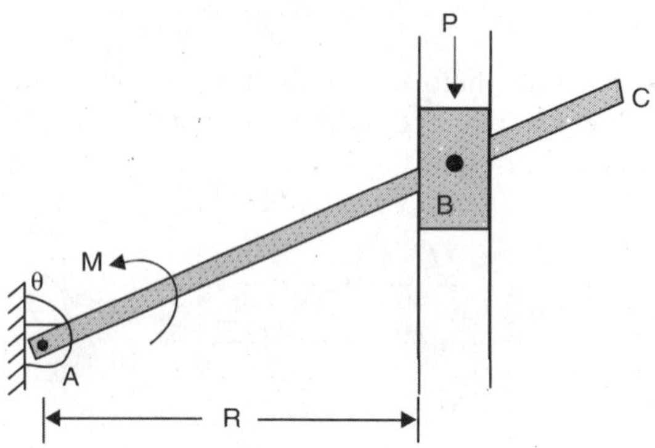

Fig. E11

Example 12

Find the force p acting on each of the two links shown in figure that will support the cylinder of weight W. Neglect weight of the links and friction. Your answer should be in terms of W, L, r and θ.

(*i*) If P is acting in horizontal direction as shown,

(*ii*) If P is acting vertically upward

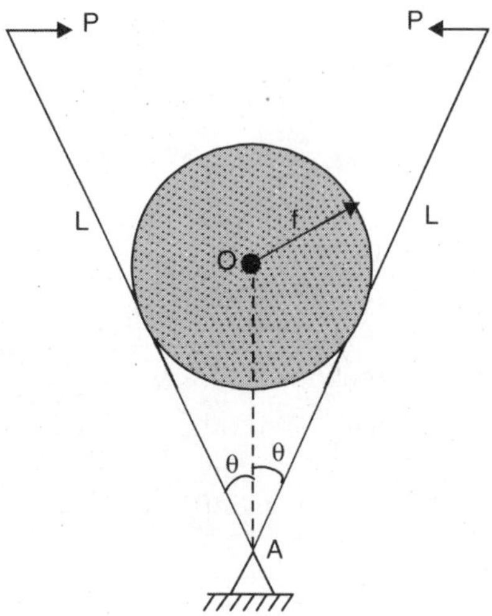

Fig. E12

Example 13

The slender rod *AB* is attached to a collar *A* and rests on a small wheel at *C* neglecting the radius of the wheel and the effect of friction, derive an expression for the magnitude of the force, required to maintain the equilibrium of the rod in terms of *P*, *a* and θ.

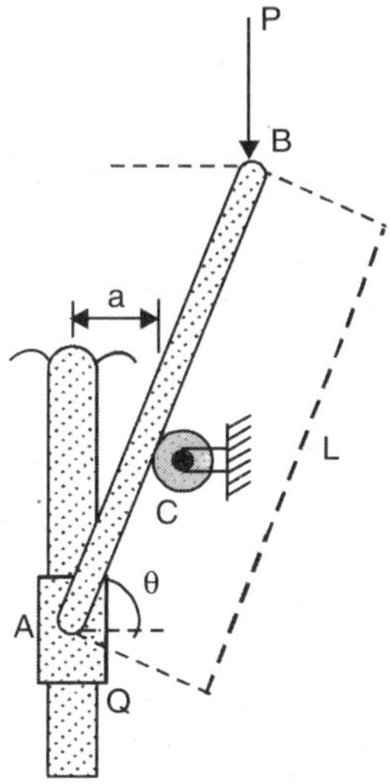

Fig. E13

Level 2

Example 1

Two beams *ABC* and *CD* are pinned together at *C* as shown in figure. A clockwise couple of 10 kNm acts at *E* and concentrated load of 60 kN at *F*. Determine reactions at *A*, *B* and *D*.

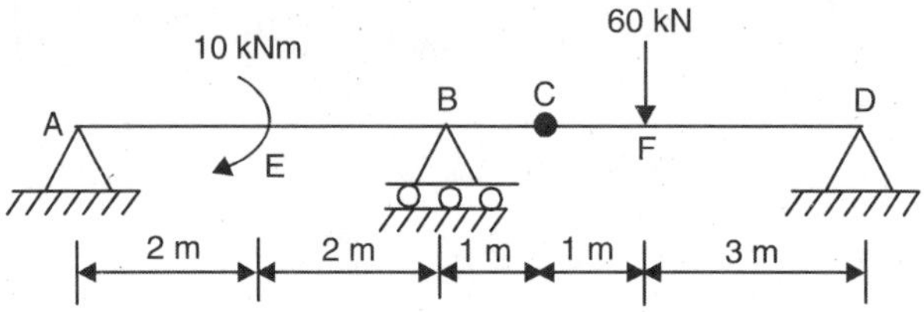

Fig. E1

Example 2

The spring is unstretched when rod *ABC* is horizontal ($\theta = 0$) and the rod is unloaded. Determine the value of θ corresponding to equilibrium when a 300-N downward load acts at *A*. The stiffness of the spring is 5 kN/m. $AC = 0.8$ m.

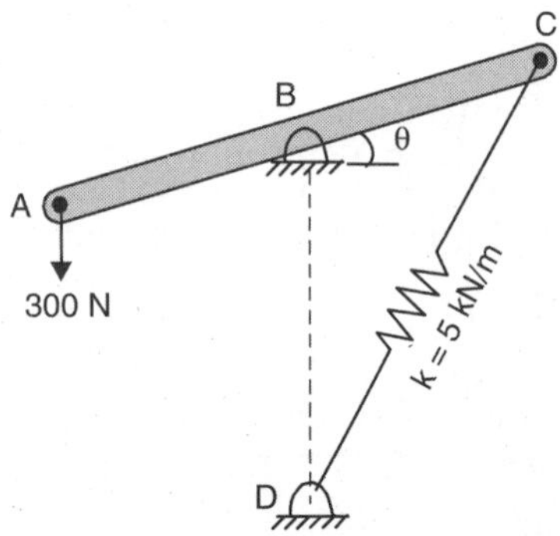

Fig. E2

Example 3

A moment '*M*' is applied to a mechanism shown in figure. A force of 50 N is applied to wheel at *C*. Wheel can rotate freely on inclined plane. Find the value of *M* to maintain equilibrium.

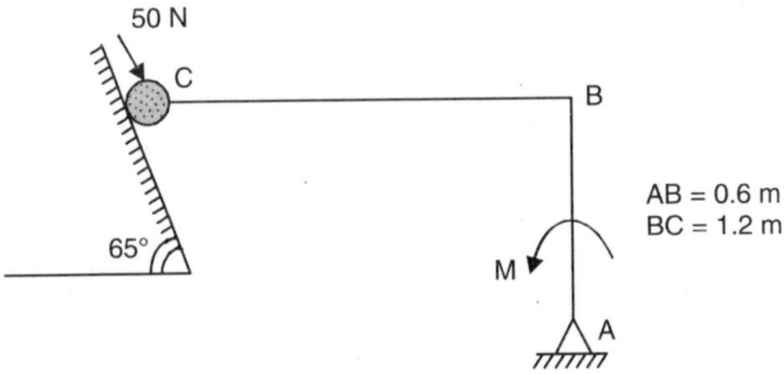

Fig. E3

Example 4

A slender rod of length 200 mm is attached to a collar at *B* and rests on a portion of a circular cylinder of radius 40 mm. Neglecting the friction, determine the value of 'θ' for the equilibrium position of the mechanism shown in the figure.

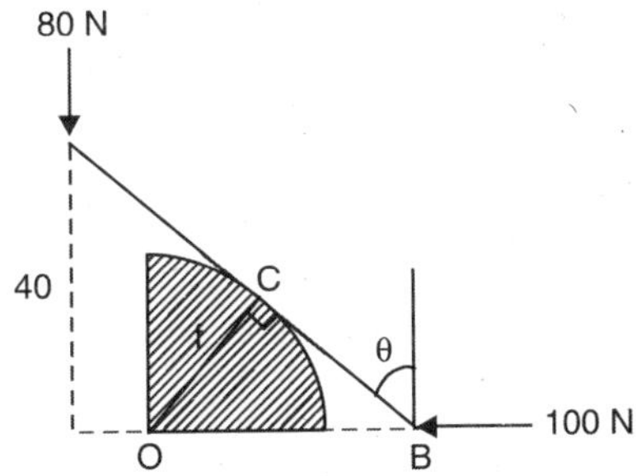

Fig. E4

Example 5

Use the method of virtual work to determine the horizontal component of reaction at *C* for the frame as shown.

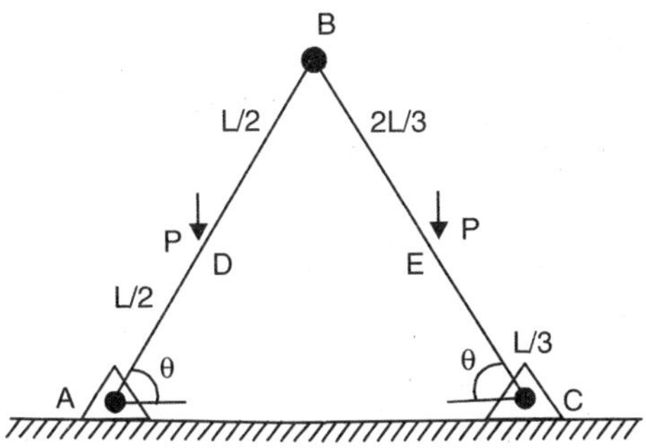

Fig. E5

Example 6

The mechanism shown is acted upon by the force *P*. Derive an expression for the magnitude of the force *q* required to maintain equilibrium.

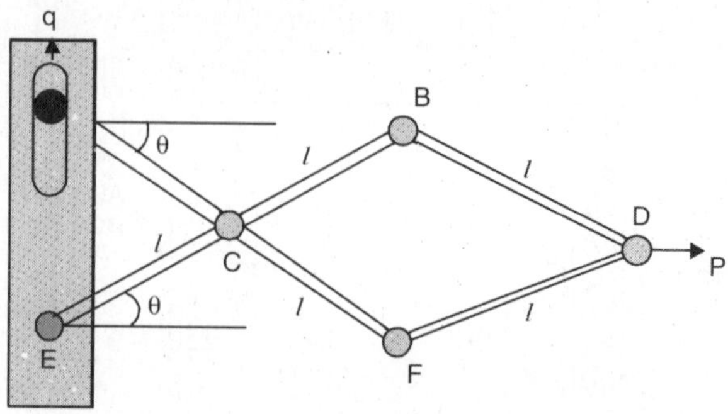

Fig. E6

Example 7

Determine the reaction components at *B* and *C*.

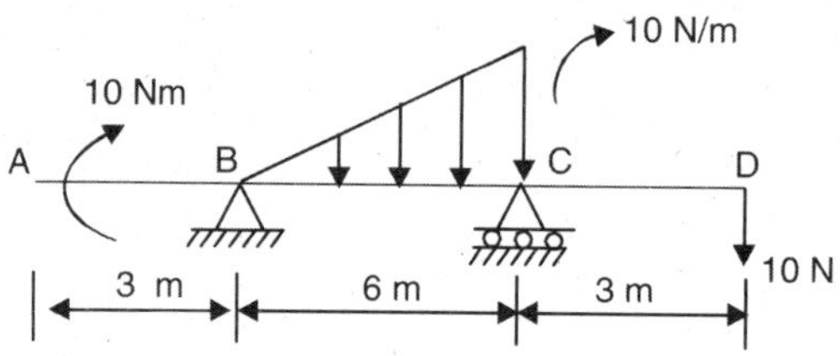

Fig. E7

Example 8

Determine the angle θ corresponding to equilibrium if $p = 4W$.

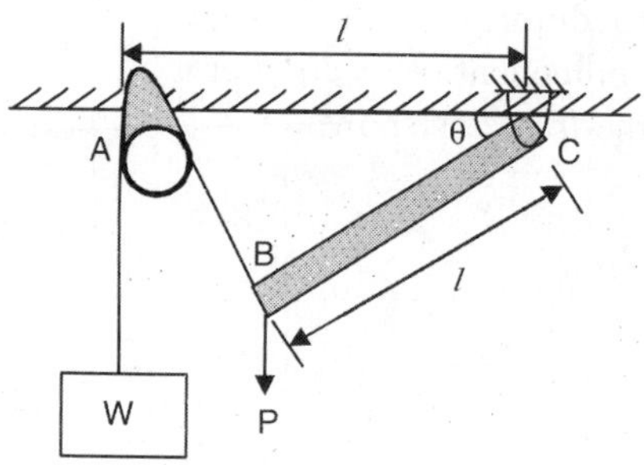

Fig. E8

Example 9

What is the compression in the piston rod of the cylinder to support the load P at a height h?

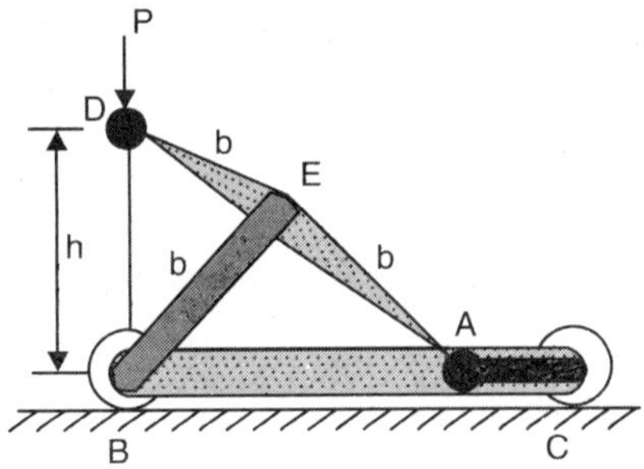

Fig. E9

Example 10

For given figure ACB, find force P to maintain equilibrium. Take $L = 0.6$ m and $\theta = 60°$.

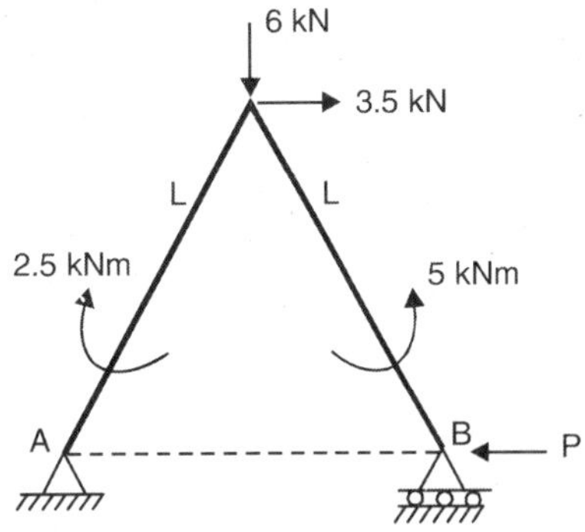

Fig. E10

Example 11

Find the reactions at support for the compound beam as shown in figure. "B" is internal hinge.

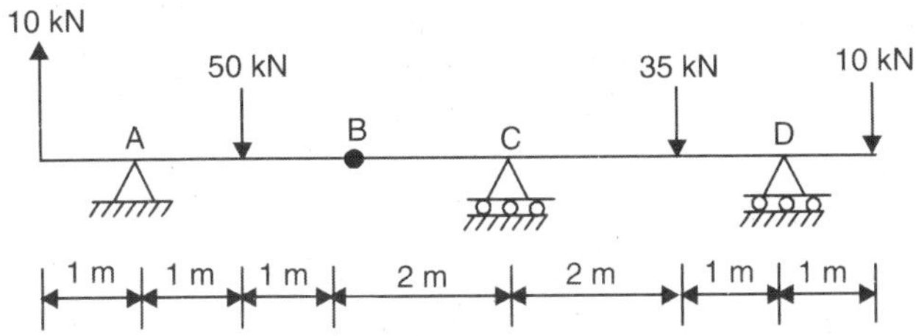

Fig. E11

Example 12

Determine the reaction components at *A*, *D* and *G*.

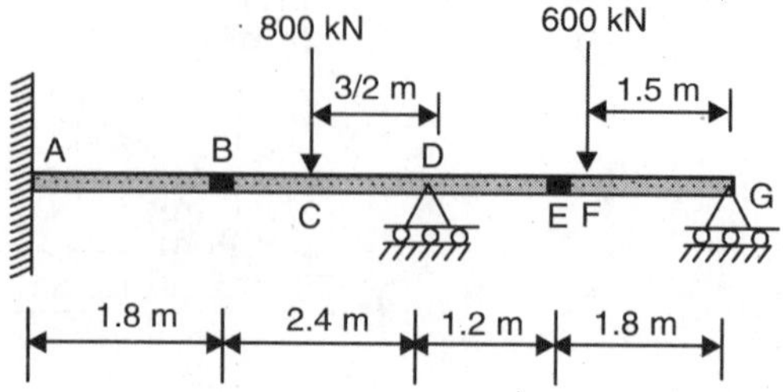

Fig. E12

Example 13

Determine the ratio of the moment of the couple *M* required to maintain equilibrium when force *P* applied at *A* is vertically downward in case first and in second case force *P* applied at *A* is horizontal and directed to the left.

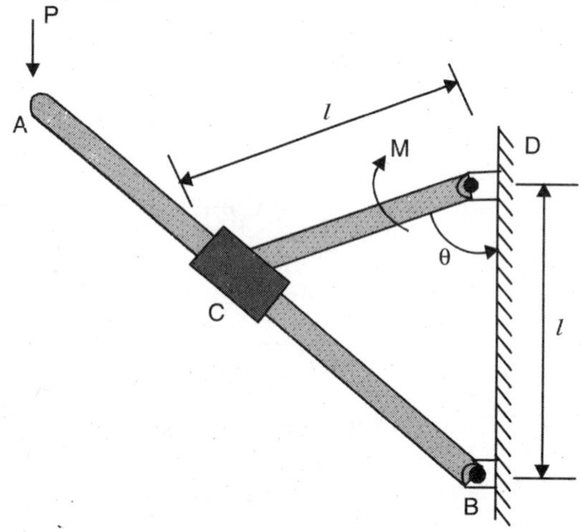

Fig. E13

Example 14

A load *W* is applied to the linkage at *B*. The spring of constant *k* is unstretched when *AB* and *BC* are horizontal. Find the value of θ for equilibrium. Given $W = 600\ N$, $K = 2500$ N/m, $l = 0.3$ m.

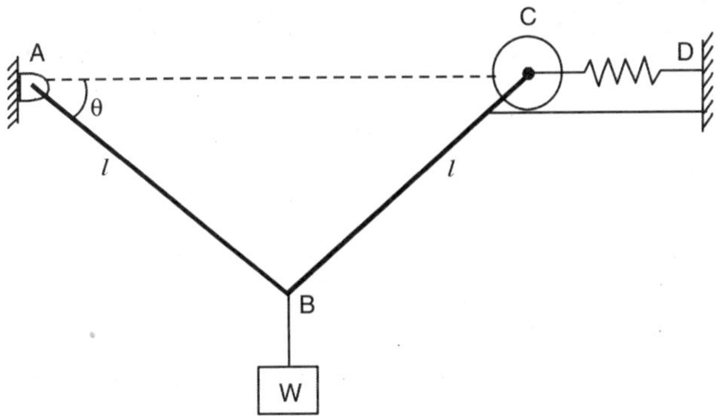

Fig. E14

Example 15

Find the reactions at support *B* which is an internal hinge.

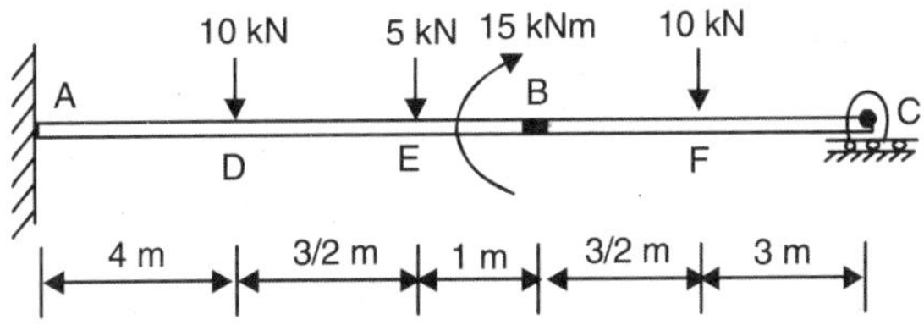

Fig. E15

Level 3

Example 1

The rod *AB* lies on two smooth planes through ends as shown in figure. Determine the inclination of the rod to the horizontal.

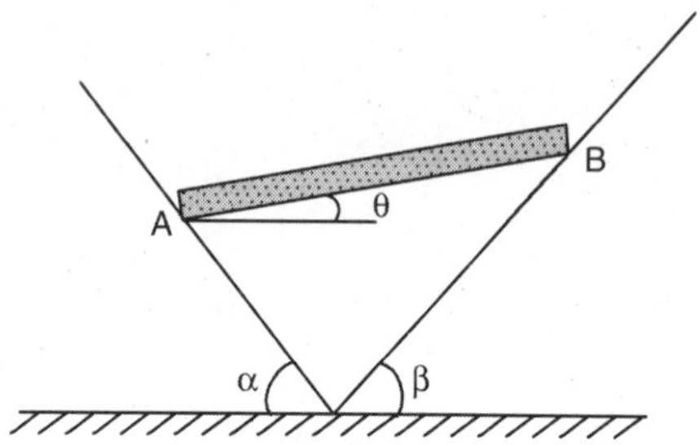

Fig. E1

Example 2

Determine the relationship between the moment *M* applied at the crank of radius *R* and the force *F* applied at the crosshead in the slider crank mechanism.

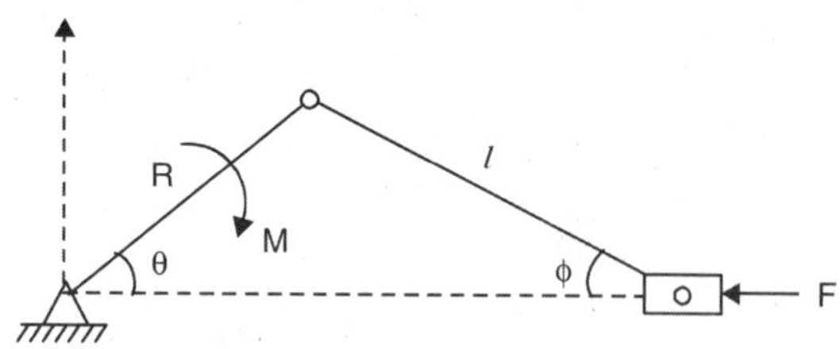

Fig. E2

Example 3

The double pendulum shown in figure consist of two links pinned smoothly together and held in position by a horizontal force *P* at point *C*. If the link *AB* is twice as long and heavy as link *BC*, find the equilibrium position as defined by angle θ_1 and θ_2.

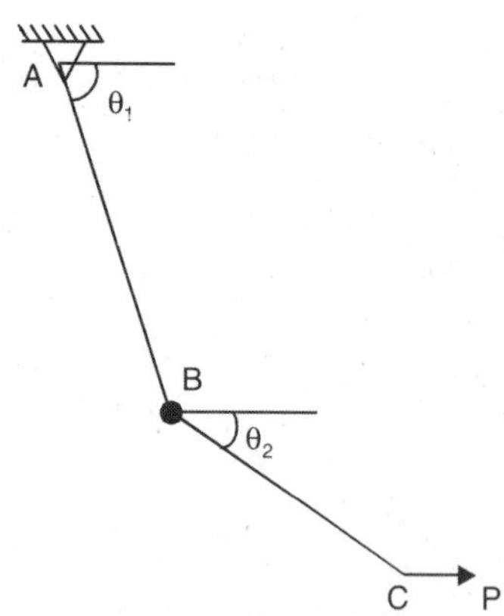

Fig. E3

Example 4

Determine the value of θ for which the structure maintain the equilibrium.

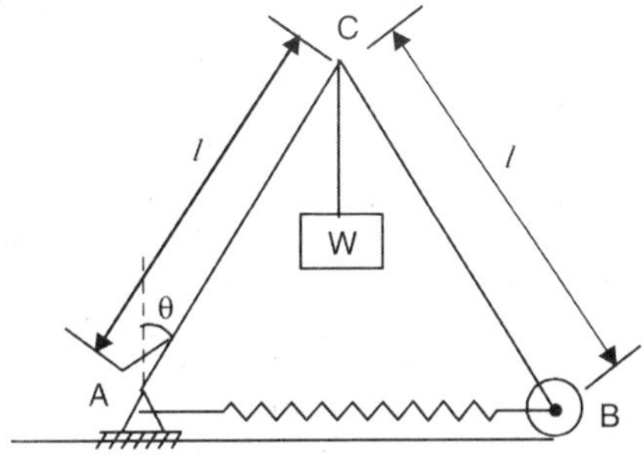

Fig. E4

Example 5

A solid hemisphere is supported by a string fixed to a point on its rim and to a point on a smooth vertical wall with which the curved surface of the hemisphere is in contact. If θ is the inclination of the string with vertical, determine the angle ϕ in terms of θ.

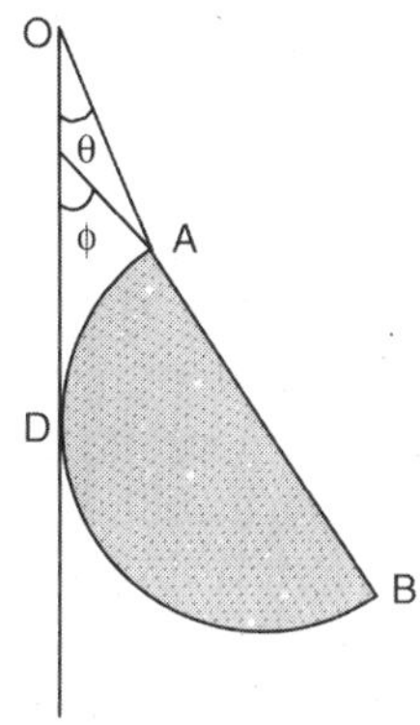

Fig. E5

Example 6

A quadrilateral *ABCD* formed of four uniform rods freely jointed to each other at their ends, the rods *AB* and *AD* begin equal and also the rods *BC*, *CD* is freely suspended from the joint *A*. A string joints *A* to *C* and is such that *ABC* is a right angle. If *W* and *W′* be the weights of a an upper and lower rod respectively, determine the tension in the string.

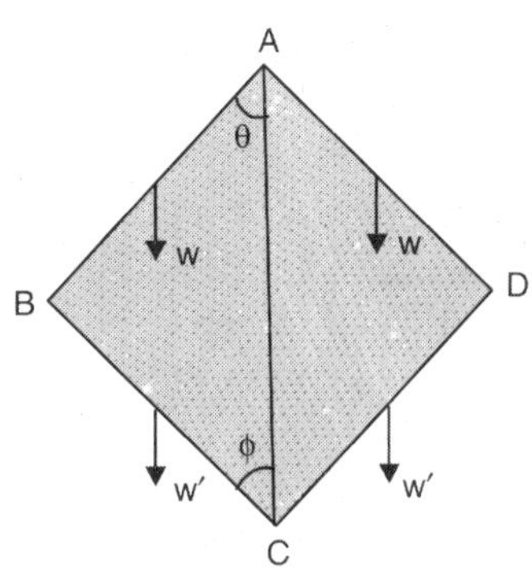

Fig. E6

Example 7

A rod *AB* is moveable about a pivot at *A*, and to *B* is attached a string whose other end is tied to a ring. The ring slides along a smooth, horizontal wire passing through *A*. Determine the value of *p* so that a rod *AB* remain in the equilibrium position as shown in figure given that *W* is weight of rod and α and β are inclinations of the rod and string to the horizontal.

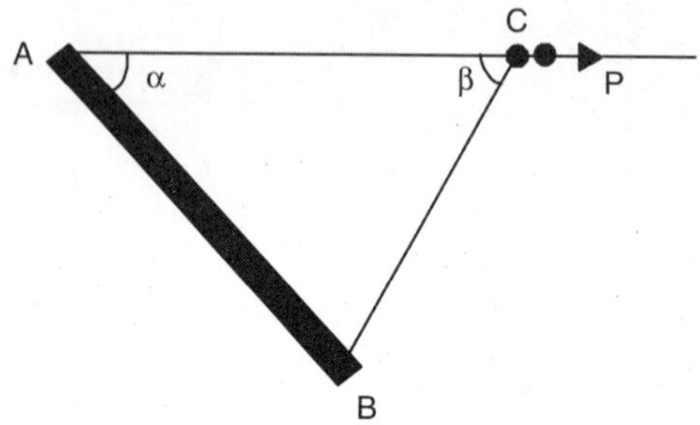

Fig. E7

Example 8

A rectangular block and a semi cylinder of the same length made of the same homogenous material are secured together and placed on a flat rough surface in equilibrium, determine the minimum radius *R* of the semi cylinder in terms of the height *H* of the block if the width of the block is *n* times the radius of the semi cylinder in order that the composite body be in stable equilibrium. Assume that the semi cylinder may roll without slip on the flat surface.

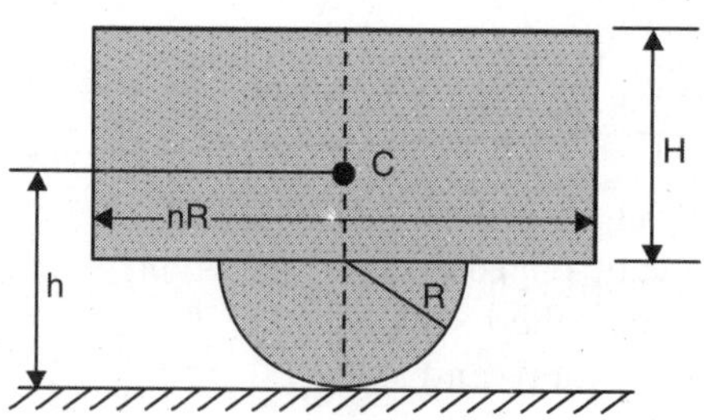

Fig. E8

Example 9

An open cylindrical can of diameter d, height h and center of gravity C rests on top of a spherical surface of radius r as shown in figure. Assuming sufficient friction to prevent slipping, find the relation between h and r consistent with stability.

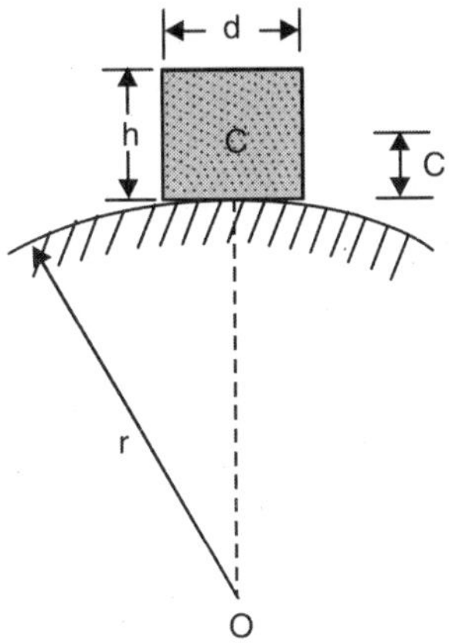

Fig. E9

5 Space Forces

Summary and Concept

In the space a force is represented by

$$\mathbf{A} = A\,(x, y, z)$$

$$|\vec{\mathbf{A}}| = \sqrt{A_x^2 + A_y^2 + A_z^2}$$

$$\cos\theta_x = \frac{A_x}{|A|} = l$$

$$\cos\theta_y = \frac{A_y}{|A|} = m$$

$$\cos\theta_z = \frac{A_z}{|A|} = n$$

$$l^2 + m^2 + n^2 = 1$$

In the space forces system apply equations of equilibrium as

$$\sum F_x = 0, \sum F_y = 0, \sum F_z = 0$$

$$\sum M_x = 0, \sum M_y = 0, \sum M_z = 0$$

These six equations may be used to find six unknowns.

Moment of a force about a point

$$\vec{M}_0 = \vec{r} \times \vec{F}$$

Where $\vec{r}$ = Position vector of point

$\vec{F}$ = Force vector

$$\vec{M}_0 = \begin{vmatrix} \hat{i} & \hat{j} & \hat{k} \\ r_x & r_y & r_z \\ F_x & F_y & F_z \end{vmatrix}$$

Moment of a force about a line

Consider a force $\vec{F} = F_x i + F_y j + F_z k$ acting on a body.

The moment about line *OL* is defined a projection of moment M_0 on to the axis *OL*.

$$\hat{e}_{OL} \Rightarrow \text{Unit vector}$$

$$M_{OL} = e_{OL} \cdot M_0$$

$$= e_{OL}\,(r \times F)$$

$$M_{OL} = \begin{vmatrix} e_x & e_y & e_z \\ x & y & z \\ F_x & F_y & F_z \end{vmatrix}$$

$$= \begin{vmatrix} \lambda_x & \lambda_y & \lambda_z \\ x & y & z \\ F_x & F_y & F_z \end{vmatrix}$$

Moment of a force about a line not passing through origin.

Consider a force

$$\vec{F} = F_x\hat{i} + F_y\hat{j} + F_z\hat{k} \text{ passing through a point } (x_2, y_2, z_2)$$

$$M_{BL} = \begin{vmatrix} \lambda_x & \lambda_y & \lambda_z \\ (x_1 - x_2) & (y_1 - y_2) & (z_1 - z_2) \\ F_x & F_y & F_z \end{vmatrix}$$

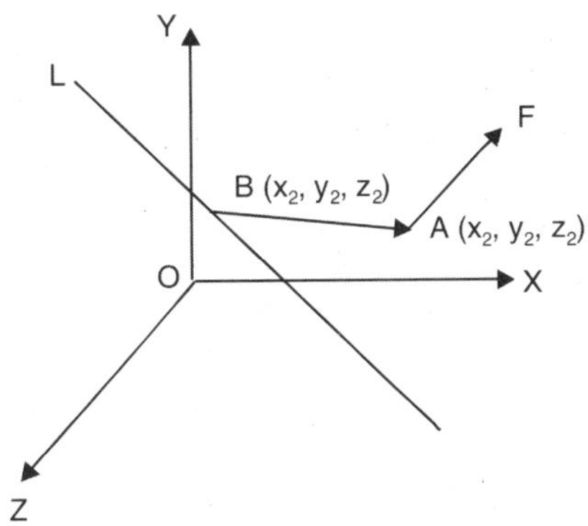

Fig. 1

Varignon's theorem

"The moment of resultant of a force system about a point is equal to the sum of the moment of various forces about the same point."

Suppose $\vec{F}_1, \vec{F}_2, \vec{F}_3 \ldots \vec{F}_n$, forces are acting on a rigid body.

Then

$$\bar{M}_R = \bar{M}_{F_1} + \bar{M}_{F_2} + \bar{M}_{F_3} + \ldots + \bar{M}_{F_n}$$

or

$$\bar{M}_R = \sum \bar{M}$$

Centralized Force System (CFS)

For CFS problems use $\Sigma F_x = 0$, $\Sigma F_y = 0$, $\Sigma F_z = 0$ after resolving the forces in the vector form.

CFS

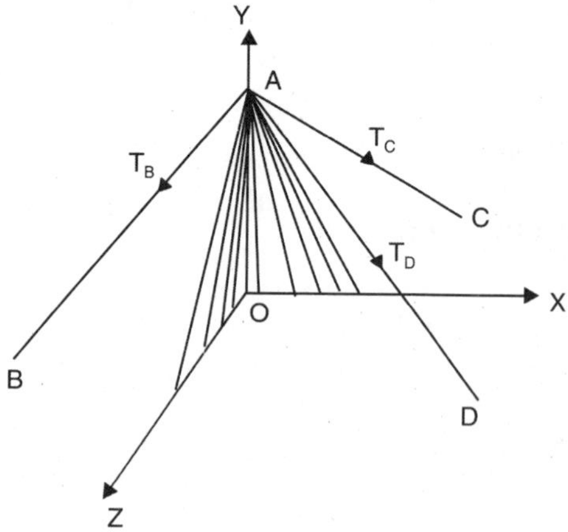

Fig. 2

Generalized Force System (GFS)

For GFS problem use

$\Sigma F_x = 0$, $\Sigma F_y = 0$, $\Sigma F_z = 0$ and $\Sigma M = 0$

Use $\Sigma F_x = 0$ or $\Sigma F_y = 0$ or $\Sigma F_z = 0$

And $\Sigma M = 0$

GFS

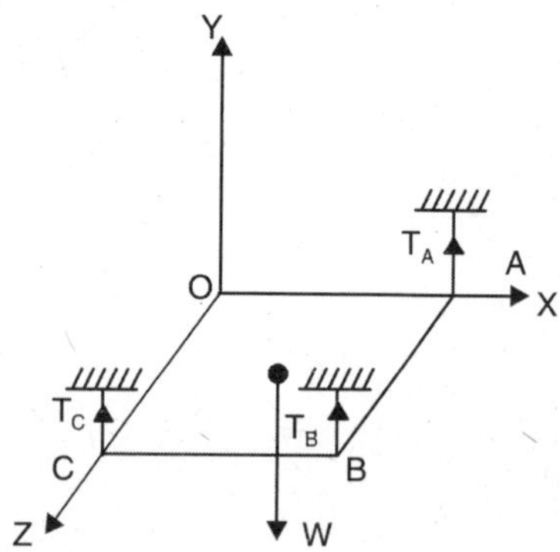

Fig. 3

Level 1

Example 1

An uniform equilateral triangular plate weighing 2.4 kN is held in a horizontal plane by means of three cable at *A*, *B* and *C*. An additional load of 1.2 kN acts at the midpoint of edge *AC* of the plate. Determine the tensions in the three cables.

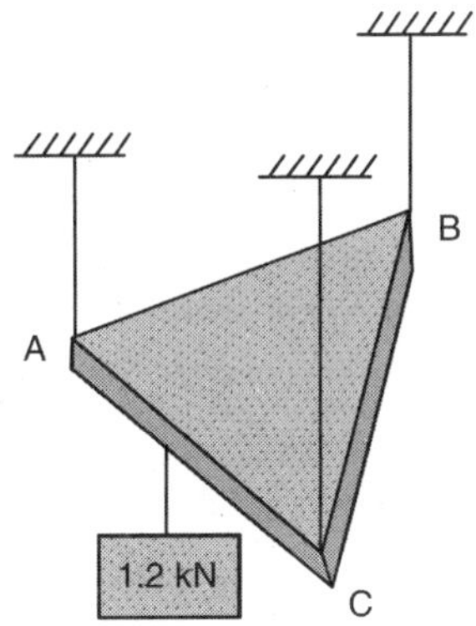

Fig. E1

Example 2

A horizontal plate (in *XZ* plane) of 3 m radius and 1.5 kN weight is supported by three vertical wires at *A*, *B*, *C*. Find tension in each of these wires.

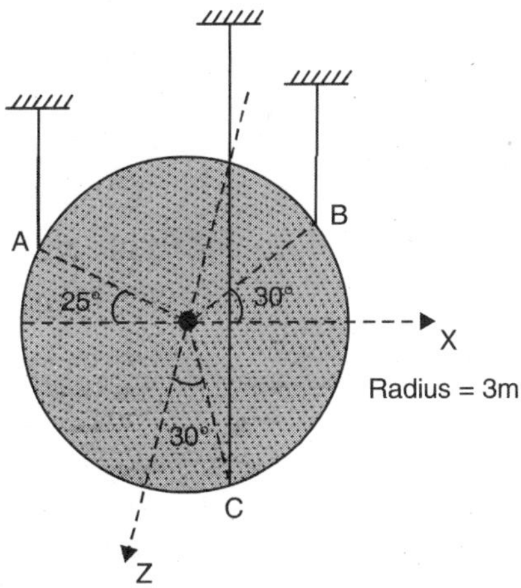

Fig. E2

Example 3

ABCDEFGH is a regular octagon of side 4 m. Figure shows positions of six columns, which carry vertical forces normal to plane of octagon and acting of its size corner pins. Locate the resultant of these forces with respect to centroid of octagon '*O*'.

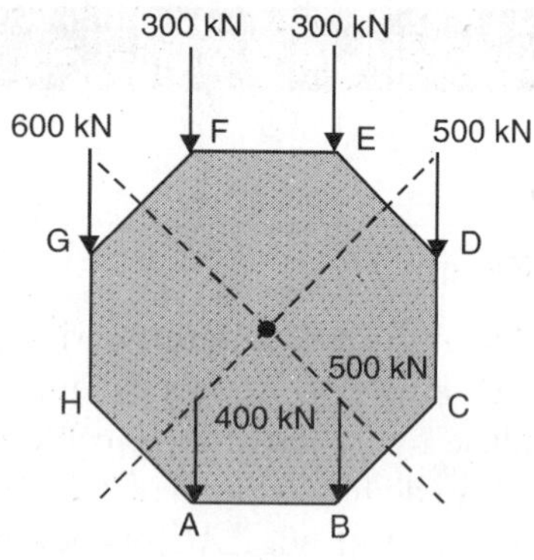

Fig. E3

Example 4

A mat supports four columns, as shown in figure. Determine the magnitude, direction and point of application of resultant of the four loads.

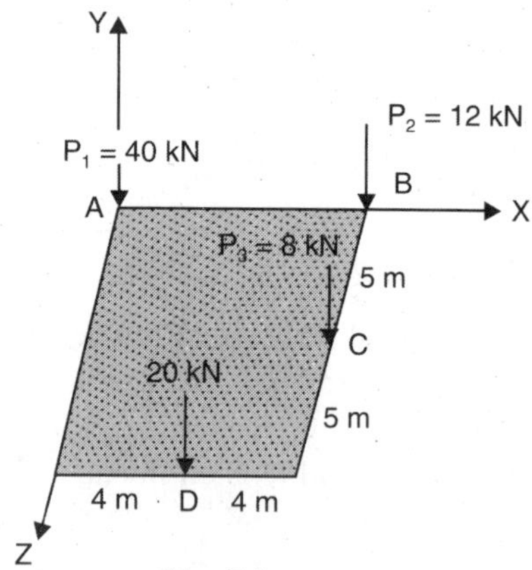

Fig. E4

Example 5

A force *P* is acting form *C* to *F*. Calculate perpendicular distance between the force *P* and the diagonal *BH*.

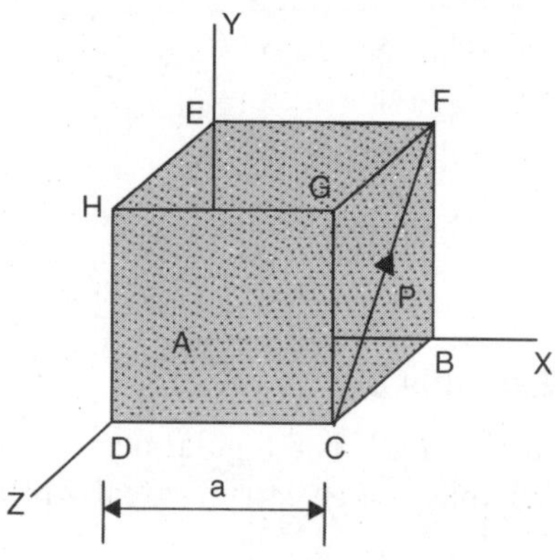

Fig. E5

Example 6

A cube of side a is acted upon by a force F along FC as shown. Determine moment of F

(*i*) About point A

(*ii*) About edge AB.

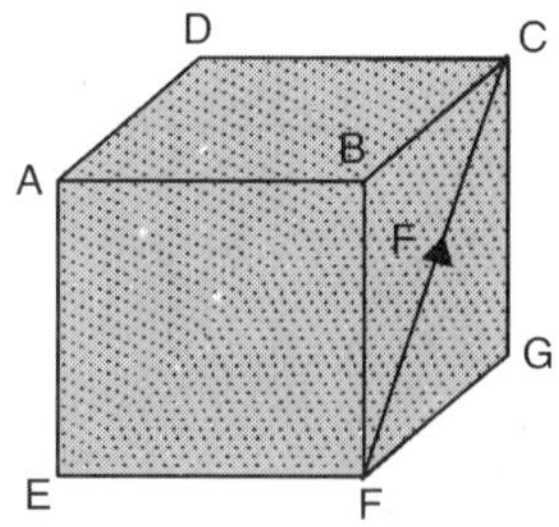

Fig. E6

Example 7

The square steel plate has a mass of 1800 kg with mass center at its center G. Calculate the tension in each of the three cables with which the plate is lift while remaining horizontal.

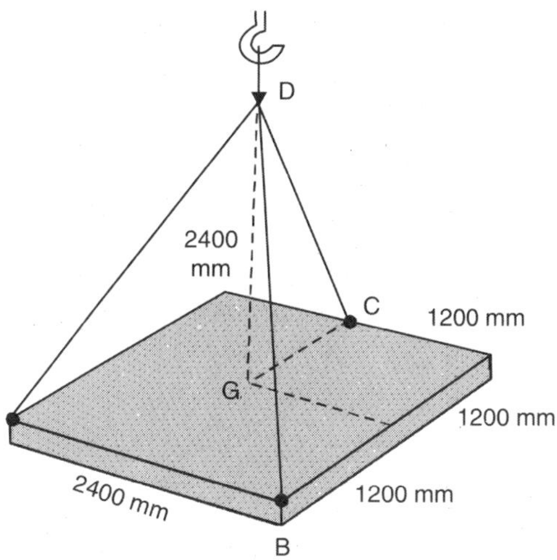

Fig. E7

Example 8

A vertical mast AB is supported at A by bell socket joint and by cables BC and DE as shown in figure. Force $\bar{F} = 500\hat{i} + (-3000)\hat{k}$ is applied at B. Find the reaction at A.

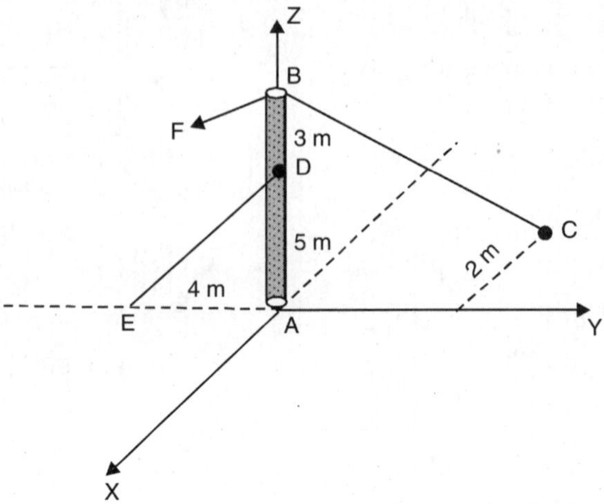

Fig. E8

Level 2

Example 1

Find the forces in the three bars 1, 2, 3 in terms of *W* for equilibrium. Assume *A* to be a smooth spherical joint (ball and socket joint). Also assume *B*, *C*, *D* as ideal ball and socket supports and bars to be weightless. Support *C* is in *x-z* plane.

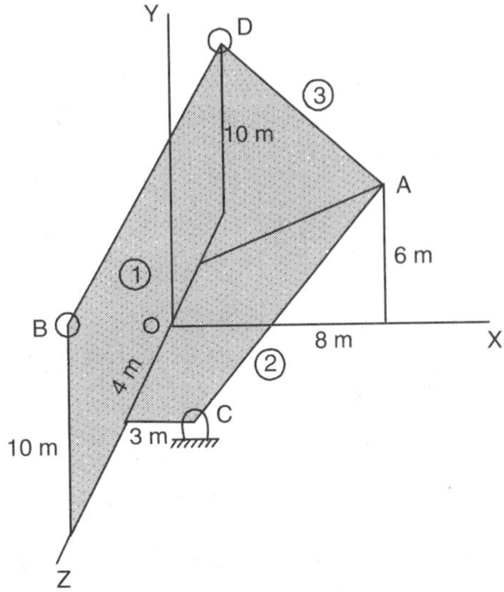

Fig. E2

Example 2

The uniform bar of length 7 m and 100 kg in mass is supported by a ball and socket joint at *A* on the horizontal floor. The ball end *B* rests against the vertical walls. Find reactions components at *A* and *B*.

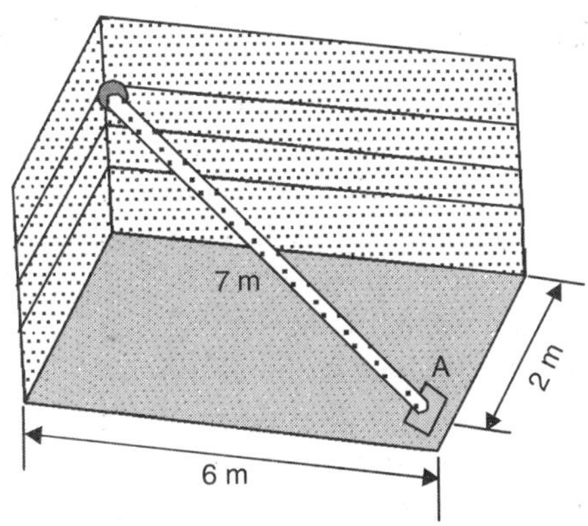

Fig. E2

Example 3

A rectangular sign board on the road has a mass of 100 kg with the center of mass in the center of the rectangle. The support against the wall at point *C* may be treated as a ball and socket joint. At center *D* support is provided in the *y*-direction only. Calculate the tensions T_1 and T_2 in the supporting wires.

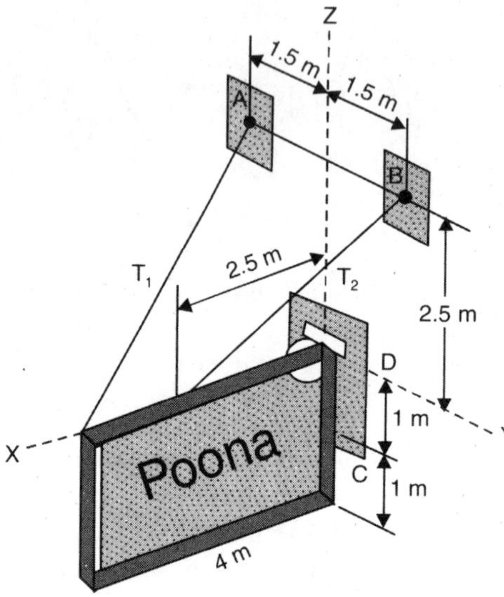

Fig. E3

Example 4

The uniform rectangular plate *ABCD* has a mass of 40 kg and is hinged at its corners A and *B* to the fixed vertical surface. A wire from *E* to *D* keeps edges *BC* and *AD* horizontal. Hinge *A* can support thrust along the hinge axis *AB*, whereas hinge *B* supports force normal to the hinge axis only. Find the tension *T* in the wire and the magnitude *R* of the force supported by hinge *B*.

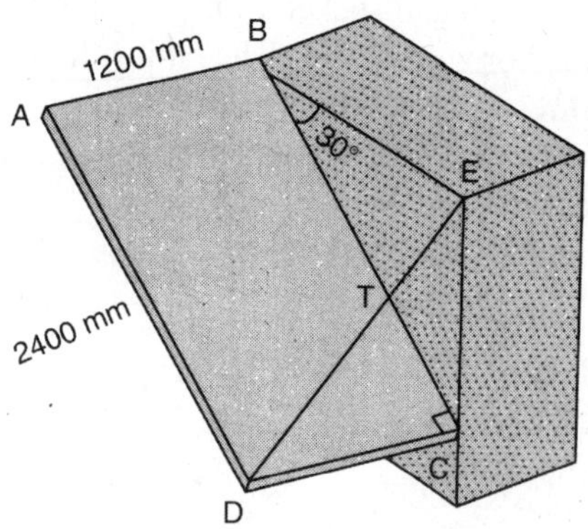

Fig. E4

Example 5

Determine the tension in cable *BC* and *BD* and reactions at the ball and socket joint *A* for the mast as shown in figure.

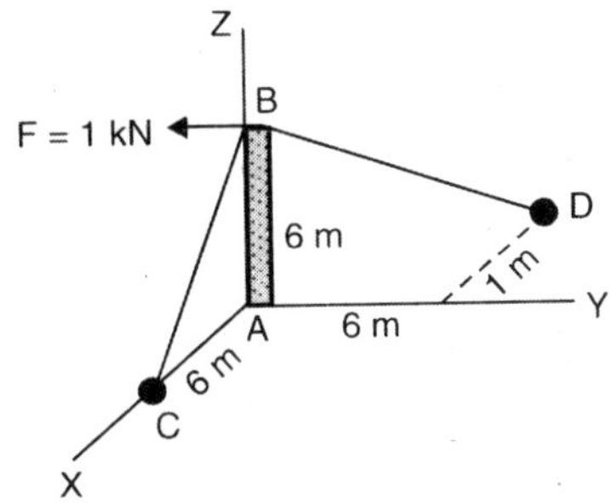

Fig. E5

Example 6

A three-legged stool or type of table is subjected to the load *L* as shown. Determine the vertical force reaction under each leg. Neglect the weight of the stool.

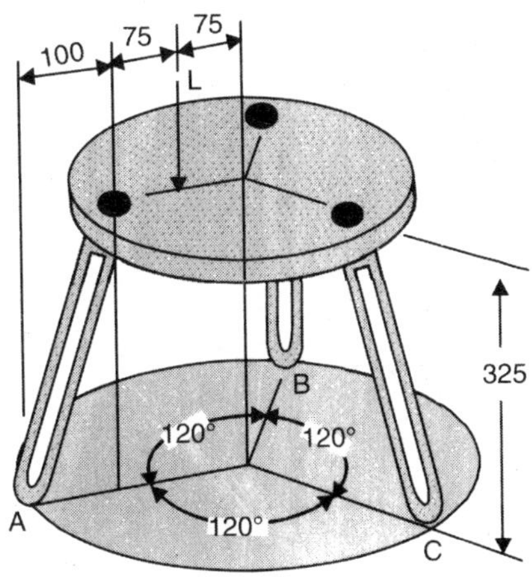

Fig. E6

Think Without Ink

1. What is the main difference between one degree of freedom and two degrees of freedom?
2. Differentiate the figures as shown below according to one-degree or two-degree.

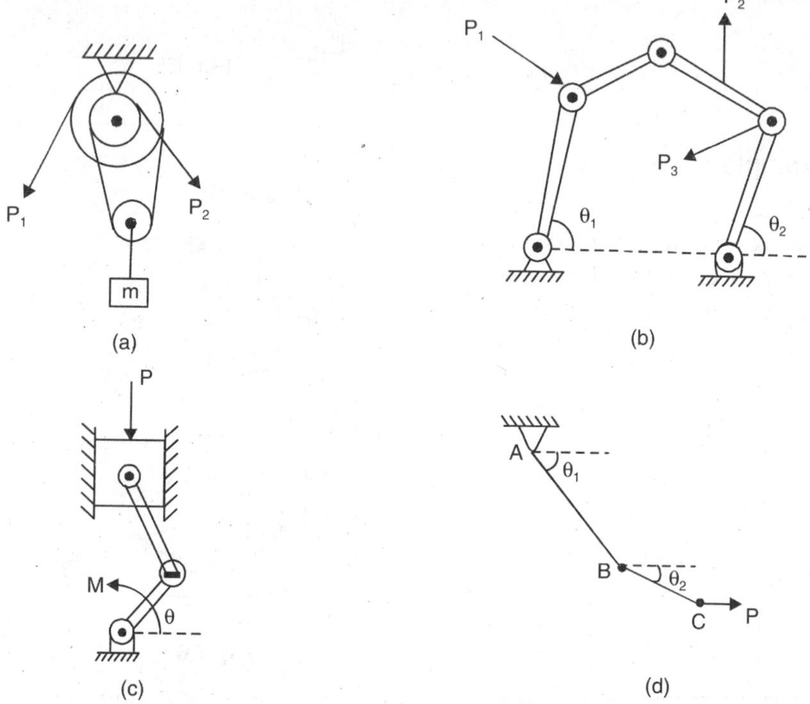

Fig. 2

3. According to the graph plotted below for a body, state the equilibrium status at various points.

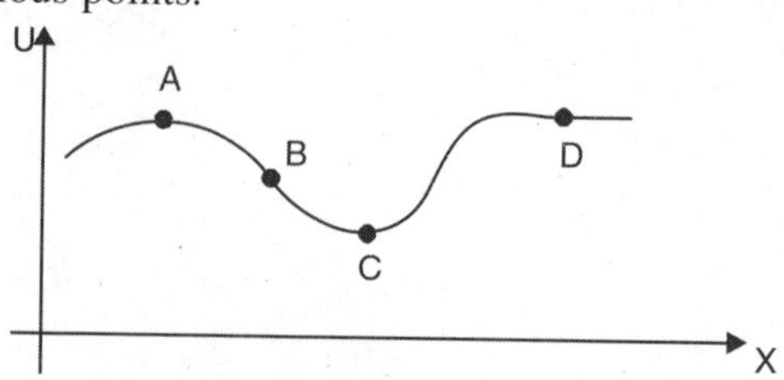

Fig. 3

4. Two blocks of mass m_1 and m_2 are connected with two rods of length L_1 and L_2 and joined it at point as shown in figure. Give the condition for states of equilibrium.

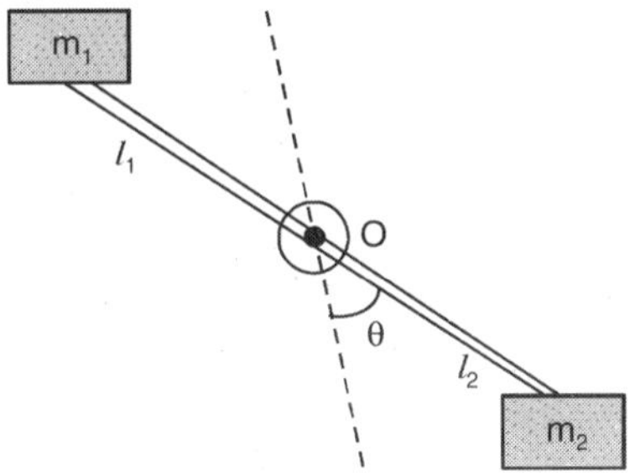

Fig. 4

5. For solving any problems in statics, how you determine the methods either virtual work method or equilibrium conditions method?
6. A beam AB of span 5 meters is carrying a point load of 2 kN at a distance 2 meters from A. Find the support reactions.

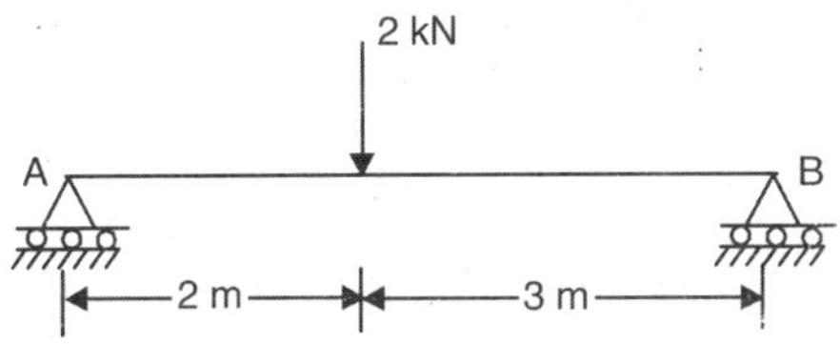

Fig. 6

7. Calculate the relation between forces P and Q for equilibrium of the bars. Length of bar = L.

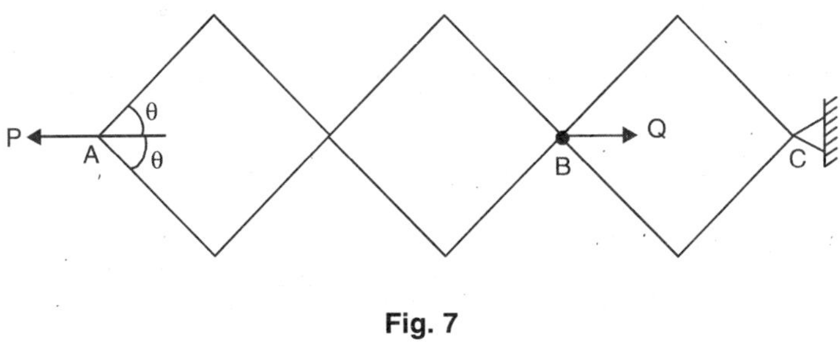

Fig. 7

8. Arrangement of three loads m_1, m_2 and m_3 are shown in figure. Neglect the friction and mass of pulleys. Find the ratio of masses of the loads at which the system is in equilibrium.

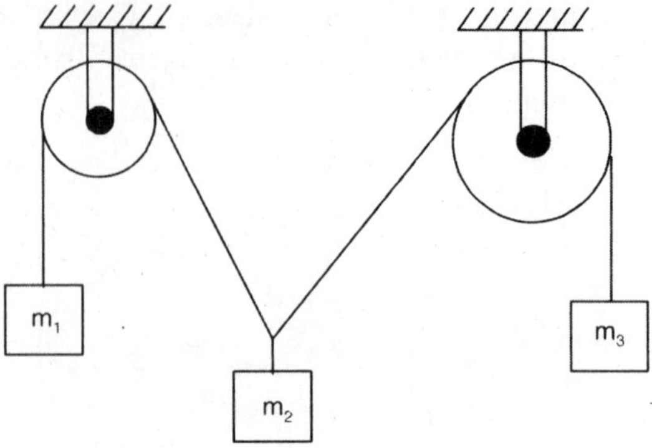

Fig. 8

9. A rod of length 6 m rests upon two supports as shown in figure. Determine range for values of load *W* for which rod will remain in equilibrium.

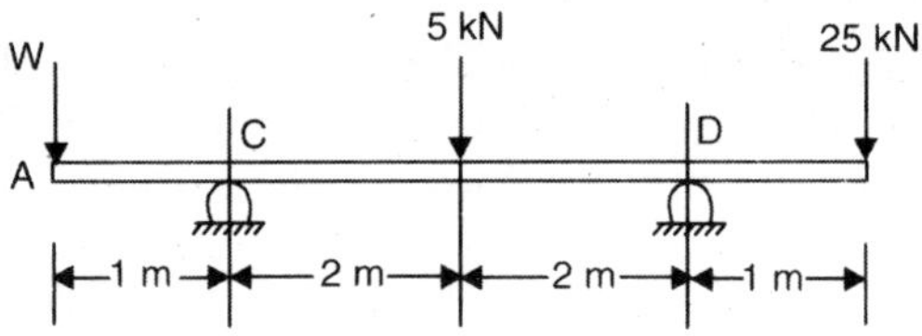

Fig. 9

10. Four figures are drawn below of a frame connecting with a load and string arrangement. In which case couple developed by the point A is maximum.

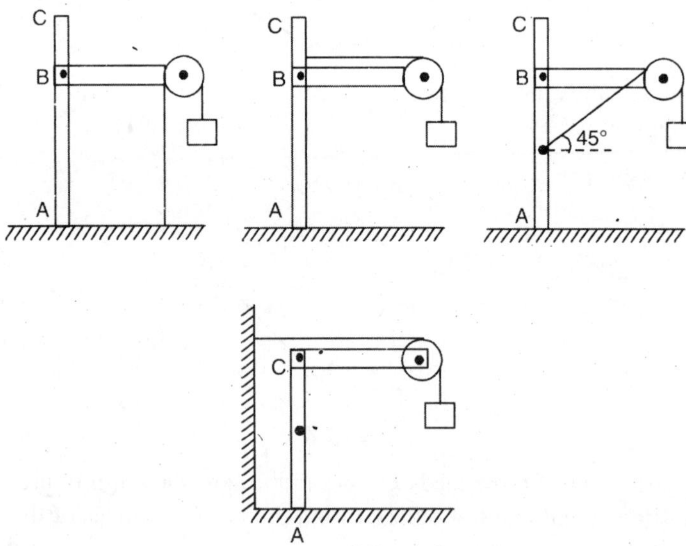

Fig. 10

11. Determine the reactions at A and B for the beam as shown in figure.

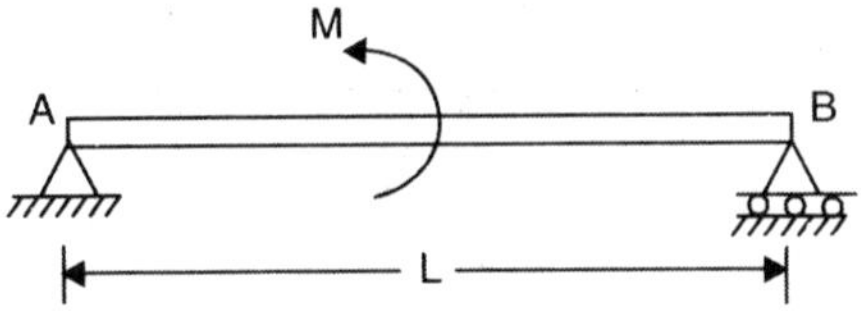

Fig. 11

12. Determine the maximum value of bending moment in the beam as shown in figure.

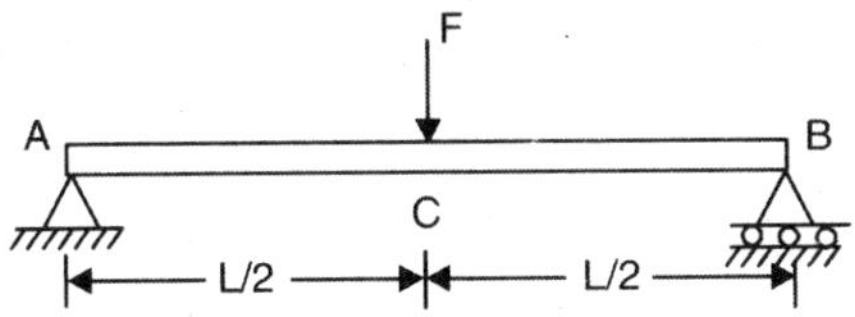

Fig. 12

13. A rope is passed round a stationary horizontal log fixed at a certain height above the ground. In order to keep a load of mass m kg suspended on one end of the rope, the maximum force T_1 should be applied to the other end of the rope. Determine the minimum force T_2 which must be applied to the rope to lift the load.
14. A block moves from rest down an inclined plane from a certain height h and then ascends along the same plane with an initial velocity such that it stops at the same height. In which case is the time of motion longer?
15. A man of mass 40 kg is managing to be at rest between two vertical walls by pressing wall A by feet and hands A and other wall B by his back. Find the normal force exerted by either wall on the man; friction coefficient is 0.5 between all contacts.

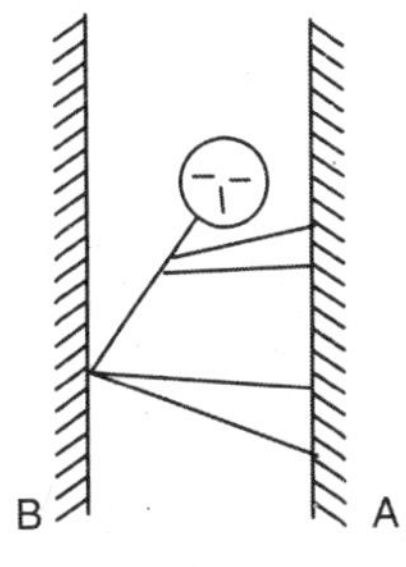

Fig. 15

16. A person of mass m_1 is applying a horizontal force F to slide a block of mass m_2. The coefficient of friction between the shoes of the person and the surface is μ_1 and that between the block and the surface is μ_2. Determine the condition so that block may not slide.
17. In which case force F applied by the person as shown in figure to lift the box will be minimum?

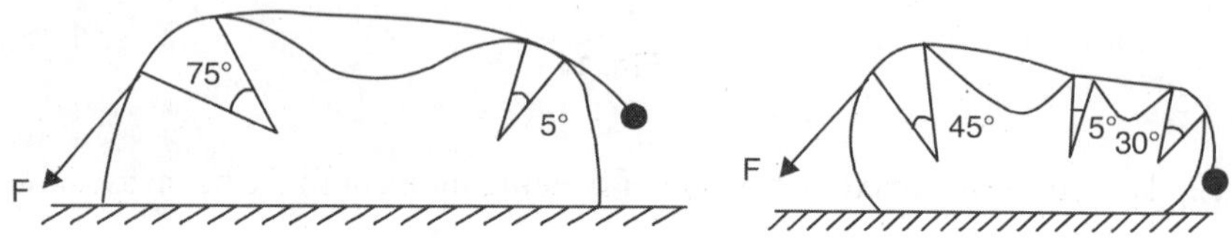

Fig. 17

18. A small sphere is kept on the bigger sphere as shown in figure. The interaction between these two spheres is 8 N when contact is smooth and 10 N when contact is rough. Determine the coefficient of friction between these two.

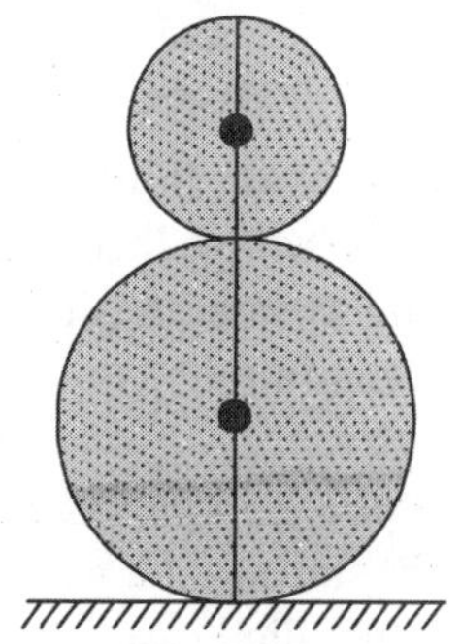

Fig. 18

19. A pair of compasses is used to describe circle on a horizontal piece of paper and has legs of equal length making an angle α with the vertical at any instant as shown. W is the vertical pressure on the handle. Determine the frictional force at one of the legs, which is rotating.

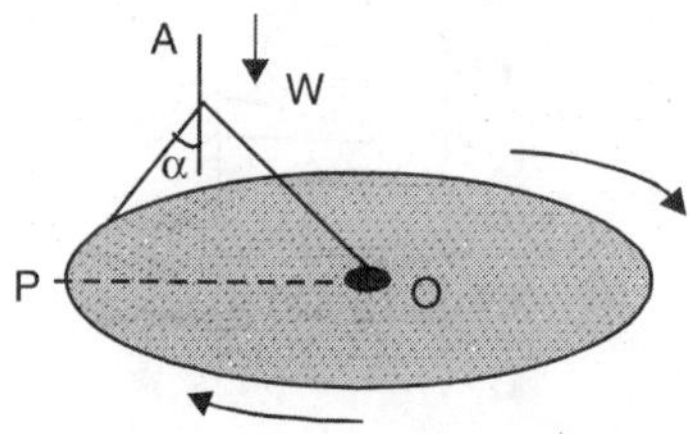

Fig. 19

Part II

Dynamics

1 Basics of Dynamics

Summary and Concept

Vector

A vector is a quantity with direction and magnitude. A vector is represented by

$$\vec{A} = |\vec{A}|\,\hat{A}$$

$|A|$ represents vector's magnitude and $\hat{A}$ gives the direction of the vector A.

Product

$$\vec{A} = A_x\hat{i} + A_y\hat{j} + A_z\hat{k}$$

$$\vec{B} = B_x\hat{i} + B_y\hat{j} + B_z\hat{k}$$

scalar product of two vector is given by

$$\vec{A}\cdot\vec{B} = |A||B|\cos\theta$$

$$= A_xB_x + A_yB_y + A_zB_z$$

where θ is the angle between two vectors A and B and also

$$\theta = \cos^{-1}\frac{A_xB_x + A_yB_y + A_zB_z}{\sqrt{A_x^2 + A_y^2 + A_z^2}\sqrt{B_x^2 + B_y^2 + B_z^2}}$$

vector product of two vectors is given by

$$\vec{A}\times\vec{B} = AB\sin\theta\,\hat{n}$$

$$= \begin{vmatrix} \hat{i} & \hat{j} & \hat{k} \\ A_x & A_y & A_z \\ B_x & B_y & B_z \end{vmatrix}$$

here $\hat{n}$ is as unit vector perpendicular to the plane which contains A and B.

Velocity and acceleration in polar coordinate systems

Velocity

$$\vec{v}(t) = \frac{d\vec{r}}{dt} = \dot{r}\hat{r} + r\dot{\theta}\hat{\theta}$$

$$V_r = \text{radial velocity} = \dot{r}$$

$$V_q = \text{transverse velocity} = r\dot{\theta}$$

$$|\vec{V}| = \sqrt{(V_r)^2 + (V_\theta)^2}$$

$$= \sqrt{(\dot{r})^2 + (r\dot{\theta})^2}$$

$$\tan \phi_1 = \left|\frac{V_r}{V_\theta}\right|$$

where ϕ_1 is an angle between velocity vector and transverse velocity vector.

Acceleration

$$\vec{a}(t) = \frac{d\vec{v}}{dt} = (\ddot{r} - r\dot{\theta}^2)\,\hat{r} + (2\dot{r}\dot{\theta} + r\ddot{\theta})\hat{\theta}$$

$$a_r = \text{radial acceleration}$$

$$= (\ddot{r} - r\dot{\theta}^2)$$

$$a_\theta = \text{transverse acceleration} = (2\dot{r}\dot{\theta} + r\ddot{\theta})$$

$$|\vec{a}| = \sqrt{a_r^2 + a_\theta^2}$$

$$= \sqrt{(\ddot{r} - r\dot{\theta}^2)^2 + (2\dot{r}\dot{\theta} + r\ddot{\theta})^2}$$

$$\tan \phi_2 = \left|\frac{a_r}{a_\theta}\right|$$

where ϕ_2 is an angle between acceleration vector and transverse acceleration vector.

Spherical coordinate system

In Cartesion coordinate system

$$x = r \sin \theta \cos \phi$$
$$y = r \sin \theta \sin \phi$$
$$z = r \cos \theta$$

In polar coordinate system

$$r = \sqrt{x^2 + y^2 + z^2}$$

$$\theta = \cos^{-1}\left(\frac{z}{\sqrt{x^2 + y^2 + z^2}}\right)$$

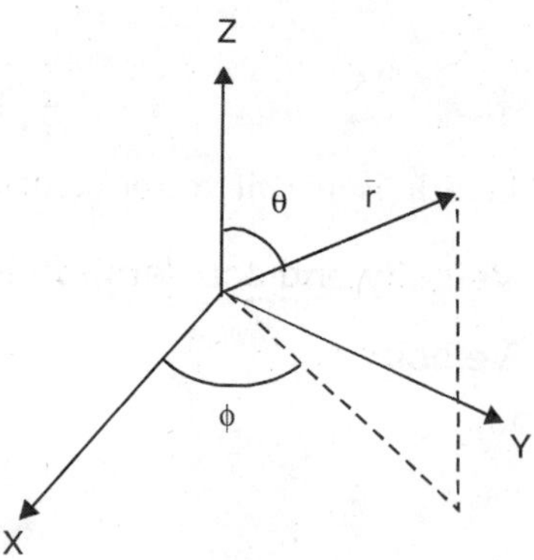

Fig. 1

$$\phi = \tan^{-1}\left(\frac{y}{x}\right)$$

Velocity and acceleration in spherical coordinate system

Velocity

$$\vec{V}(t) = \frac{d\vec{r}}{dt} = \dot{r}\hat{r} + r\dot{\theta}\hat{\theta} + r\sin\theta\,\hat{\phi}$$

Acceleration

$$\vec{a}(t) = \frac{d\vec{v}}{dt} = \left(\ddot{r} - r\dot{\theta}^2 - r\sin^2\theta\dot{\phi}^2\right) + \left(\frac{1}{r}\frac{d\left(r^2\dot{\theta}^2\right)}{dt} - r\sin\theta\cos\theta\,\dot{\phi}^2\right) +$$

$$\left(\frac{1}{r\sin\theta}\frac{d}{dt}\left(r^2\sin^2\dot{\phi}\right)\right)\hat{\phi}$$

In cylindrical coordinate system

Velocity

$$\vec{V}(t) = \frac{d\vec{r}}{dt} = (\dot{p})\hat{p} + (p\dot{\phi})\hat{\phi} + (\dot{z})\hat{z}$$

$$\vec{a}(t) = \frac{d\vec{v}}{dt} = \left(\ddot{p} - (\phi)^2\right)\hat{p} + \left(2\dot{p}\dot{\phi} + p\ddot{\phi}\right)\hat{\phi} + \ddot{z}\hat{z}$$

Radius of trajectory $y = f(x)$ at the point (x, y)

$$\rho = \frac{[1 + (dy/dx)^2]^{3/2}}{[dy/dx^2]}$$

ρ represents radius of curvature.

Velocity of approach and relative angular velocity

If two points A and B are separated by distance l at a certain moment of time move with velocities $\vec{V}_1$ and $\vec{V}_2$ at angles θ_1 and θ_2 with the direction A to B.

Then

$$V_1\cos\theta_1 - V_2\cos\theta_2 = -\frac{dl}{dt}$$

$$\left|\vec{V}_{12x}\right| = \frac{-dl}{dt}$$

$|\vec{V}_{12x}|$ represents velocity of approach. In case of rigid body distance between two points is constant so

$$|\vec{V}_{12x}| = 0$$

$$V_1 \cos \theta_1 = V_2 \cos q_2$$

And

The angular velocity of *B* relative to *A*

$$= \frac{V_2 \sin\theta_2 + V_1 \sin\theta_1}{l} = \frac{|\vec{V}_{21x}|}{1}$$

$$|\vec{\omega}_{AB}| = |\vec{\omega}_{BA}| = \frac{|\vec{V}_{21x}|}{l} = \frac{|\vec{V}_{12x}|}{l}$$

Instantaneous angular velocity and angular acceleration of a rigid body

$$\vec{\omega} = \frac{d\vec{\theta}}{dt}, \ \vec{\alpha} = \frac{d\vec{w}}{dt}$$

Resultant angular velocity $\vec{\omega}$ of a rigid body rotating simultaneously about two intersecting axes:

$$\vec{\omega} = \vec{\omega}_1 + \vec{\omega}_2 + \ldots$$

Equation of trajectory of a projectile

$$y = x(\tan\alpha) - \frac{gx^2}{2u^2 \cos^2 \alpha}$$

Here α = Angle of projection

u = Projection velocity

Relative motion between two particles

$$\vec{r}_{BA} = \vec{r}_B - \vec{r}_A$$

$$\vec{V}_{BA} = \vec{V}_B - \vec{V}_A \text{ and}$$

$$\vec{a}_{BA} = \vec{a}_B - \vec{A}_A$$

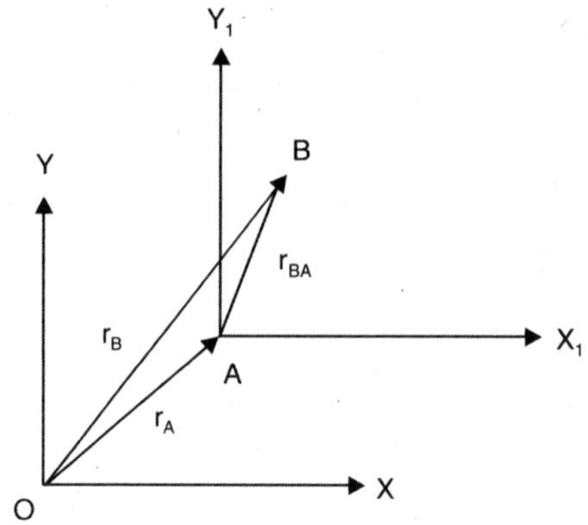

Fig. 2

Level 1

Example 1

The V-S graph for a vehicle movement is as shown. Determine a cceleration of vehicle at $S = 50$ m and $S = 150$ m. Draw a-s graph.

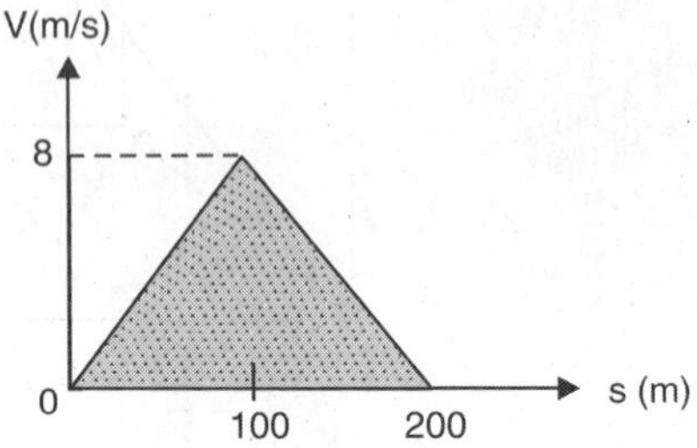

Fig. E1

Example 2

Two cars A and B start from rest at the origin $x = 0$ and move along a straight line such that $a_A = (6\,t - 3)$ m/s^2 and $a_B = (12\,t^2 - 8)$ m/s^2 where t is in sec. Determine the distance between them when $t = 4$ sec and the total distance each has traveled in $t = 4$ sec.

Example 3

Derive relation between V_A and V_B in terms of b and y.

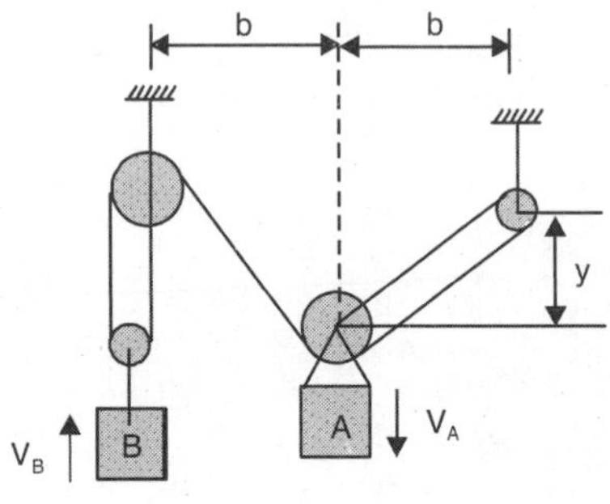

Fig. E3

Example 4

The motion of a particle is defined by the equations $x = \cos(\pi t)$ and $y = 1 - 2\cos(2\pi t)$ where x and y in meter and t is in sec. Determine velocity and acceleration when $t = 0$.

Example 5

The slotted arm AB drives the pin P through the spiral $r = 0.2\ e^{\theta}$ where r is in meter and θ in rad. If $\dot{\theta} = 3$ rad/s and constant, determine acceleration of pin when $\theta = 120°$.

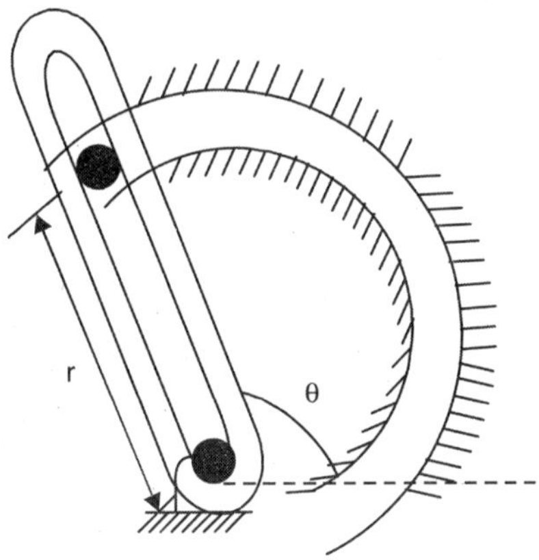

Fig. E5

Example 6

At a certain instant, cylinder A has a downward velocity of 0.8 m/s and an upward acceleration of 2 m/s^2. Determine the corresponding velocity and acceleration of cylinder B.

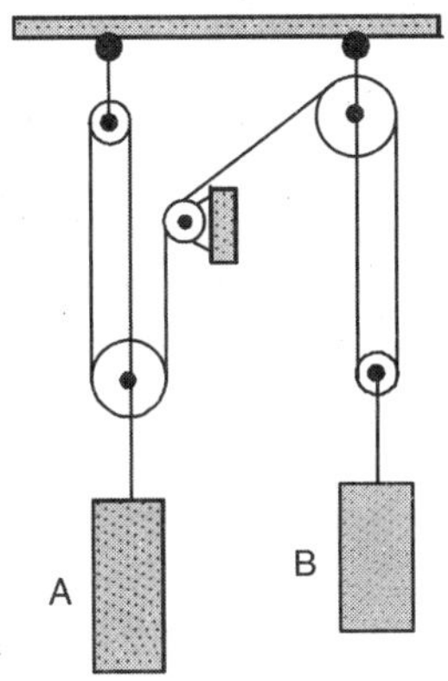

Fig. E6

Example 7

At a certain instant a particle which moves on a curved path has co-ordinates $x = 4$ m, $y = 3$ m, $\dot{x} = -3$ m/s, $\dot{y} = 4$m/s. Also $a_r = 10$ m/s^2 and $a_\theta = 15$ m/s^2, for this instant determine $\dot{r}$, $\ddot{r}$, $\dot{\theta}$ and $\ddot{\theta}$ and radius of curvature.

Example 8

Two particles are simultaneously thrown from roofs of two high buildings, as shown in Fig. Their velocities of projection are 2 ms^{-1} and 14 ms^{-1} respectively. Horizontal and vertical separation between points A and B is 22 m and 9 m respectively. Calculate minimum separation between the particles in the process of their motion.

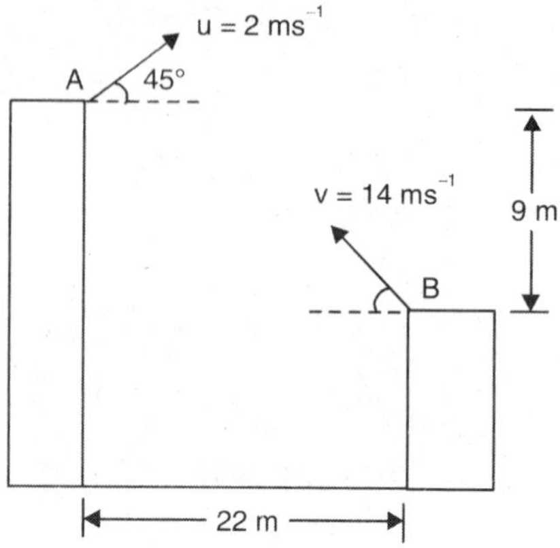

Fig. E8

Example 9

A fighter plane flies at a velocity of 300 m/s. On the fighter plane there is a gun, which shoots at a rate of 40 rounds per second with a muzzle velocity of 1200 m/s. The shots are aimed at another fighter plane flying at a velocity of 200 m/s. Find the rate at which the projectiles hit the target plane:

1. When the two planes move towards one another.
2. When the two planes move away from one another.

Example 10

The acceleration for a car traveling along a straight line is as shown. Given that $x = 0$ and $v = 60$ m/s when $t = 0$. Determine

1. The velocity and position of plane at $t = 20$ sec.
2. Its average velocity during the interval $6 < t < 14$ sec.

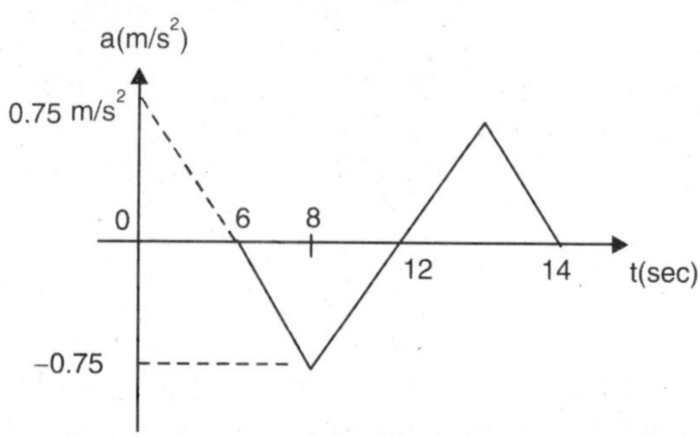

Fig. E10

Example 11

Passengers in the jet transport *A* flying east at a speed of 800 km/h observe a second jet plane *B* that passes under the transport in horizontal flight. Although the nose of *B* is pointed in the 45° north east direction plane *B* appears to the passengers in *A* to be moving away from the transport at the 60° angle as shown. Determine the true velocity of *B*.

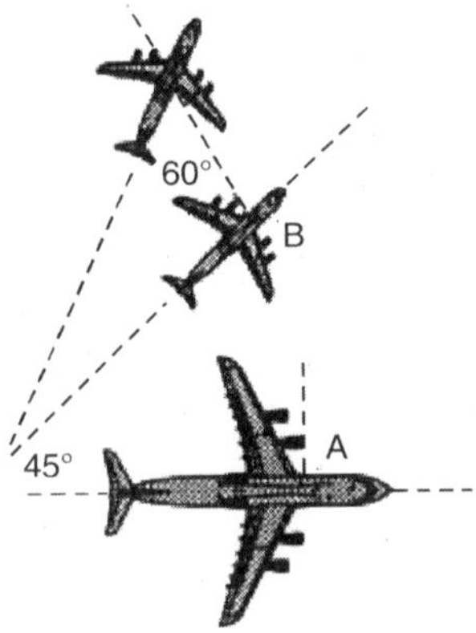

Fig. E11

Example 12

Determine the constraint equation which relates the accelerations of bodies *A* and *B*. Assume that the upper surface of *A* remains horizontal.

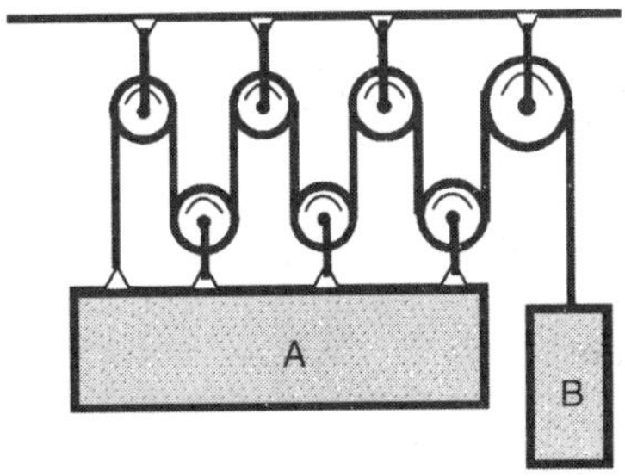

Fig. E12

Example 13

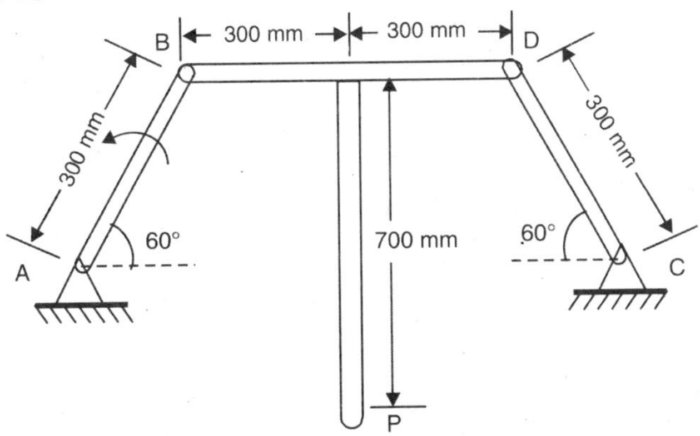

Fig. E13

The similar links *AB* and *CD* rotate about the fixed pins at *A* and *C*. If *AB* has an angular velocity $\omega_{AB} = 8$ rad/s, determine the angular velocity of *BDP* and the velocity of point *P*.

Example 14

Rod *AB* is attached to a collar at *A* and is fitted with a small wheel at *B* which moves on a circular track of 300 mm radius. *A* has a velocity of 250 mm/s upward. Determine the magnitude and direction of the velocity of point *B*. Rod *AB* is 200 mm inclined.

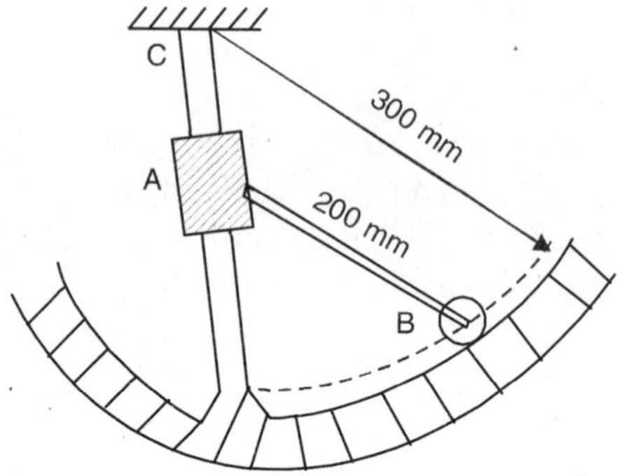

Fig. E14

Example 15

Under the action of force *P*, the constant acceleration of block *B* is 3 m/s^2 to the right. At the instant when the velocity of *B* is 2 m/s to the right, determine the velocity of *B* relative to *A*, the acceleration of *B* relative to *A*, and the absolute velocity of point *C* of the cable.

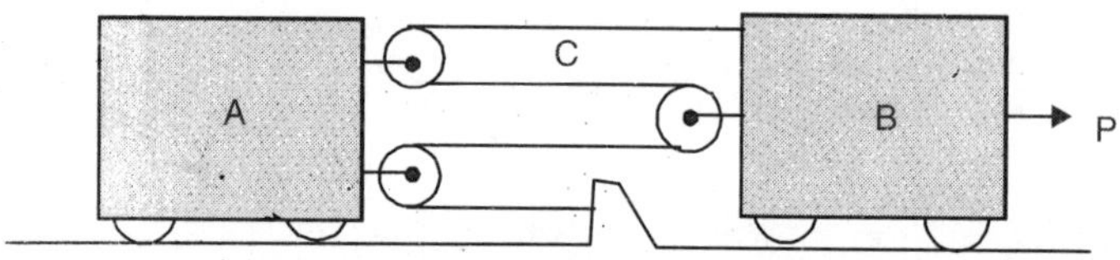

Fig. E15

Example 16

A rocket starts vertically up from the ground with a velocity of 120 m/s. Due to the aerodynamic drag and the gravitational acceleration, the net downward acceleration of the rocket is given by $a = g + 0.005\, V^2$ where V is instantaneous velocity upwards. At the top most

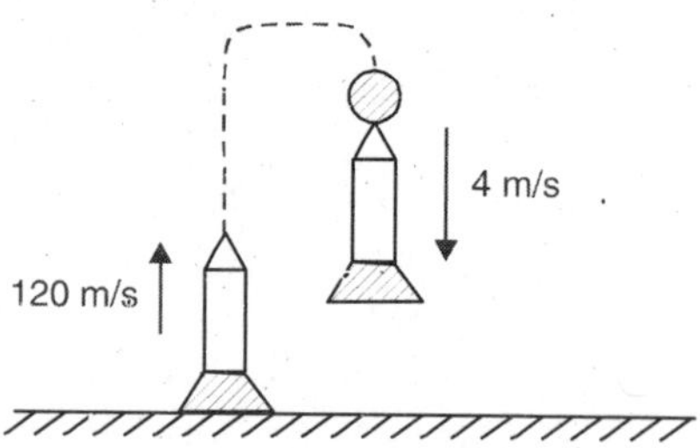

Fig. E16

point, a parachute ejects out of the rocket quickly acquires a constant downward velocity of 4 m/s. Find the total flight time of the rocket, i.e. time elapsed from take off till landing.

Example 17

The small block P starts from rest at time $t = 0$ at point A and moves up the incline with constant acceleration α. Determine $\dot{r}$ as a function of time.

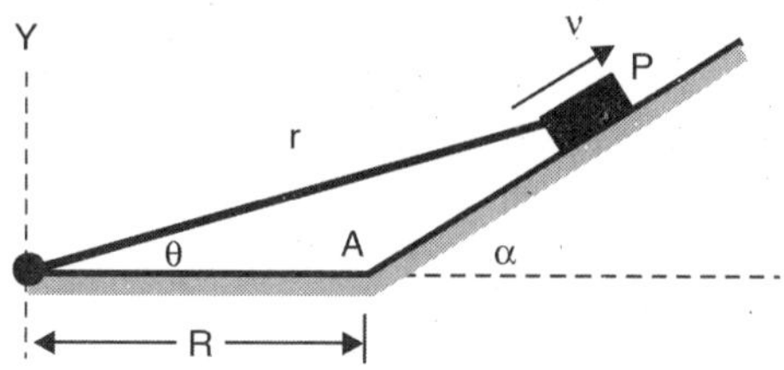

Fig. E17

Example 18

A rocket has expanded all its fuel when it reaches position A, where it has a velocity u at an angle θ with respect to the horizontal. It then begins unpowered flight and attains a maximum added height h at position B after traveling a horizontal distance s from A. Determine the expressions for h and s, the time of flight from A to B, and the equation of the path.

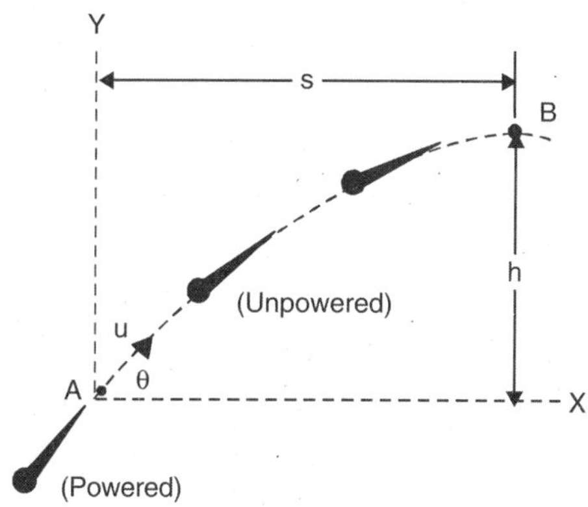

Fig. E 18

Example 19

For the pulley system shown each of the cables at A and B is given a velocity of 2 m/s in the direction of the arrow. Determine the upward velocity v of the load.

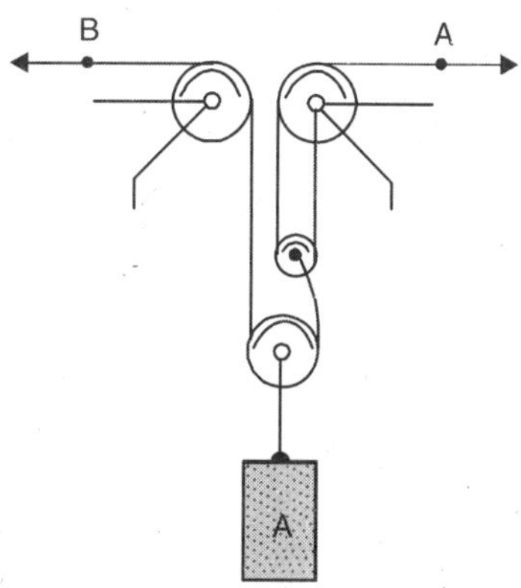

Fig. E19

Example 20

The wind appears to blow from south when it is observed from a ship moving due east at 9 km/h. After the ship has changed course and speed, it has started moving due north at 6 km/h. The wind appears to blow from southwest.

Assuming that the direction and velocity of the wind is constant during the entire period of observation, compute the magnitude and direction of true wind velocity.

Level 2

Example 1

A train passes a station *A* at 40 km per hour and maintains this speed for 7 km, and is then uniformly retarded, stopping at *B* which is 8 km from *A*. A second train starts from A the instant the first train passes and being uniformly accelerated for part of the journey and uniformly retarded for the rest stops at B at the same time as the first train. What is its greatest speed on the journey?

Example 2

For given case determine acceleration a_1 and a_2 during both intervals.

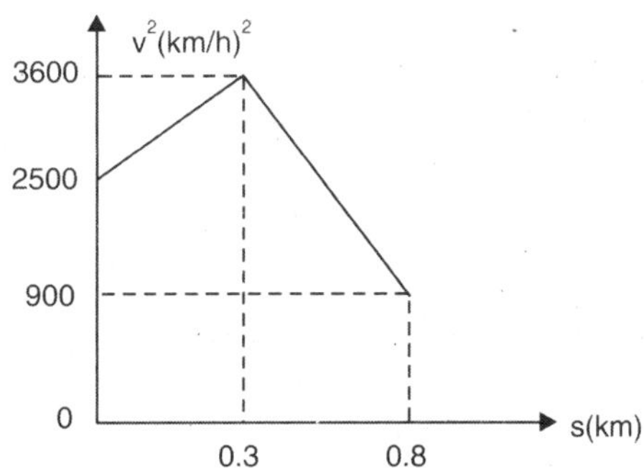

Fig. E2

Example 3

A man standing at *A* is following the movement of a car *B* which is traveling around a curved track at a constant speed of 25 m/s. Determine the angular rate $\dot{\theta}$ at which the man must turn in order to keep the eyes directed on the car at the instant $\theta = 45°$.

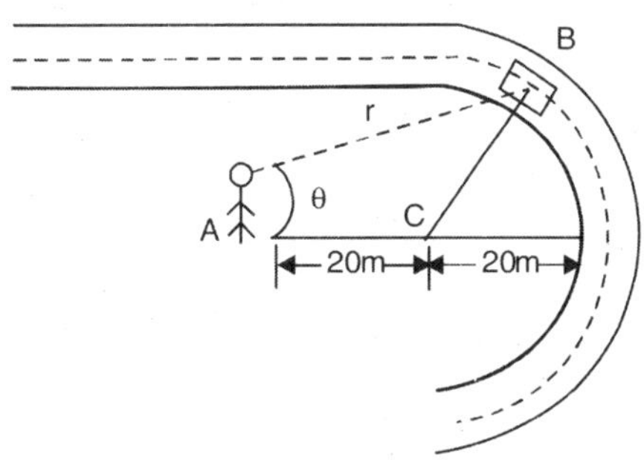

Fig. E3

Example 4

The position coordinate of a particle which is confined to move along a straight line is given by $s = 2t^3 - 24t + 6$, where s is measured in meters from a convenient origin and t is in seconds. Determine (*a*) the time required for the particle to reach a velocity of 72 m/s from its initial condition at $t = 0$, (*b*) the acceleration of the particle when $v = 30$ m/s.

Example 5

In the pulley configuration shown, cylinder A has a downward velocity of 0.3 m/s. Determine the velocity of *B*.

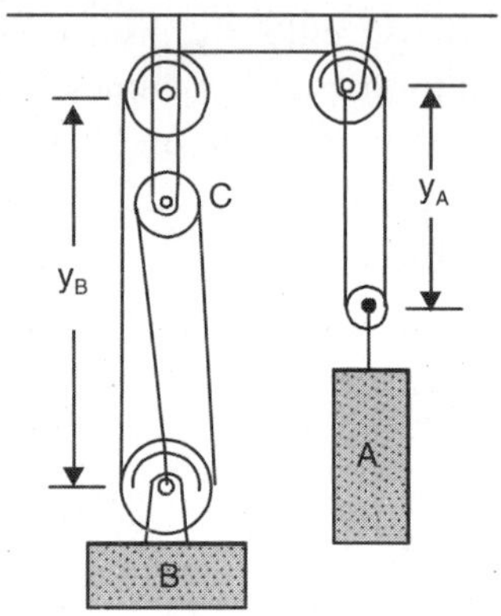

Fig. E5

Example 6

Collar *A* and block *B* are connected by a cable passing over three pulleys *C*, *D* and *E* as shown in Fig. E6. Pulleys *C* and *E* are fixed while *D* is attached to a collar which is pulled down with constant velocity 75 mm/s at $t = 0$, collar *A* starts moving downward from position *K* with a constant acceleration and no initial velocity. Knowing that the velocity of collar *A* is 300 mm/sec, as it passes through point *L*, determine the change in elevation, the velocity and acceleration of block *B* when collar *A* passes through *L*.

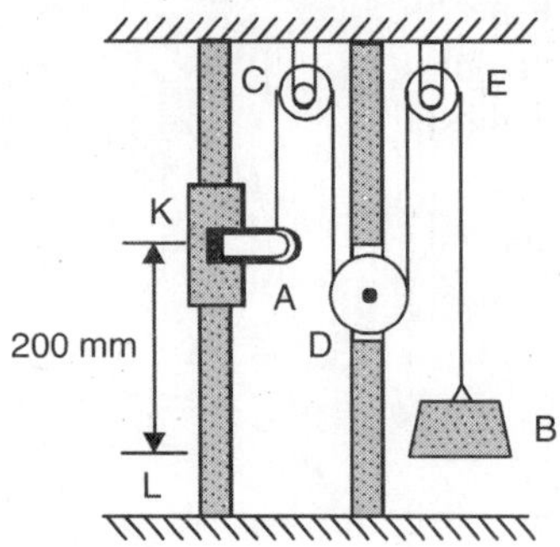

Fig. E6

Example 7

The motor at C and the motor at D draw the cable with a constant velocity of 4 m/s and a constant acceleration of 8 m/s^2 respectively. When $h = 3$ m, the motor at D is switched on. Find the time needed for $h = 0$, i.e. when crates A and B will be at the same level refer Fig. E7.

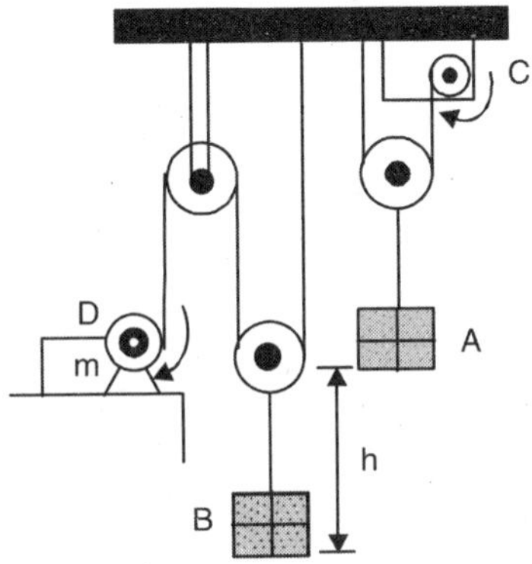

Fig. E7

Example 8

A glass of milk is poured into a glass of height 140 mm and inside diameter 66 mm. If the initial velocity of milk is 1.2 m/s at an angle of 40° with the horizontal, determine the range of values of height 'h' for which the milk will enter the glass. Refer Fig. E8.

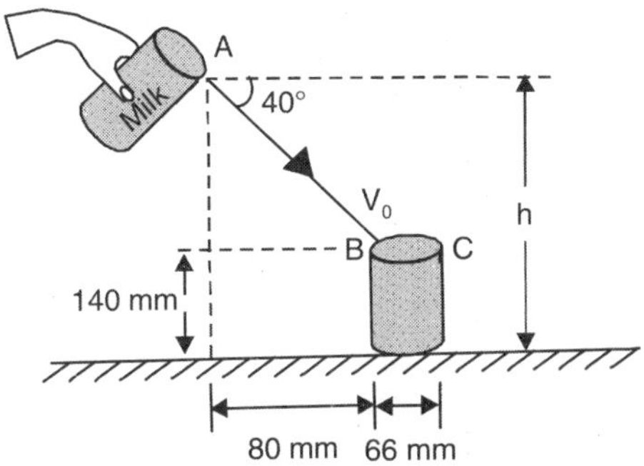

Fig. E8

Example 9

A particle moves along the x-axis with an initial velocity $v_x = 50$ m/s passing the origin when $t = 0$. For the first 4 seconds it has no acceleration, and thereafter it is acted on by a retarding force which gives it a constant acceleration $a_x = -10$ m/s^2. Calculate the velocity and the x coordinate of the particle for the conditions of $t = 8$ s and $t = 12$ s and find the maximum positive x-coordinate reached by the particle.

Example 10

The double gear shown rolls on the stationary lower rack the velocity of its center A is 1.2 m/s directed to the right. Determine (a) the angular velocity of the gear, (b) the velocities of the upper rack R and of point D of the gear. $R_1 = 150$ mm, $R_2 = 100$ mm.

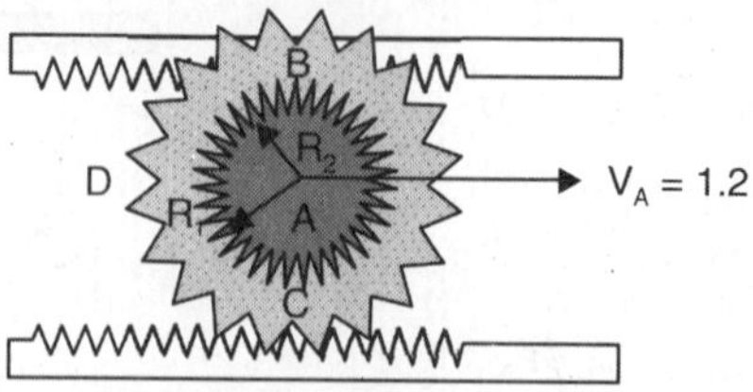

Fig. E10

Example 11

A gun is aimed at point A and fired. Calculate the distance y below A to the point 'B' where the bullet strikes. The velocity of bullet is 800 m/s.

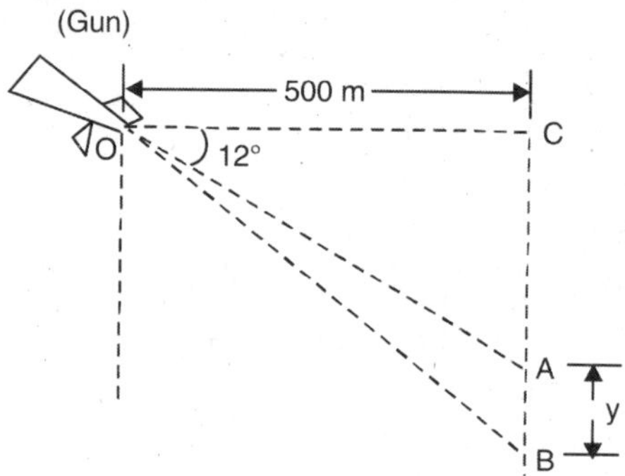

Fig. E11

Example 12

A wheel of radius r rolls without slip along the x-axis with constant speed V_0. Derive the equation of velocity of point A on the rim of wheel in terms of V_0, r and t.

Example 13

A particle travels along branch of a hyperbola $x^2/a^2 - y^2/b^2 = 1$. If the particle is moving with constant speed V_0. Determine the acceleration of particle when it is at 'A'.

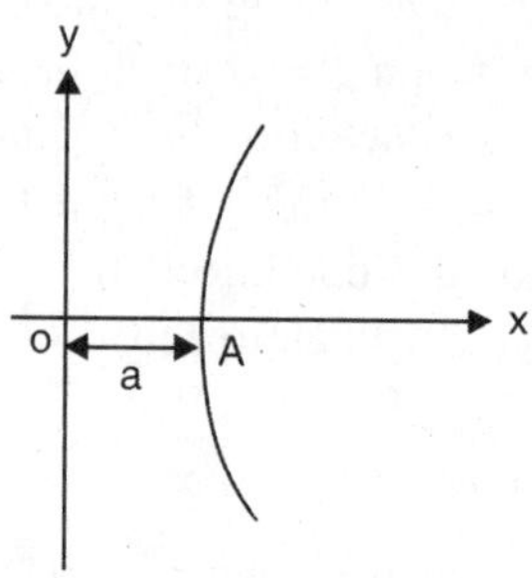

Fig. E13

Example 14

A ball is to be thrown from a height of 2 m from the floor of a horizontal tunnel of 10 m diameter. If velocity V_0 = 100 kmph, find maximum possible range inside the tunnel.

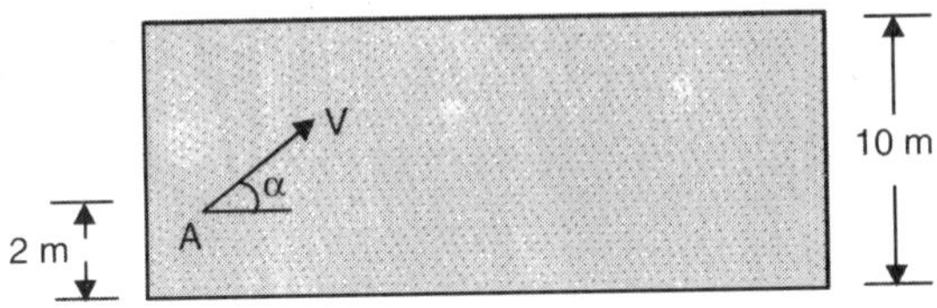

Fig. E14

Example 15

A ball is thrown horizontally from the top of a 50 m in cliff at A with a speed of 15 m/s and lands at point C. Because of a strong horizontal wind the ball has a constant acceleration in the negative x direction. Determine the radius of curvature ρ of the path of the ball at B where its trajectory makes an angle of 45° with the horizontal. Neglect any effect of air resistance in the vertical.

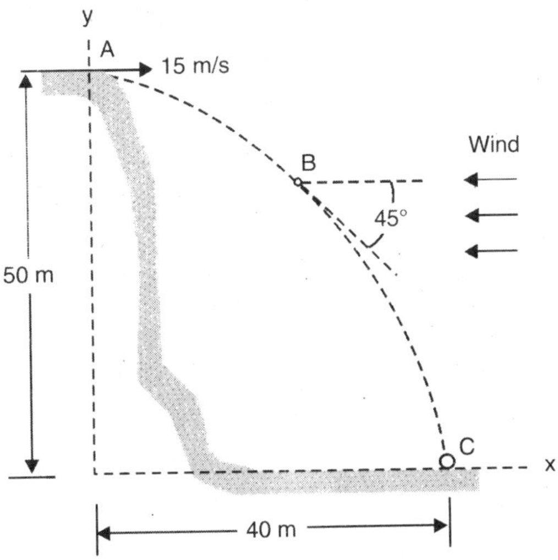

Fig. E15

Example 16

Find V_B in terms of y.

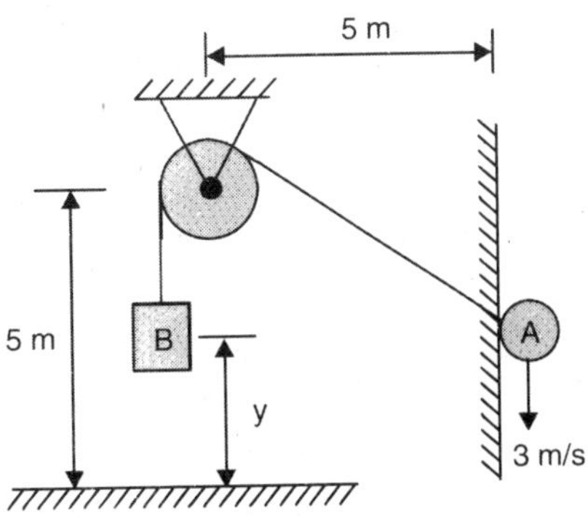

Fig. E16

Example 17

For a car moving along a straight-line *s-t* diagram is as shown in Fig. Plot *v-t* and *a-t* diagrams for time period $0 \le t \le 15$ sec.

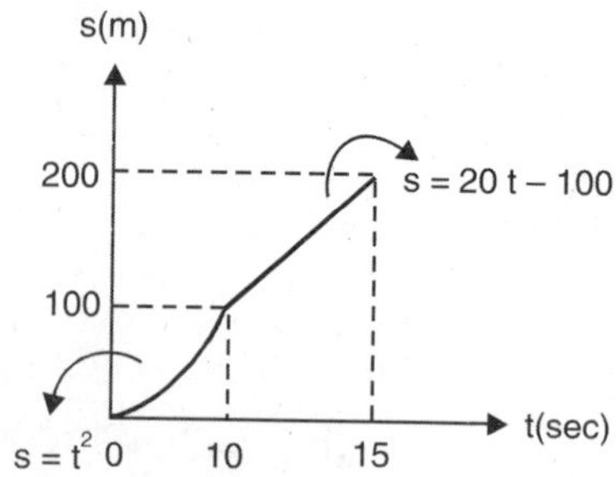

Fig. E17

Example 18

A circle *A* of radius a turns around its center with uniform angular velocity ω. *A* circle *B* of radius *b* rolls on the circle *A* and its uniform angular velocity is ω′. Find the time taken

(*i*) For the point of contact to make a complete circle of *A*.

(*ii*) For the center of *B* to return to a former position.

Determine the accelerations of the common point of the two circles and the greatest acceleration of a point on the circle *B*.

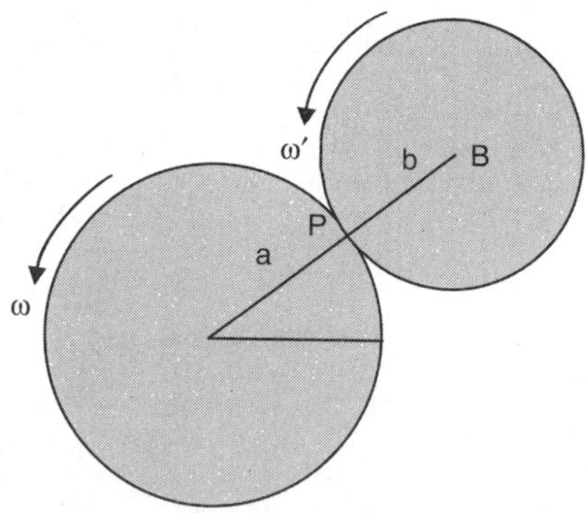

Fig. E18

Example 19

A particle is moving along a vertical circle of radius $R = 20$ m with a constant speed $V = 31.4\ \text{ms}^{-1}$ as shown in Fig. E19. Straight line *ABC* is horizontal and passes through the center of the circle. A shell is fired from point A at the instant when particle is at *C*. If distance *AB* is $20\sqrt{3}$ m and shell collide with the particle at *B,* calculate

(*i*) Smallest possible value of the angle θ of projection,

(*ii*) Corresponding velocity *u* of projection.

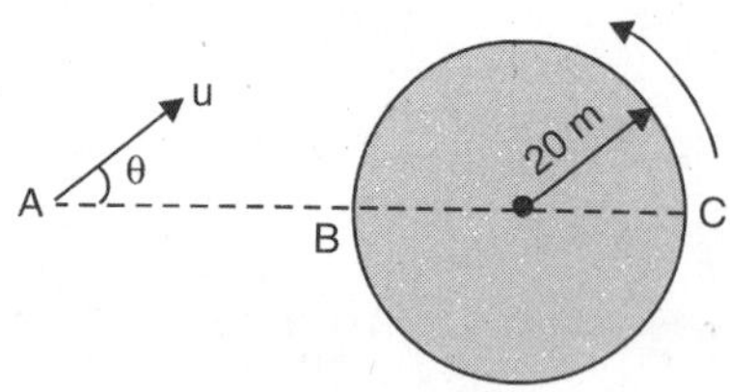

Fig. E19

Example 20

A cannon is placed on an inclined plane, as shown in Fig. A projectile is fired with an initial velocity v_0, forming an angle θ relative to the inclined plane.

1. Calculate the range R on the inclined plane between the firing and the hitting points.
2. Calculate the angle θ, which maximizes the range.

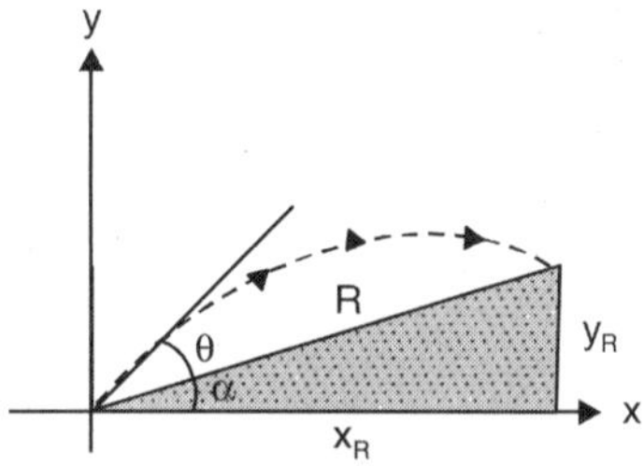

Fig. E20

Example 21

A thin smooth pipe, rotated by an engine, rotates around one corner of a horizontal table with a constant angular velocity ω_0. At $t = 0$, when the pipe creates an angle $\theta = 0$, a small mass m is shot through its edge at the center of the circle, with velocity v_0. The mass moves in the pipe without friction. The angular velocity of the pipe remains unchanged.

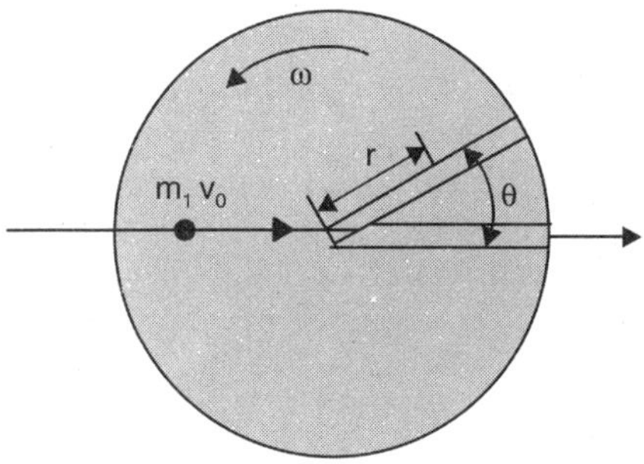

Fig. E21

Draw the path of the particle before it leaves the pipe and after it leaves the table.

Example 22

A particle travels along a parabolic path with a constant speed. Find the maximum acceleration if the path is $y = kx^2$ where k is a constant.

Example 23

The tractor A is used to hoist the cable B with the pulley arrangement shown. If A has a forward velocity v_A, determine an expression for the upward velocity v_B of the cable in terms of x.

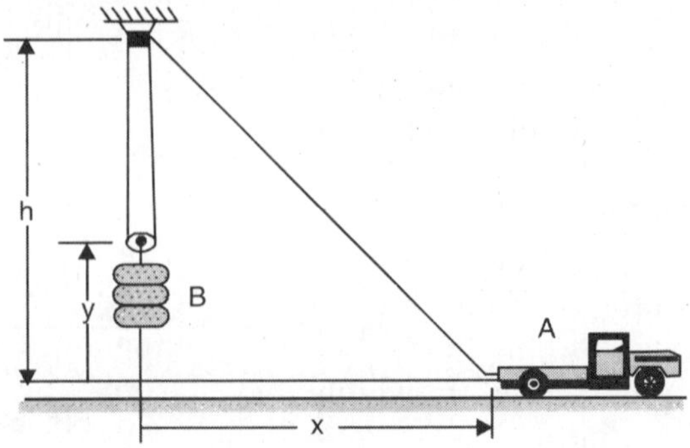

Fig. E23

Example 24

A rod *OA* is hinged at *O* has a pin welded to it. The distance *OP* is 100 mm. The pin panes thrown the slot of a slotted horizontal bar which is moving upwards with constant velocity 100 mm/sec. At the instant when $y_p = 50$ mm, calculate the *x*-component of velocity and acceleration of *PL*.

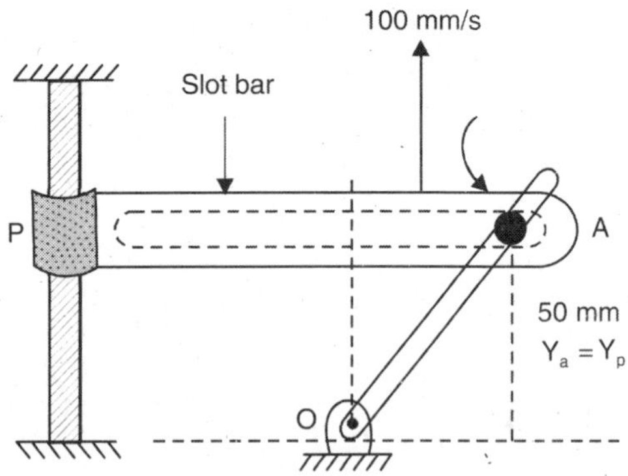

Fig. E24

Example 25

Slider block *A* as shown in figure moves downward at a speed of 0.5 m/s, the velocity with respect to *A* of the portion of the belt B between the pulleys *C* and *D* is $V_{CD/A}$ = 2 m/s. Determine the velocity of portion CD of the belt when (*a*) θ = 45° (*b*) θ = 60°.

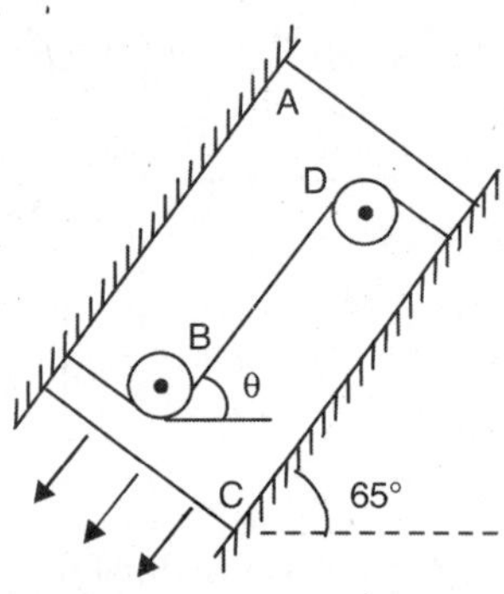

Fig. E26

Example 26

On a ground, an experiment is going on by two friends Mickey and Buttu, both are separated by distance d to each other. Wind is blowing from Mickey to Buttu. Mickey calls Buttu at any instant but she listens after time t_1 but when both interchange the position she listens after time t_2.

Calculate the velocity of sound in still air and the velocity of wind.

Example 27

A disc of radius r rolls without slipping with velocity and acceleration to the left shown in figure. Determine the acceleration of point A and C on the disc.

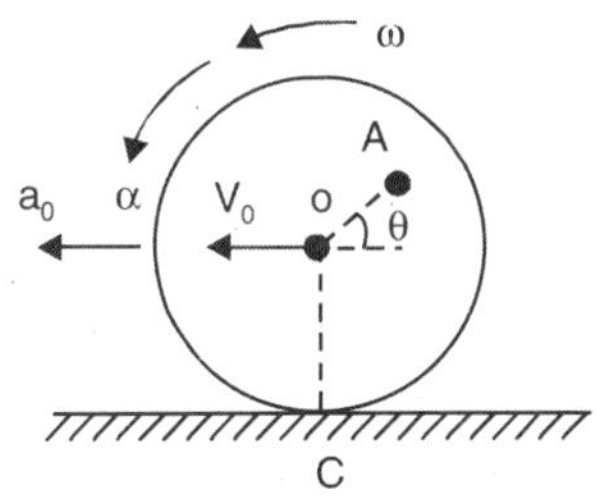

Fig. E 27

Example 28

Two points describe concentric circles with velocities varying inversely as the square roots of the radii of the circles; R_1 and R_2. Find positions in which the relative angular velocity vanishes.

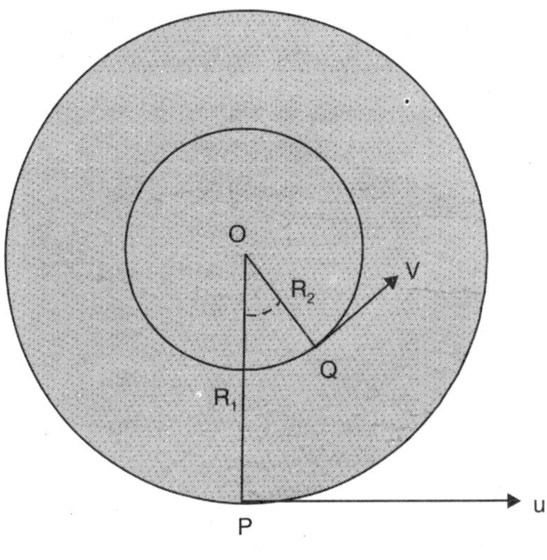

Fig. E28

Example 29

A freighter is moving at a speed of 8 knots when its engines are suddenly stopped. It takes 10 minutes for the freighter to reduce its speed to 4 knots, determine and plot the distance s in nautical miles moved by the ship and its speed v in knots as functions of the time t during this interval. The deceleration of the ship is proportional to the square of its speed so that $a = -kv^2$.

Example 30

Collars A and B slides along the fixed right-angled rods and are connected by a cord of length L. Determine the acceleration a of collar B as function of y if collar A is given a constant upward velocity v_A.

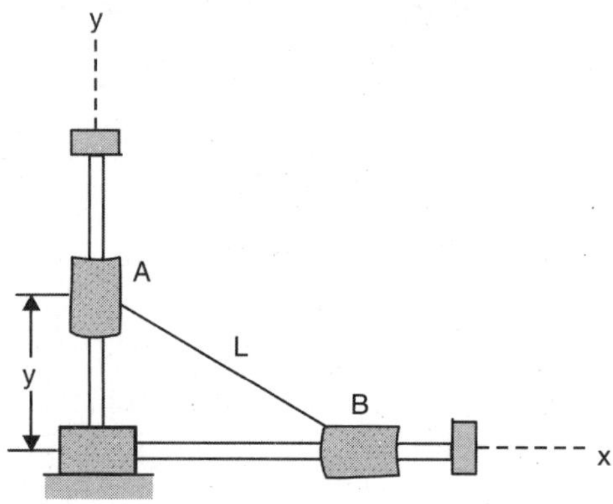

Fig. E30

Example 31

A particle 'P' moves along the spiral path $\gamma = \left(\frac{10}{\theta}\right)$m, where θ is in radians. If it maintains a constant speed of $v = 6$ m/s. Determine the magnitude of v_r and v_θ as a function of θ and evaluate each at $\theta = 1$ radian.

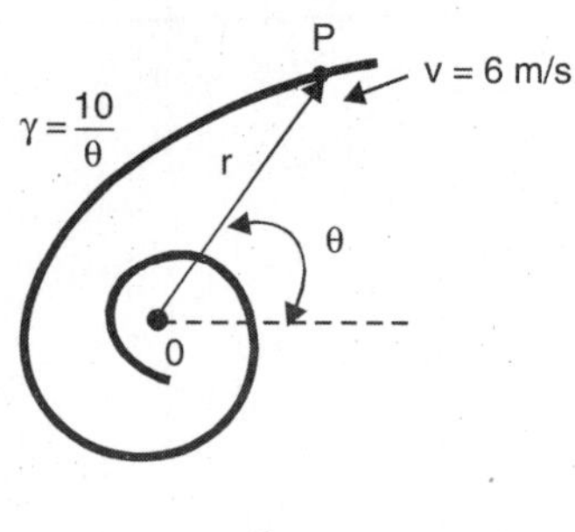

Fig. E31

Example 32

Three boys Jack, Sam and Mikey are sitting on the ground with forming an equilateral triangle of length l. Each of the boys moves with constant speed v_0. Jack always has its velocity along Sam, Sam along Mikey, and Mikey along Jack. At what time will all boys meet each other?

Example 33

The water sinker is attached with the spool as shown in figure. If water sinker rotates with angular velocity 20 rad/s and spool rotates with angular velocity 40 rad/s, simultaneously assembly rotates about the vertical axis with an angular velocity 20 m/s. Determine the magnitude ω of the total angular velocity of the water sinker.

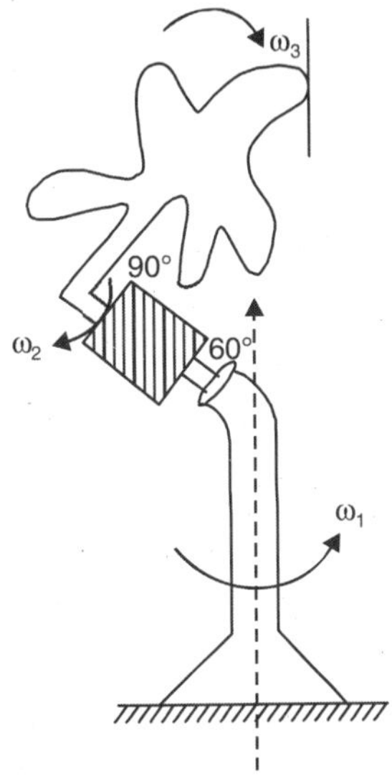

Fig. E33

Example 34

Neglect the diameter of the small pulley attached to body A and determine the magnitude of the total velocity of V in terms of the velocity v_A, which body A, has to the right. Assume that the cable between B and the pulley remains vertical and solve for a given value of x.

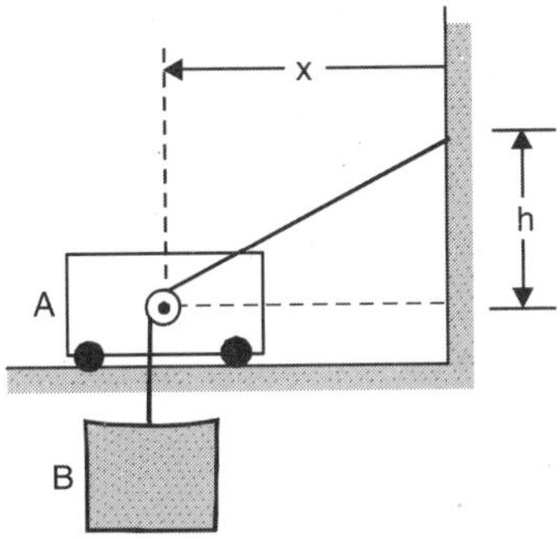

Fig. E34

Level 3

Example 1

If velocity of block B with respect to block A is $V_{B/A} = 5.6$ m/s at 70° with x-axis in 1st quadrant, find the velocities of A and B.

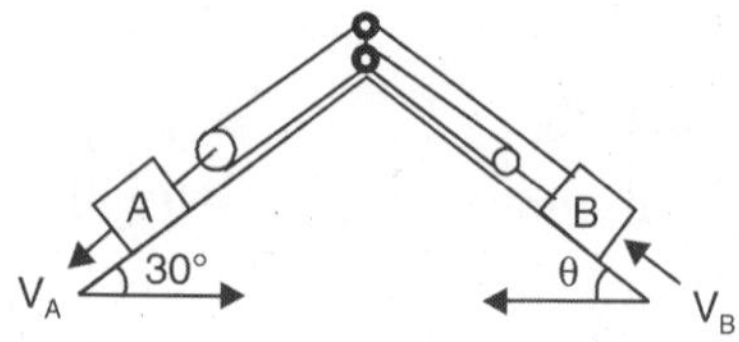

Fig. E1

Example 2

A magnetic tape is wound on an empty spool rotating at a constant angular velocity the final radius r_f of the winding was found to be three times as large as the initial radius r_i. The winding time of the tape is t_1. Determine the time required for winding a tape whose thickness is half that of the initial tape.

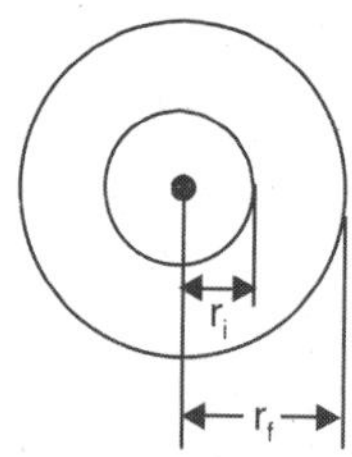

Fig. E2

Example 3

A sitting rat in a garden suddenly sees a cat. To save its life the rat runs away in a straight line with constant velocity u. Without any time lag the cat starts with a constant speed $v > u$ to catch the rat. At the initial moment directions of velocity of cat and rat is perpendicular and l is the separation between them. If the cat always heads towards the rat, find the time after which the cat catches the rat.

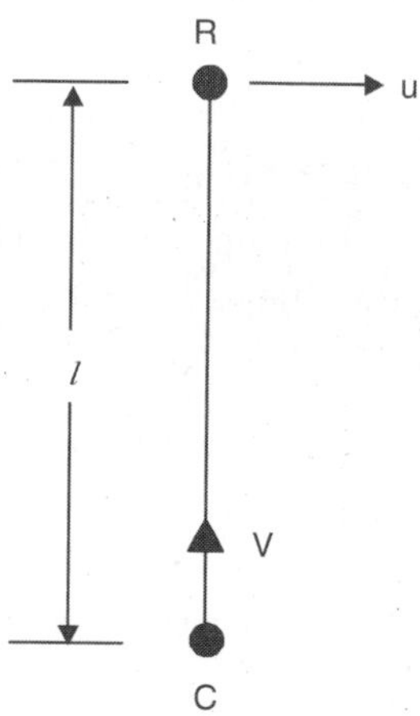

Fig. E3

Example 4

A weightless inextensible rope rests on a stationary wedge forming an angle θ. One end of the rope is fixed to the wall at point A. A small load is attached to the rope at point B. The wedge starts moving to the right with a constant acceleration a. Determine the acceleration of the load when it is still on the wedge.

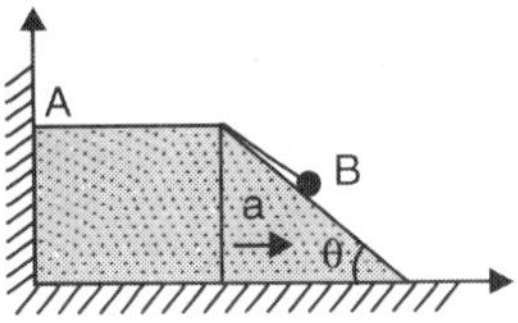

Fig. E4

Example 5

A cylinder of radius r_3 rolls inside a hollow cylinder of radius r_1 and r_2 (outer and inner). A cord is connected at point A i.e. top of the hollow cylinder is drawn by velocity v_0 as shown in figure. Determine the angular velocity of the cylinder.

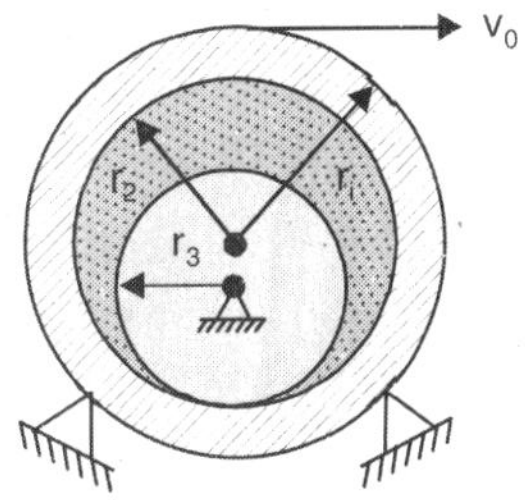

Fig. E5

Example 6

Two boys move with velocities v_1 and v_2 at an angle α and β respectively as shown in figure on horizontal plane. Determine the minimum distance between boys and corresponding time.

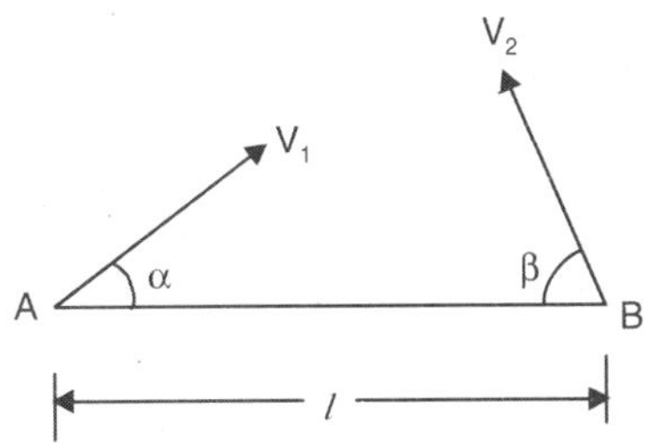

Fig. E6

Example 7

Magnetic tape runs over the idler pulley in a computer as shown. If the total acceleration of a point P on the tape in contact with the pulley makes an angle of 4° with the tangent to the tape at time $t = 0$ when the velocity v of the tape is 4 m/s, determine the time t require to bring the pulley to stop with constant deceleration. Assume no slipping between the pulley and the tape.

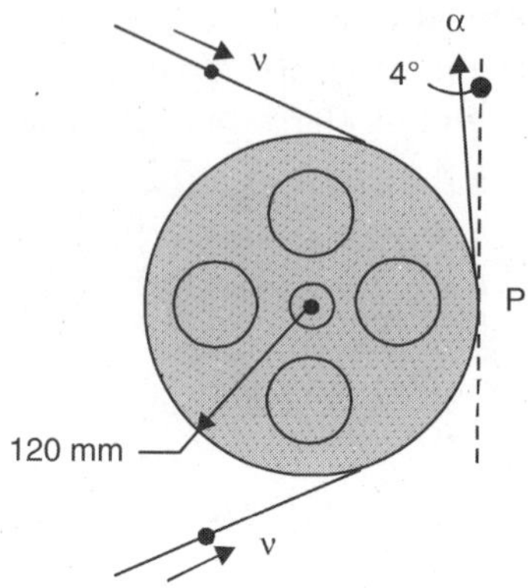

Fig. E7

Example 8

In the Geneva mechanism disk D rotates with a constant counter-clockwise angular velocity ω of 10 rad/s. At the instant when $\phi = 150°$ determine the angular acceleration of disk.

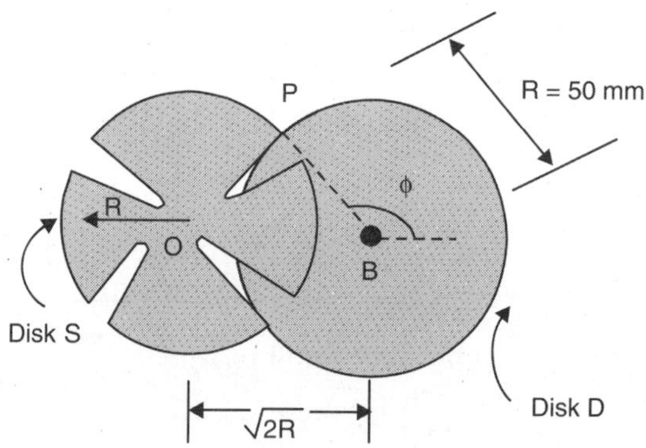

Fig. E8

Example 9

A projectile is ejected into an experimental fluid at time $t = 0$. This initial speed is v_0 and the angle to the horizontal is θ. The drag on the projectile results in an acceleration term $a_D = -kv$, where k is a constant and v is the velocity of the projectile. Determine the x-and y-component of both the velocity and displacement as functions of time. What is the terminal velocity? Include the effects of gravitational acceleration.

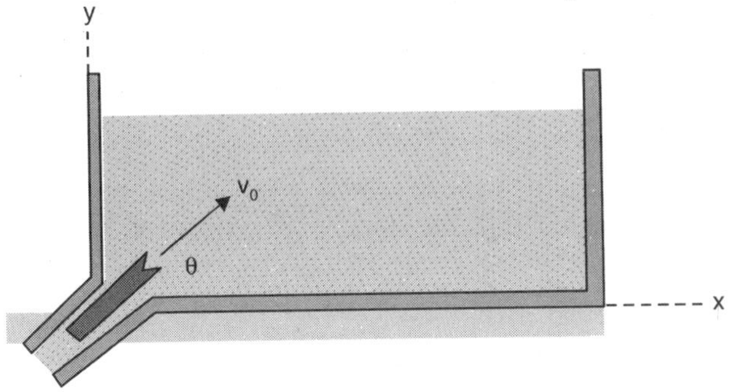

Fig. E9

Example 10

The point A on the rod CA is confined to move in the rotating slot of rod OD. The mechanism shown in the figure. Angular velocity of the rod OD is 2 rad/s for the position shown in figure. Determine the angular acceleration of AC and acceleration of A relative to the rotating slot in rod OD.

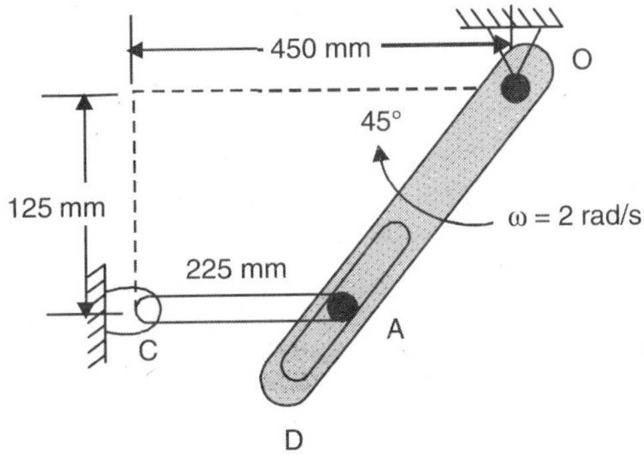

Fig. E10

Example 11

The two particles A and B starts at the origin O and travel in opposite direction along the circular path at constant speeds $V_A = 0.7$ m/s and $V_B = 1.5$ m/s respectively as shown. Determine in $t = 2$ s (a) the displacement along the path of each particle, (b) the position vector to each particle, and (c) the shortest distance between the particles.

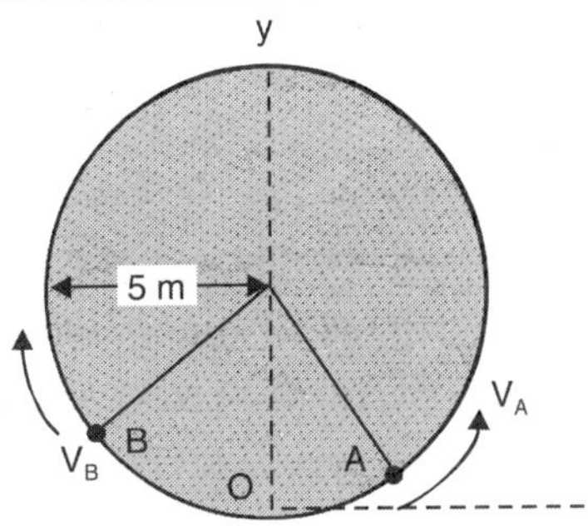

Fig. E11

Example 12

A spool or bobbin or yo-yo rolls without slipping on a horizontal surface so that the velocity v of the end of the string is directed along the horizontal. A board hinged at B leans against the spool. Determine the angular velocity ω of the board as function of α.

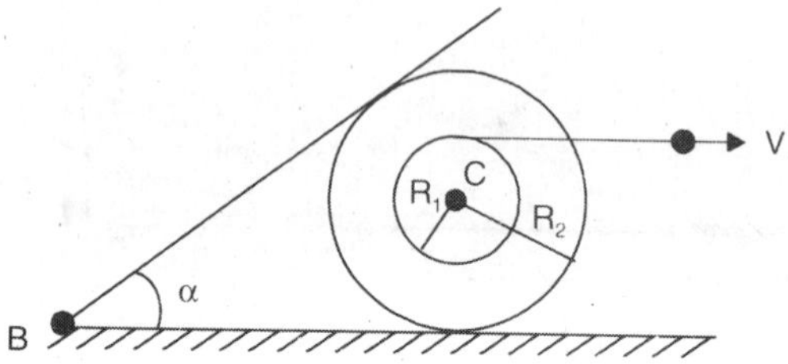

Fig. E12

Example 13

A rocket is fired with velocity 50 m/s from point B simultaneously a car starts its motion with velocity 20 m/s as shown in figure. After 4 sec rocket deflects its direction of motion making an angle 30° with vertical without changing its speed at that instant. Calculate the angular acceleration of line of sight observed by car observer during the motion of rocket when it deflects at point C and its topmost point of trajectory.

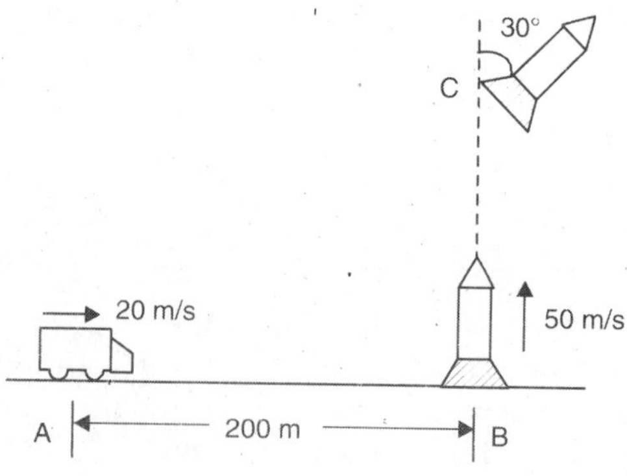

Fig. E13

2 Newton's Laws

Summary and Concept

Newton's second law

In an inertial frame (k)

$$\sum \vec{F} = m\vec{a}$$

in projection form,

$$\sum F_x = ma_x$$

$$\sum F_y = ma_y$$

$$\sum F_z = ma_z$$

in non-inertial frame (k') which rotates with a constant angular velocity ω about an axis translating with an acceleration a_0.

$$\vec{F} - m\vec{a}_0 + m\omega^2\vec{\rho} + 2m(\vec{v}' \times \vec{\omega}) = m\vec{a}'$$

where $\vec{\rho}$ is the position vector of the particle relative to the rotation axis and v' is the velocity of body m relative to the rotating frame.

For the translational motion of a system of bodies

$$\sum \vec{F}_{ext} = \sum m_i \, \vec{a}_{ic}$$

here $\vec{F}_{ext}$ represent net external forces acting on the system and $\vec{a}_{ic}$ is the linear acceleration of i-th

constituent of the system whose mass is m_i.

Newton's second law in rotation

$$\sum \vec{\tau}_i = I\vec{\alpha}$$

here $\sum \vec{\tau}$ represents net torque on the system about an axis and I and α are moment of inertia and angular acceleration about an axis.

And $$\vec{\tau} = \vec{r} \times \vec{F}$$

An inertial frame is a system in which the laws of physics are valid. An inertial frame is usually defined either as a system which moves at a constant velocity or as a system at rest.

If K is an inertial frame, then k' which is rotated relative to k and k'' which moves at a constant velocity relative to k are also inertial systems.

$$\vec{F} = m\vec{a}$$

in non-intertial reference frame

$$\sum \vec{F}' = m\vec{\alpha}'$$

here k' accelerates at a rate a_0 relative to k.
Newton's law in the accelerating frame at reference.

$$\sum \vec{F}' = \sum \vec{F} - m\vec{a}_0$$

k' accelerates at an acceleration a_0 along a straight line relative to k.

$$\sum \vec{F}' = ma' = \sum F - ma_0$$

– ma_0 is nothing but D'Alembert's force

k' rotates around a direction $\hat{n}$ at an angular velocity $\vec{\omega} = \omega\hat{n}$, relative to an inertial frame k.

$$\vec{a}_R = \vec{a} - 2\vec{\omega} \times \vec{V}_R - \vec{\omega} \times \left(\vec{\omega} \times \vec{V}_R\right) - \left[\vec{\omega} \times (\vec{\omega} \times \vec{r})\right]$$

$$m\vec{a}_R = m\vec{a} - 2m\left(\vec{\omega} \times \vec{V}_R\right) - m\left[\vec{\omega} \times (\vec{\omega} \times \vec{r})\right]$$

where

$\vec{V}_R$ = Particle's velocity in k′ frame

$\vec{r}$ = Particle's position

ω = Angular velocity

$\vec{a}_R$ = Acceleration of particle in k′ frame

$\vec{a}$ = Acceleration of particle in k frame

$-2\left(\vec{\omega} \times \vec{V}_R\right)$ is called the coriolis acceleration

$-\left(\vec{\omega} \times (\vec{\omega} \times \vec{r})\right)$ is called the centrifugal acceleration

$m\vec{a}_R$ is force acting in the k' frame.

Level 1

Example 1

A rubber string of mass m and rigidity k is suspended at one end. Determine the elongation Δl of the string.

Example 2

A smooth incline of angle α is accelerated at a rate a. A block of mass m is placed on the incline. At $t = 0$ the block is released, and begins moving.

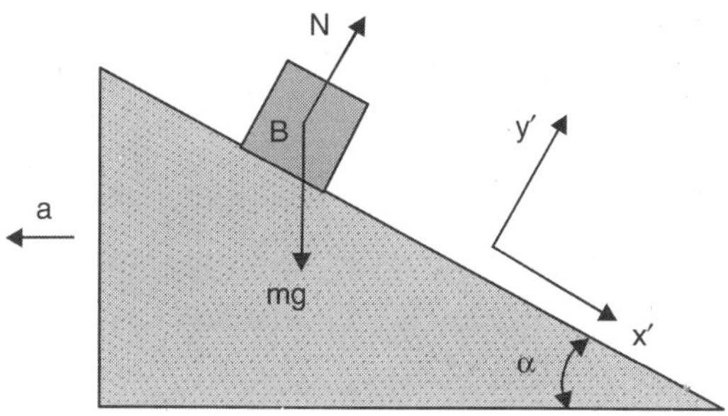

Fig. E2

How much time is required for the block to slide a distance L along the incline?

Example 3

All the surfaces are frictionless and the strings and the pulleys are light. Find the acceleration of mass m_1.

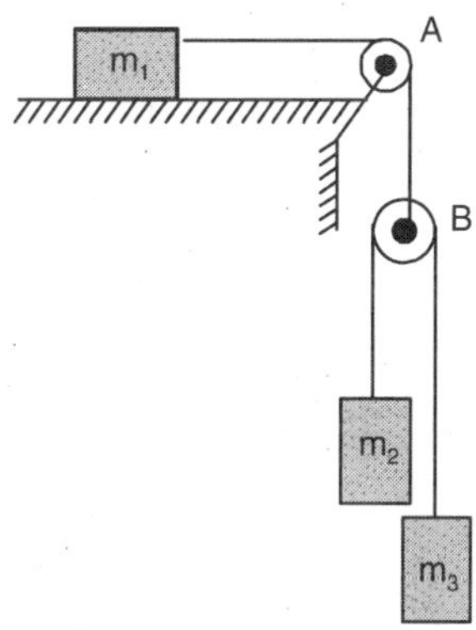

Fig. E3

Example 4

The roller coaster cars shown are traveling at a speed of v = 50 km/h when they pass over the top of hill. If the radius of curvature is 21 m, determine the acceleration of the cars as they pass over the top of hill.

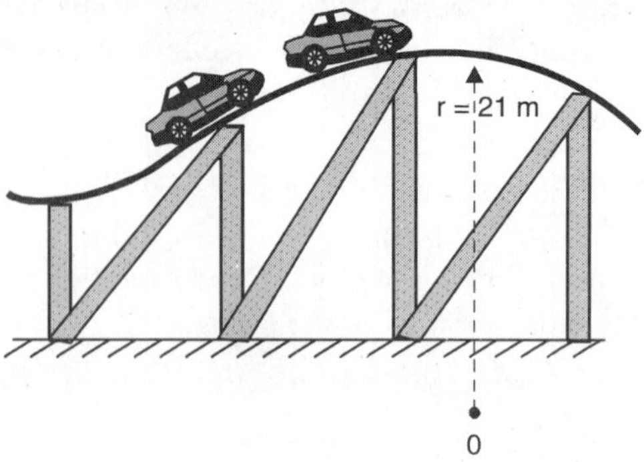

Fig. E4

Example 5

Determine the maximum value of mass m_3 in order that m_2 accelerates over the table without m_1 slipping over m_2. The masses of blocks m_1 and m_2 are 2 kg and 4 kg respectively and coefficient of static friction between m_1 and m_2 is 0.4 and the coefficient of kinetic friction between m_2 and the table is 0.2. Pulley is light and frictionless.

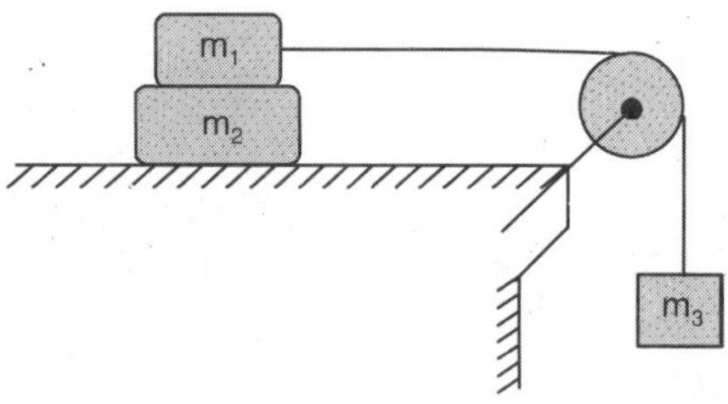

Fig. E5

Example 6

Each of the three plates has a mass of 10 kg. If the coefficient of friction at surface of contact are $\mu_s = 0.3$ and $\mu_k = 0.2$ respectively, determine the acceleration of each plate when the three horizontal forces are applied as shown in figure.

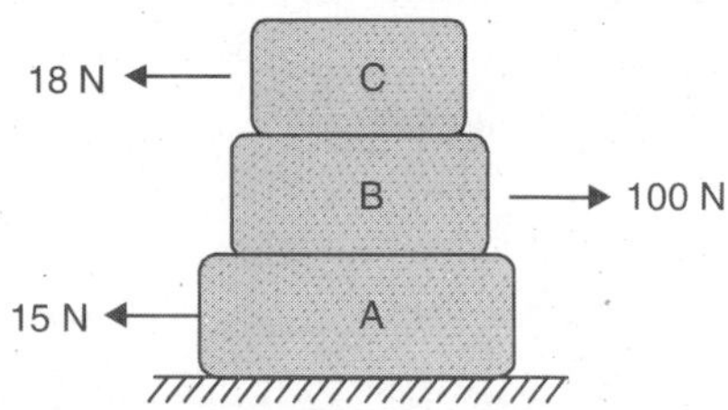

Fig. E6

Example 7

A water container undergoes constant acceleration down on an inclined plane as shown in figure. Determine the angle made by the water level with the inclined surface. (g = 10 m/s^2).

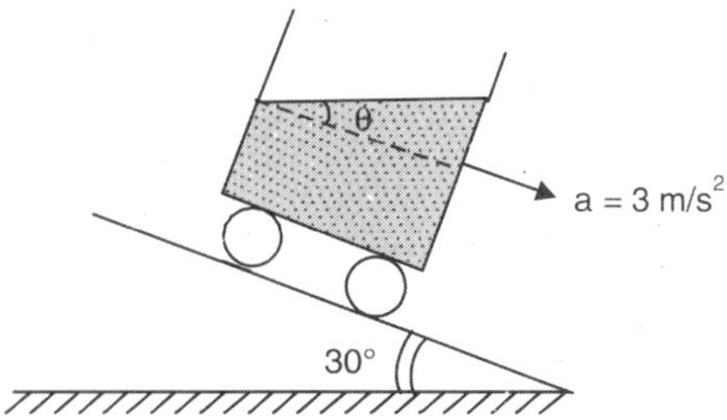

Fig. E7

Example 8

Two blocks A and B having masses 50 kg and 100 kg respectively are connected by a string which passes over frictionless pulley as shown in figure. Coefficient of friction between block and surface is 0.2 for both A and B. Determine force P if

(*a*) The system is prevented to move towards left.

(*b*) The system is just on the point of moving towards right.

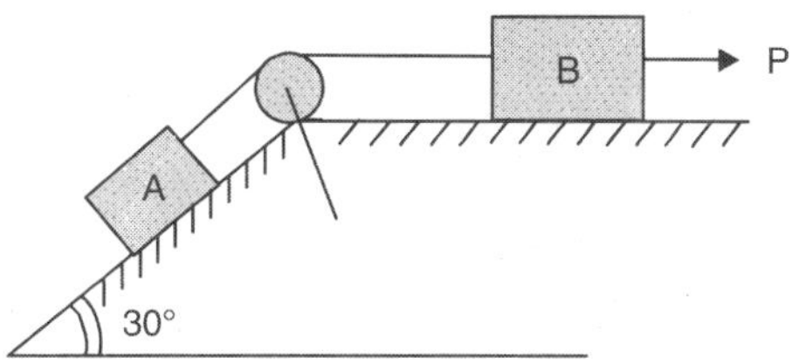

Fig. E8

Example 9

The speed of a car increases uniformly with time from 50 kmph at A to 100 kmph at B during 10 seconds. The radius of curvature of the hump at A is 40 m. The magnitude of total acceleration of the car's mass center is the same at B as at A. Compute the radius of curvature R_B of the dip of the road at B. The mass center of the car is 0.6 m from the road as shown in figure. If mass of the car is 250 kg compute the reaction offered at position A and B.

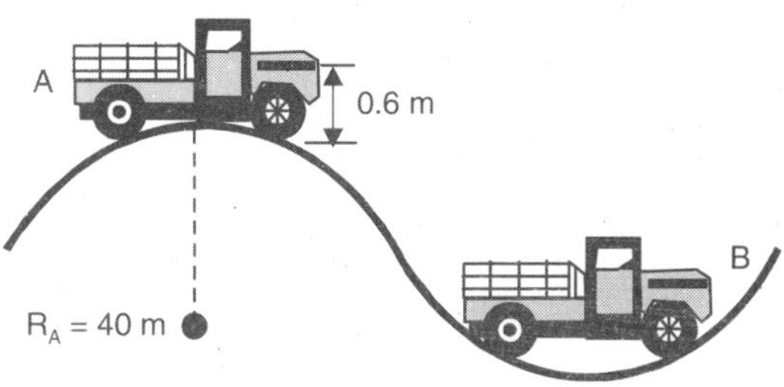

Fig. E9

Example 10

Five identical balls each of mass m are joined with right equiangular frame of negligible mass. The assembly rests on smooth horizontal plane. *A* force F is applied on the topmost ball. Determine the instantaneous acceleration of the assembly.

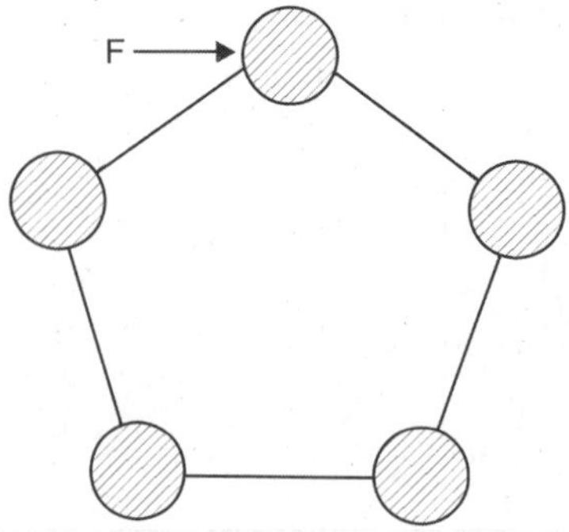

Fig. E10

Example 11

A sphere falls freely under gravity from a height of 19.6 m and then enters water. Determine the time it will take to come again to the water surface. Neglect air resistance and viscosity. Density of sphere is half that of water.

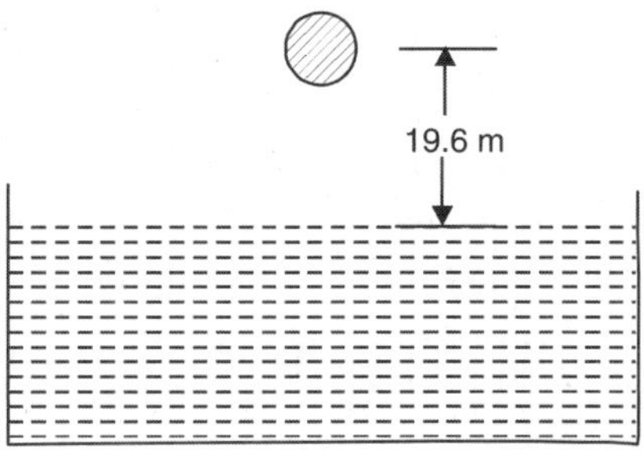

Fig. E11

Example 12

A particle of mass 20 gm is attached to a 0.2 m long string fixed to the apex of a cone with rough surface as shown in figure. At time $t = 0$, the cone is given a constant angular acceleration $\alpha = 0.5$ rad/s^2 by a simple motor mechanism about its vertical axis to lift the particle from the surface. There is no relative motion with respect to surface of the cone. Determine the time after which particle will leave contact with the surface and also find tension in the string at that instant.

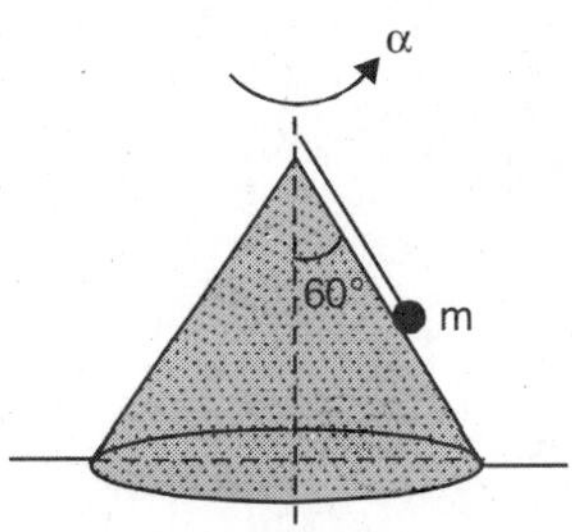

Fig. E12

Example 13

The bracket shown is used to transport cylindrical drum from one elevation to another. If co-efficient of static friction between drum and bracket is $\mu = 0.3$, determine the smallest ratio h/d for which the drum will tip before it slides. Given $g = 10 \text{ m/s}^2$, $a = 2.5 \text{ m/s}^2$.

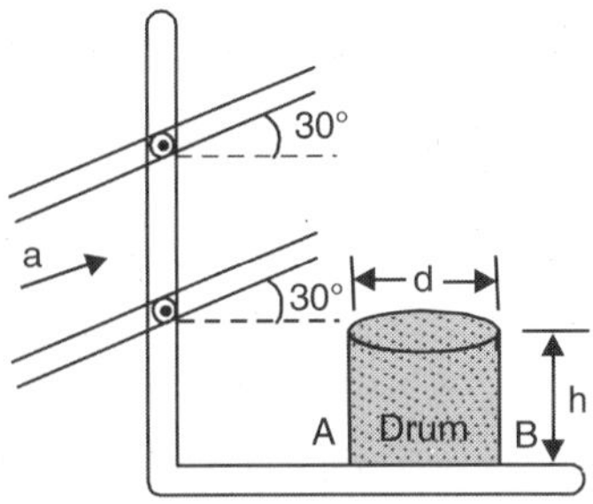

Fig. E13

Example 14

The sports car having a mass of 1700 kg is traveling horizontally along a 20° banked track which is circular and has a radius of curvature of 100 m. If the coefficient of static friction between the tyres and road is $\mu_s = 0.2$, determine the maximum constant speed at which car can travel without sliding up the slope. Neglect the size of car.

Example 15

A uniform rectangular block 0.6 m × 1.2 m and weight 1000 N is placed on rough inclined plane ($\mu_k = 0.2$). It is subjected to force $P = 1500$ N applied parallel to the plane as shown. Calculate maximum and minimum values of 'd' so that block does not overturn as it slides up the plane whose slope is 3/4.

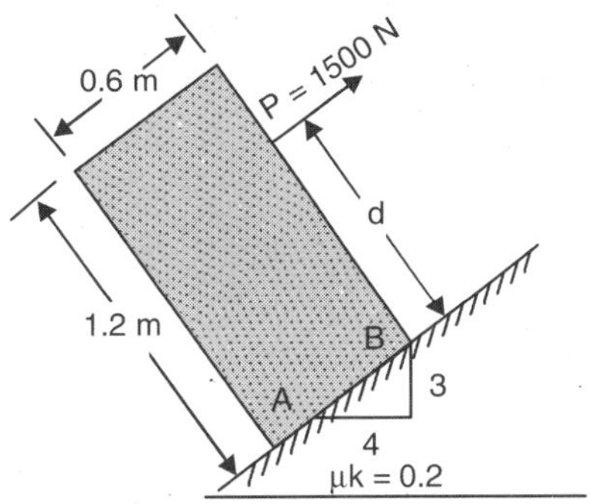

Fig. E15

Example 16

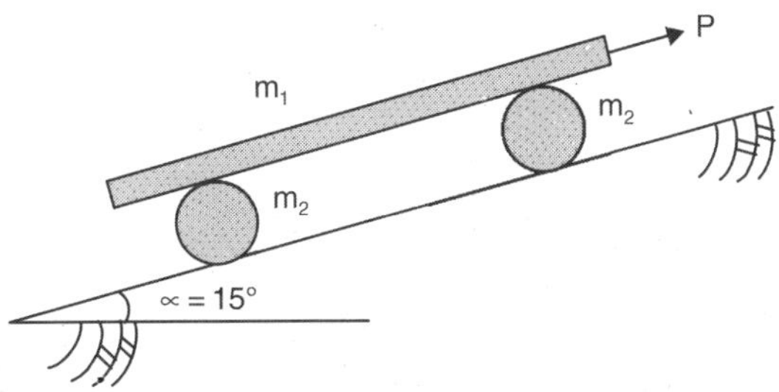

Fig. E16

A beam of mass $m_1 = 100$ kg, supported on two cylindrical rollers each of mass $m_2 = 20$ kg and radius $r = 100$ m, is moving up the plane by a force P applied parallel to the plane as shown in Fig. E16. Find the force P if the beam is moving up with acceleration of 0.1 m/s^2. Assume no slip at the contacts.

Example 17

The coefficient of friction between the blocks is μ. Find the minimum and the maximum force F that can be applied in order to keep the smaller blocks at rest.

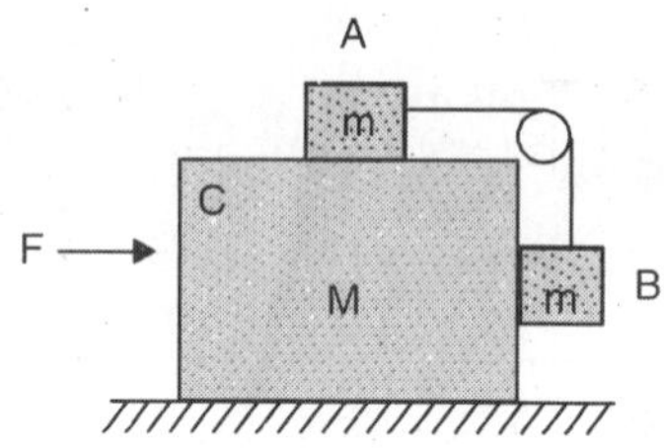

Fig. E17

Example 18

Find the acceleration of the smaller block with respect to the wedge. Assume all surfaces are smooth.

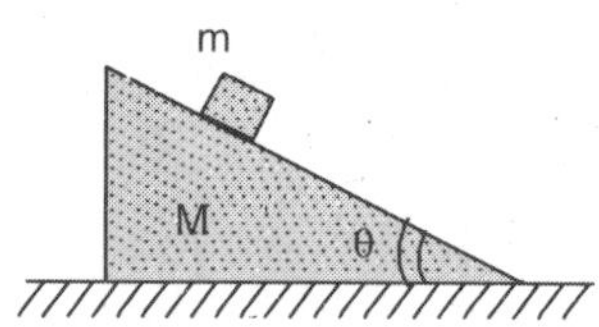

Fig. E18

Example 19

Two blocks of mass M_1 and M_2 are suspended from a rigid support as shown in figure. The length of each cable is 1 meter. The upper wire has no mass and the lower cable has a uniform mass of λ kg/m. The whole system of blocks, cables and support have an upward acceleration of 0.2 m/s^2. Determine the tension in the midpoint of the lower wire and mid-point of the upper wire. Given $M_1 = 2.9$ kg, $M_2 = 1.9$ kg, $\lambda = 0.2$

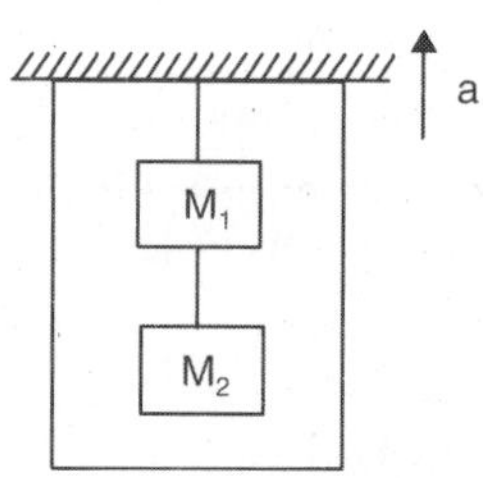

Fig. E19

Example 20

A small object of mass m is released from rest at A and slide down the smooth circular surface of radius R. A conveyor belt is joined at end point B. The radius of pulley of conveyor belt is r. Determine the normal contact force as a function of θ and also determine the minimum value of angular velocity ω of the conveyor pulley so that object can safely transfer on the conveyor without sliding, compare the value of ω when the *AB* track will be like as shown in figure.

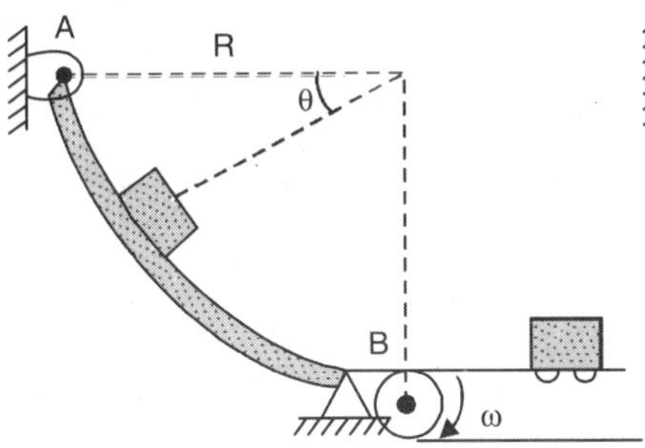

Fig. E20

Example 21

Determine the acceleration of the mass center of the assembly as shown in figure. Each block has mass m and there is no friction at contact.

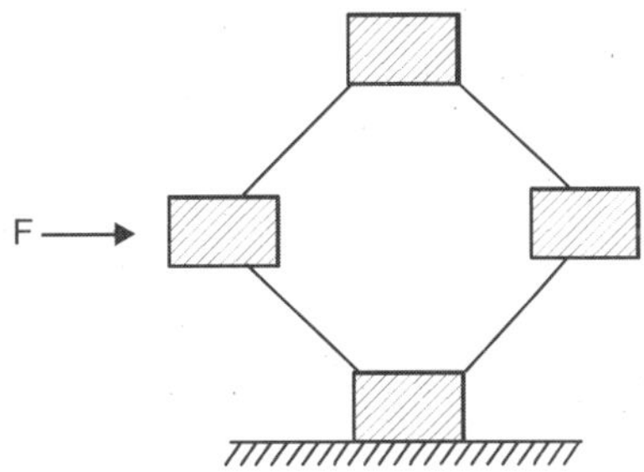

Fig. E 21

Example 22

A particle slides from top of a smooth inclined plane having inclination θ, fixed in an elevator going up with an acceleration a. Find the time taken by the particle to reach the bottom.

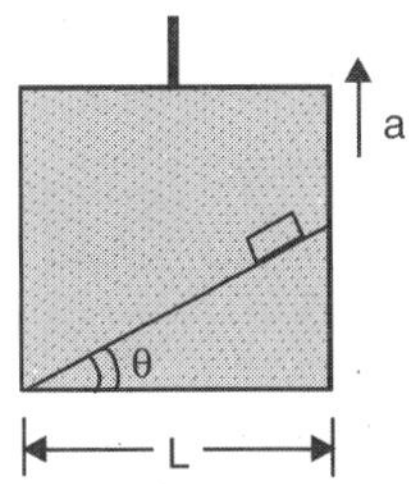

Fig. E22

Example 23

A boy of mass M kg standing in a suspended cage of mass m kg has arranged to pass the rope over a fixed pulley. The boy pulls the rope with acceleration in order to rise. He exerts an effective weight F of the cage. Determine the acceleration of the boy and tension in the string.

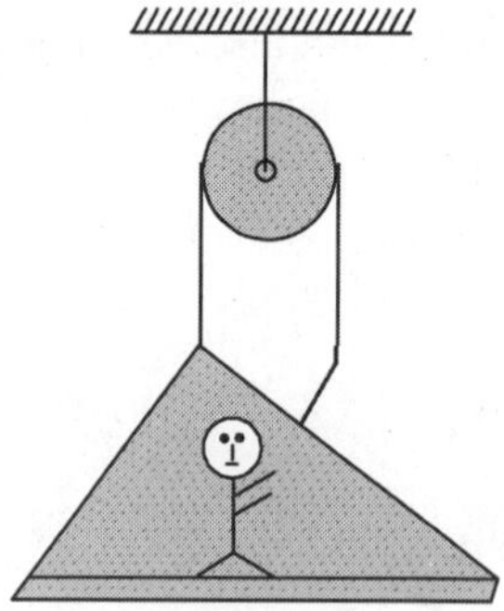

Fig. E23

Example 24

Determine the bank angle θ for the airplane flying at 600 km/h and making a turn of 3 km radius. Assume that the force exerted by air is normal to the wing surface.

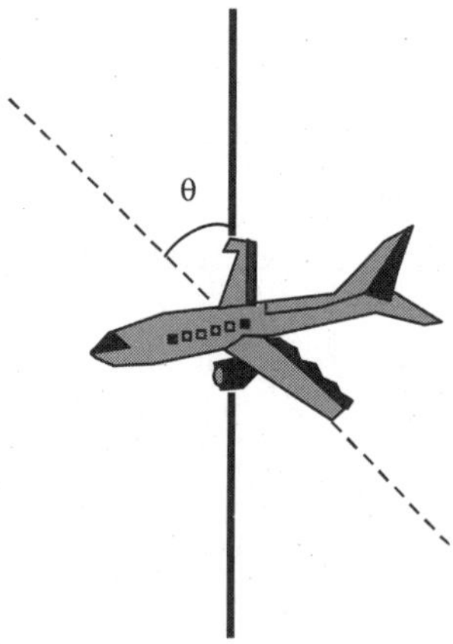

Fig. E24

Level 2

Example 1

A liquid is kept in a cylindrical vessel, which is rotated along its axis. The liquid rises at the sides of vessel. The speed of rotation is 2 rev/s and radius of vessel is 0.05 m. Find the difference between maximum and minimum height of the liquid level.

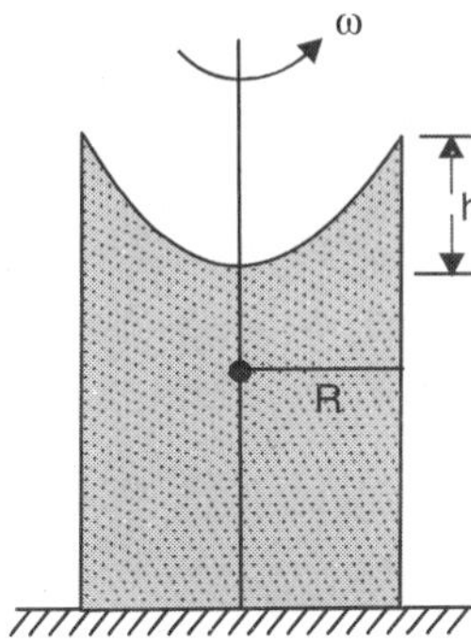

Fig. E1

Example 2

A series of tile bundles. A is transported to the roof-top using a motor driven lift consisting of a horizontal platform *BC* that rides on rolls attached to the sides of a ladder. The lift starts from rest and initially moves with a constant acceleration a_1. The lift then decelerates at a constant rate of a_2 and comes to rest at *D* near the top of the ladder. If $\mu_s = 0.3$ and $\mu_k = 0.2$ between the bundles and the horizontal platform determine the largest allowable a_1 and a_2 if the bundles are not allowed to slide on the platform.

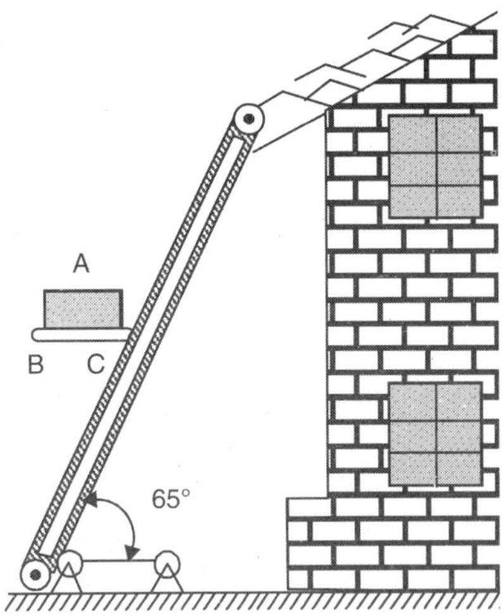

Fig. E2

Example 3

A slider C can slide on a semi-circular rod of radius 0.6 m, which is rotating about the vertical axis AB with constant angular velocity of 6 rad/s as shown. Determine the minimum coefficient of static friction between the slide and the rod if the slider is not to slide when (*a*) $\theta = 70°$, (*b*) $\theta = 30°$. In each case indicate the direction of impending motion.

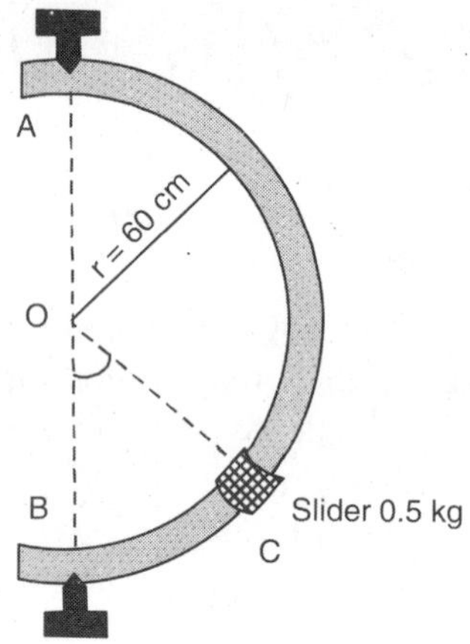

Fig. E3

Example 4

A force P of magnitude 3 N is applied to a tape wrapped around a uniform disc of mass 2.4 kg. If the disc rest on a frictionless horizontal surface, determine the acceleration of (*a*) point A, and (*b*) point B.

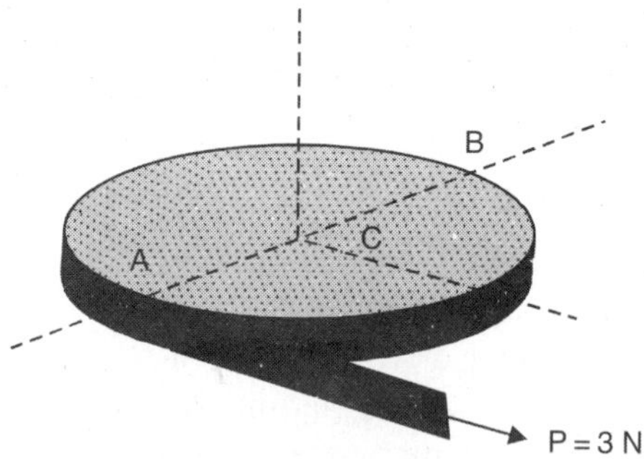

Fig. E4

Example 5

At what uniform speed of rotation around the vertical axis AB will the bells C and D of equal weights $W_1 = 10$ N begin to lift the weight of the device $W_2 = 20$ N as shown. Neglect all friction and the weights of four hinged bars. The length of each bar is 250 m.

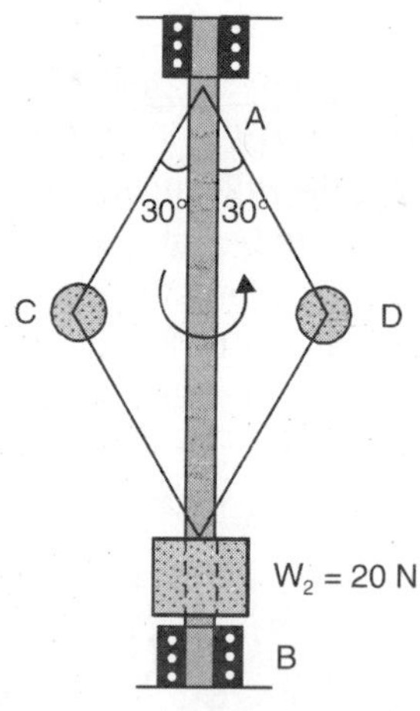

Fig. E5

Example 6

A uniform rectangular plate has a mass of 5 kg. It is held in position by three ropes as shown in figure. Rope is cut. At this instant determine.

(*a*) Acceleration of the plate.

(*b*) Tensions in ropes *AE* and *BD*. Take $\theta = 30°$.

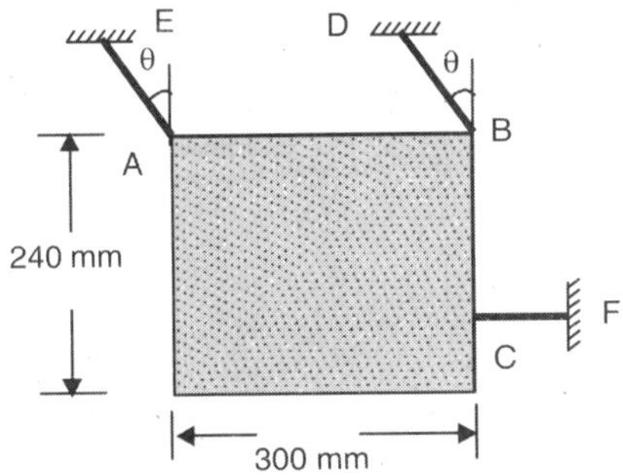

Fig. E6

Example 7

A block of mass 1 kg is held stationary with the help of a string as shown in figure. The tank is accelerating vertically upwards with an acceleration $a = 1$ m/s^2. Determine the tension in the string and find the acceleration of the block when string is cut. Density of water = 10^3 kg/m^3 and density of block = 0.8×10^3 kg/m^3.

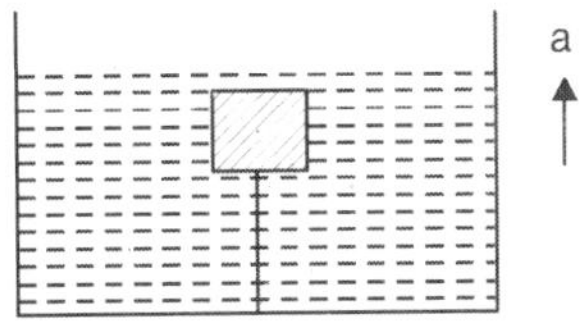

Fig. E7

Example 8

A drum has a weight of 100 N and can be treated as a solid cylinder. If the block has a weight of 40 N, determine the angular acceleration α_D of the drum if the block is allowed to fall freely. Compare this value of α_D with that determined by removing the block and applying a force of 40 N to the cord.

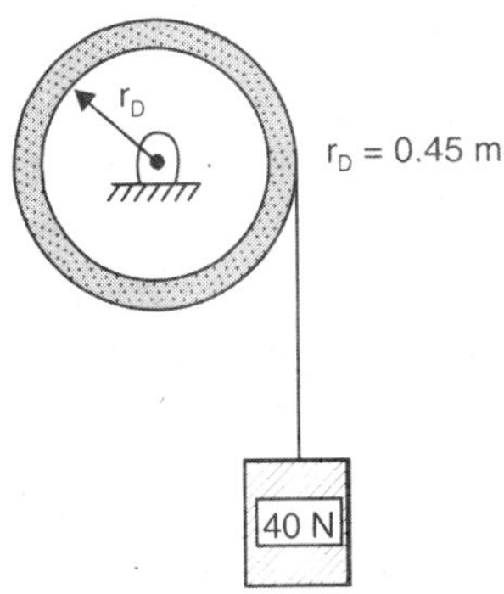

Fig. E8

Example 9

A uniform slender rod of mass 20 kg is pivoted at *A* and swings freely in the vertical plane. If the bar is released from rest in the horizontal position as shown in figure, determine the reaction at *A* at an instant after release.

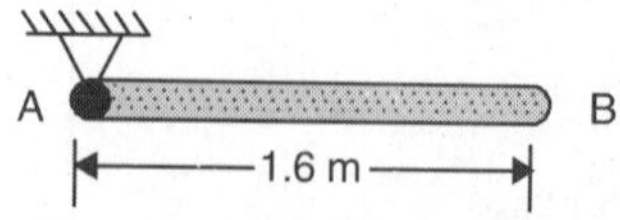

Fig. E9

Example 10

The disk has a mass *M* and radius *R*. If the block of mass *m* is attached to the cord, determine the angular acceleration of the disk when the block is released from rest. Also, what is the velocity of the block after it falls a distance 2*R* starting from rest.

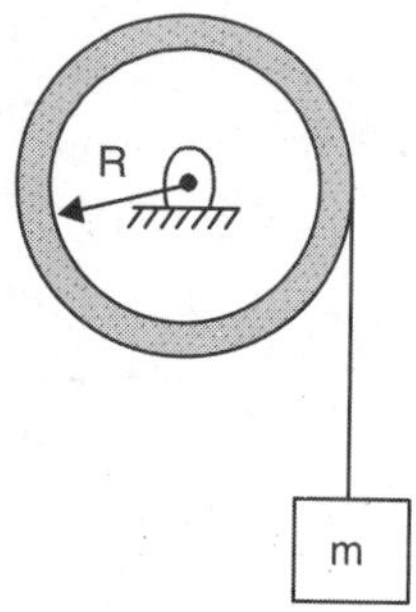

Fig. E10

Example 11

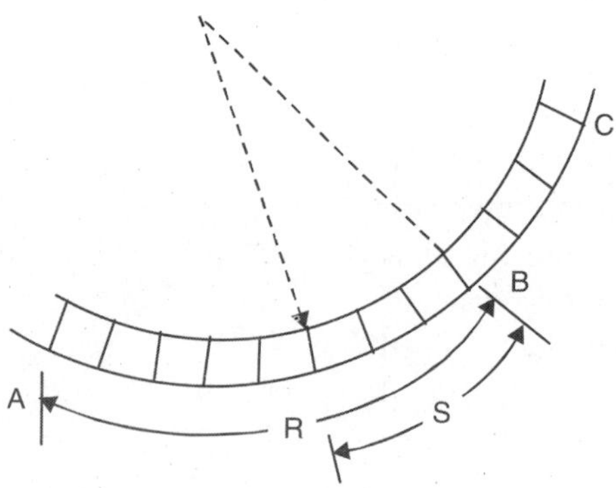

Fig. E11

The portion of railroad track as shown in figure consists of a tangent *CB* and a spiral BA having length *R* and radius of curvature $\rho = \dfrac{R^2}{S}$ where *S* is measured from B towards A. A locomotive of weight w = 600 kN starts from rest at A and increases its speed along AB with constant tangential acceleration of g/10. Find the maximum lateral thrust on the outer rail during the motion from A to B. Where does this occur?

Example 12

The slender rod AB of weight $2W$ is held in equilibrium by two counter weights each of weight W. If the wire at B is cut, determine the acceleration at that instant of
(a) Point A, (b) Point B

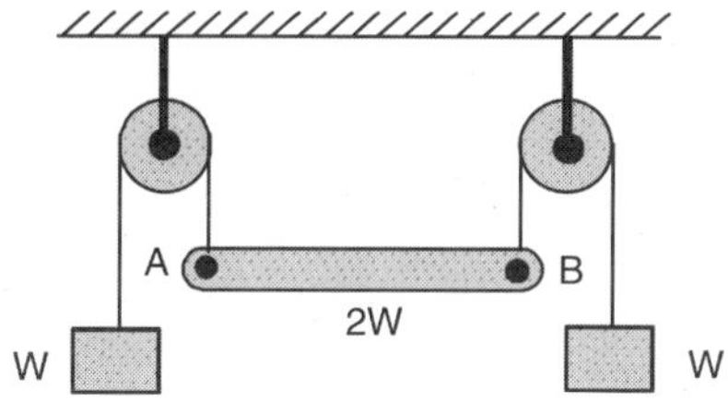

Fig. E12

Example 13

A bob of mass m is connected by string of length and attached to the top most point of the hook of the small tower. The whole arrangement is kept on the car. The car moves with an acceleration a_0 in the direction shown in the figure, initially at the position $\theta = 0$ bob is released freely. Determine the expression for the tension T in the string as a function of θ and find the magnitude of tension at $\theta = \pi/2$.

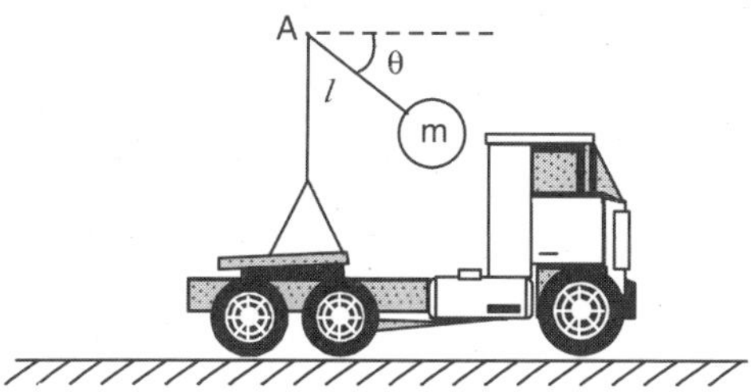

Fig. E13

Example 14

A chain AB of length l is kept in a smooth horizontal tube so that its fraction of length h hangs freely and touches the surface of the table with its end B as shown in the figure. At any instant the end A of the chain is set free. With what velocity will this end of the chain slip out of the tube.

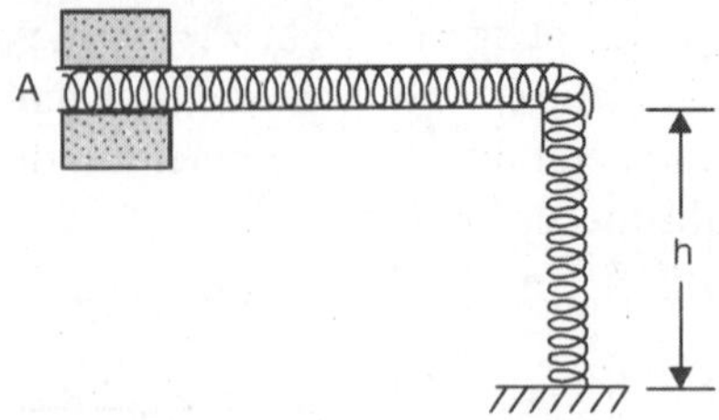

Fig. E14

Example 15

Consider a very thin, uniform chain with length L and mass M, which falls onto a table from an initial state in which the bottom edge of the chain is at height d above the table.

1. Calculate the force applied on the table
   (*a*) As a function of time,
   (*b*) When the whole chain has fallen onto the table.
2. Repeat your calculations of the previous section, for the particular case, $d = L$.

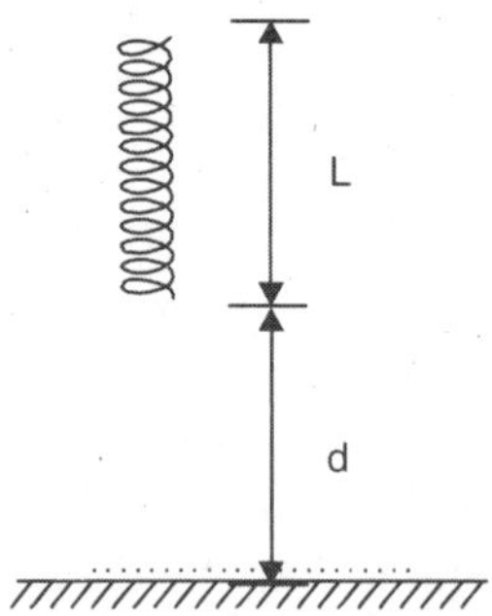

Fig. E15

Example 16

A circle of radius $R = 2$ m is marked on upper surface of a horizontal board, initially at rest. A man starts from rest along the circle with a tangential acceleration $a_t = 0.25$ ms^{-2}. At the same instant board accelerates upwards with acceleration $a_2 = 2.5$ ms^{-2}. If the coefficient of friction between board and man is $\mu = 0.1$, what distance will the man travel on the board without sliding?

Example 17

For the system at rest shown, determine the accelerations of all the loads immediately after the lower thread keeping the system in equilibrium has been cut. Assume that the threads are weightless and inextensible, the springs are weightless, the mass of the pulley is negligibly small, and there is no friction at the point of suspension.

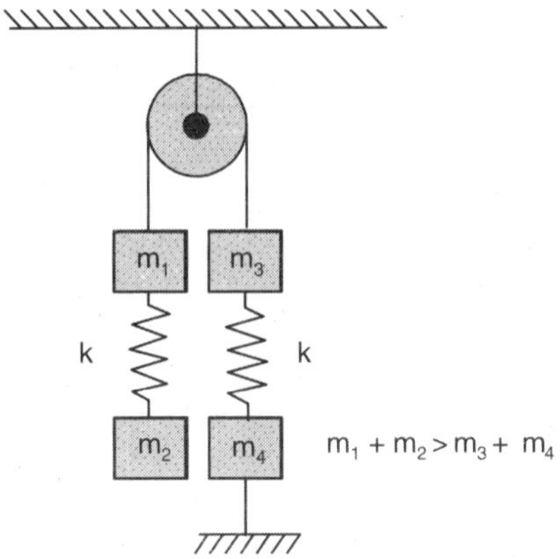

Fig. E17

Example 18

Block M with another block m over it, is placed on a fixed inclined plane of inclination α. If all the contacts are smooth, find the acceleration of m.

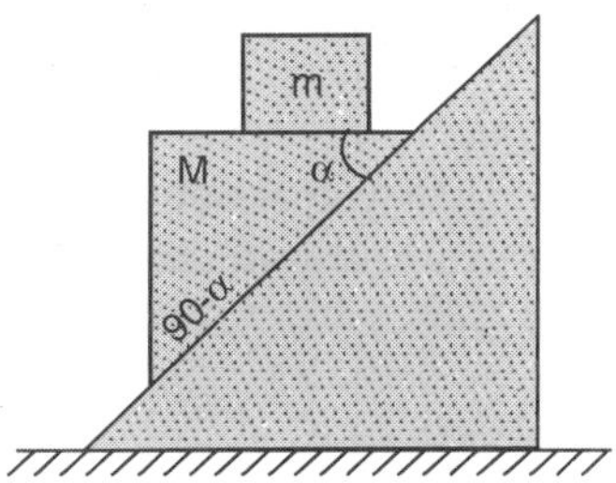

Fig. E18

Example 19

A heap of chain is lying on a horizontal table as shown in figure. A small section of the chain is released through a hole in the table. Calculate the velocity of chain as a function of x hanging vertically.

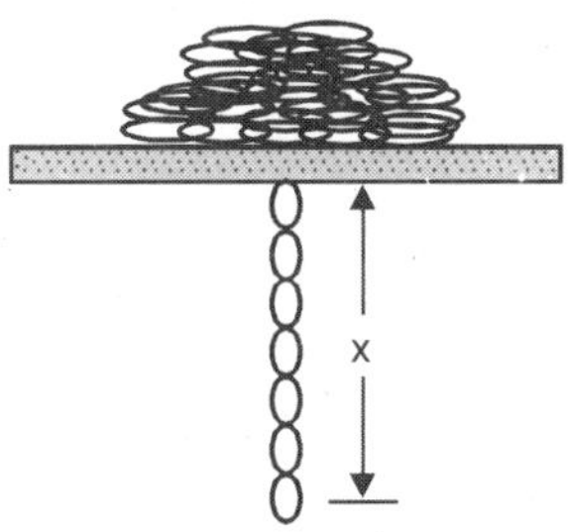

Fig. E19

Example 20

A horizontal weightless rod of length $3l$ is suspended on two vertical strings. Two loads of mass m_1 and m_2 are in equilibrium at equal distances from each other and from the ends of the strings.

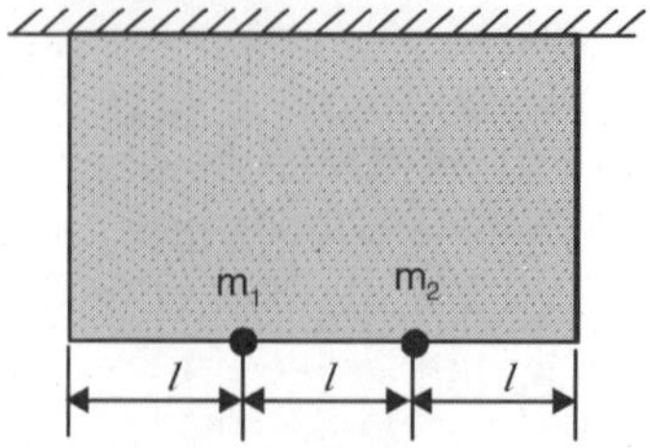

Fig. E20

Determine the tension T of the left string at the instant when the right string snaps.

Example 21

Three identical blocks A, B and C, each of mass $m = 7$ kg are connected with each other by light and inextensible strings, as shown in Fig. E21. Strings pass over light and frictionless pulleys fixed to the edges, trolley of mass $M = 21$ kg. If co-efficient of friction between blocks and troll surfaces is $\mu = 0.6$, calculate maximum possible value of angle θ so that block B remains stationary relative to the trolley.

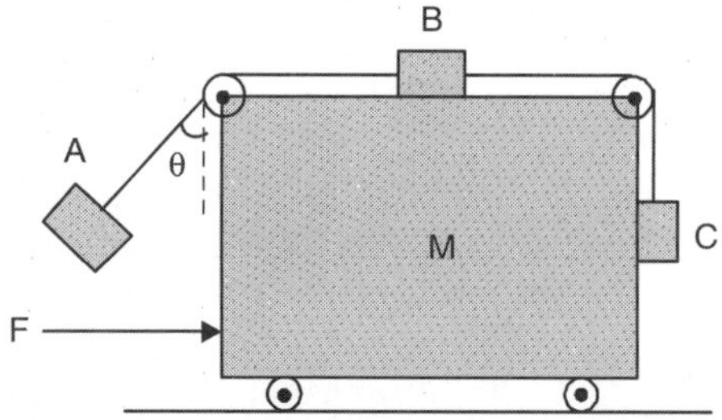

Fig. E21

Calculate the force F to be applied horizontally on the trolley. ($g = 10$ ms^{-1})

Level 3

Example 1

As an arrangement shown in figure. Write the equations of motions of each blocks m_1, m_2 and m_3.

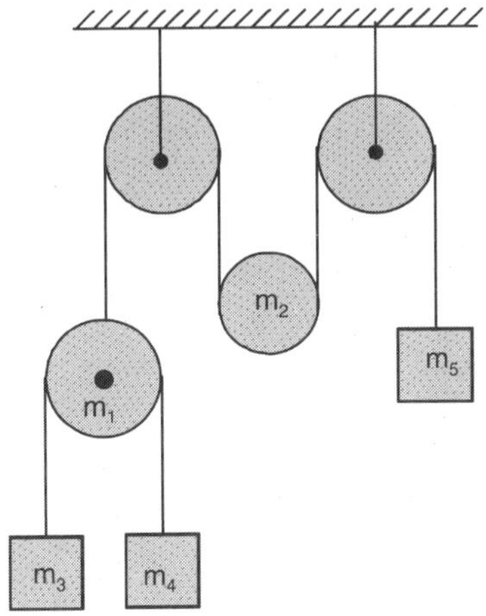

Fig. E1

Example 2

In the arrangement shown, mass of blocks A, B and C is 7.5 kg, 6 kg and 1 kg respectively. The pulley is solid circular disc of mass 0.5 kg, radius 20 cm and thickness 1 cm. Thread between block A and pulley is horizontal and that between pulley and block C is vertical. The pulley is free to rotate about axis O without friction and thread does not slip over its curved surface. Neglecting friction between blocks B and C and that between blocks and the floor, calculate resultant acceleration of block C when the system is released.

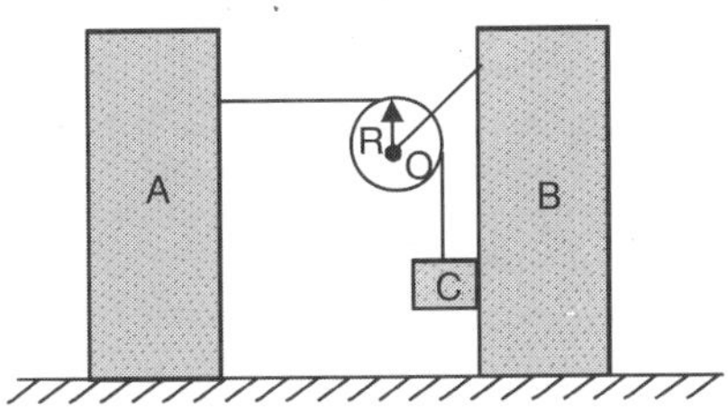

Fig. E2

Example 3

A cylinder and a wedge with a vertical face, touching each other, move along two smooth inclined planes forming the same angle α with the horizontal. The masses of the cylinder and the wedge are m_1 and m_2 respectively. Determine the force of normal pressure exerted by the wedge on the cylinder neglecting the friction between them.

Fig. E3

Example 4

A chain of length L and mass per unit length λ is given a constant upward velocity v_1 as shown in figure write expressions for F, the force R supports the fixed end and the tension T in the chain at the loop in terms of x.

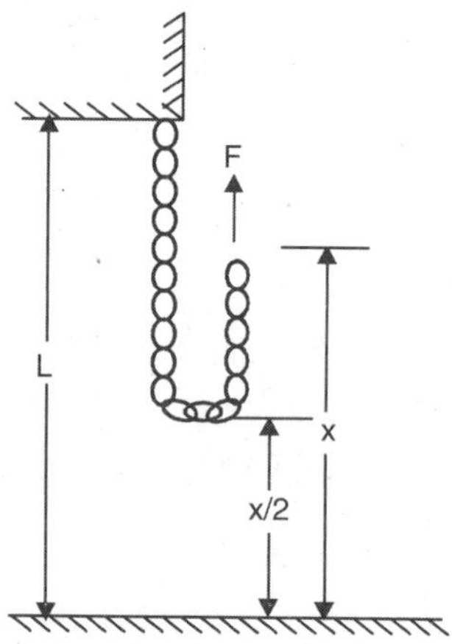

Fig. E4

Example 5

A uniform flexible chain of length 2.5 m rests on a fixed smooth sphere of radius $R = 4/\pi$ m such that one end A of chain is at top of sphere while the other end B is hanging freely. Chain is held stationary by a horizontal thread PA shown. Calculate acceleration of chain when the thread is burnt. ($g = 10 \text{ ms}^{-2}$)

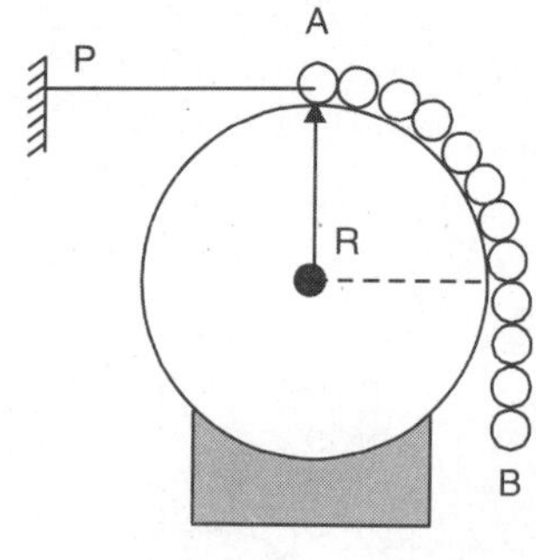

Fig. E5

Example 6

A board is fixed to the floor of an elevator such that the board forms angle $\theta = 37°$ with horizontal floor of the elevator accelerating upwards. A block is placed on point A of the board as shown. When motion with velocity $v_1 = 4\sqrt{2}$ ms^{-1} is given to the block, it comes to rest after moving a distance $l = 1.6$ m relative to the board. Its velocity was $v_2 = 4$ ms^{-1} down the board when it returns to point A. Calculate acceleration a of elevator and coefficient of friction μ between the board and the block. ($g = 10$ ms^{-1})

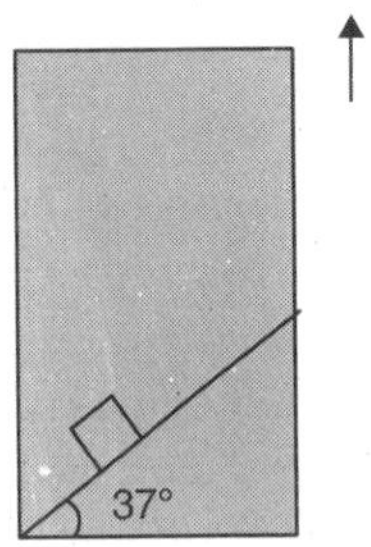

Fig. E6

Example 7

A heavy chain is initially at rest on the rough surface with $x = 0$. A constant force F is applied along a horizontal surface as shown in figure when all part of chain remain on the rough surface, i.e. $x = 0$. Determine the velocity V of the chain when $x = L$. λ is the mass per unit length and μ_k is the coefficient of kinetic friction between the rough surface and the chain.

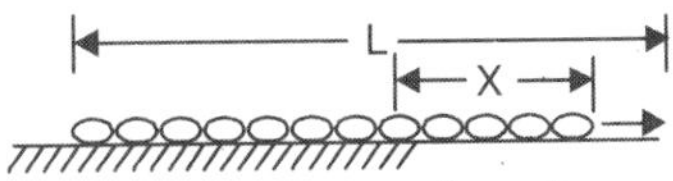

Fig. E7

Example 8

In the arrangement shown, mass of blocks A, B and C is 20 kg, 10 kg and 2.5 kg respectively. Bottom surface of A is smooth, while coefficient of friction between B and floor is 0.2 and that between blocks A and C is 1/3. System is released from rest at $t = 0$ and pulleys are light and frictionless.

Calculate acceleration of block C.

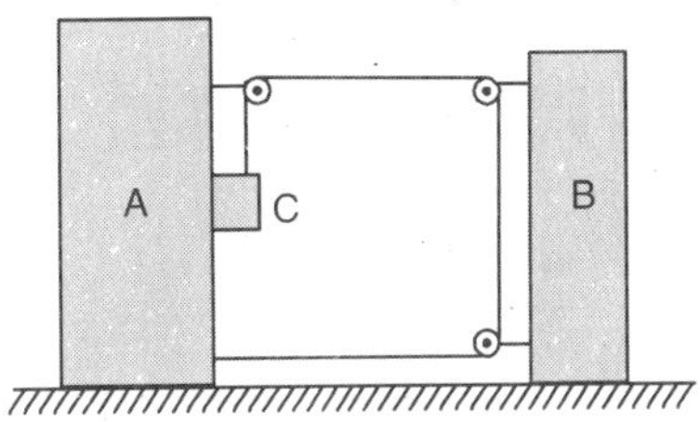

Fig. E8

Example 9

A uniform chain $A'B'$ of length $2l$ having mass λ per unit length is hanging from ceiling of elevator by two light, inextensible threads AA' and BB' of equal length as shown in Fig. E9. Distance AB is very small. At a certain instant, elevator starts ascending with constant acceleration a. Two seconds after the beginning of motion, thread BB' is burnt. Assuming that instant to be $t = 0$, calculate tension in thread AA' at time t.

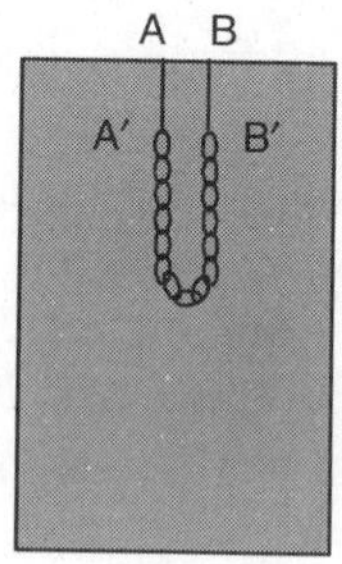

Fig. E9

Example 10

The system of bodies shown in the figure is released from rest. Assuming all the contacts to be smooth, find the acceleration of each block (m_A = 14 kg, m_B = 11 kg, and m_C = 50 kg).

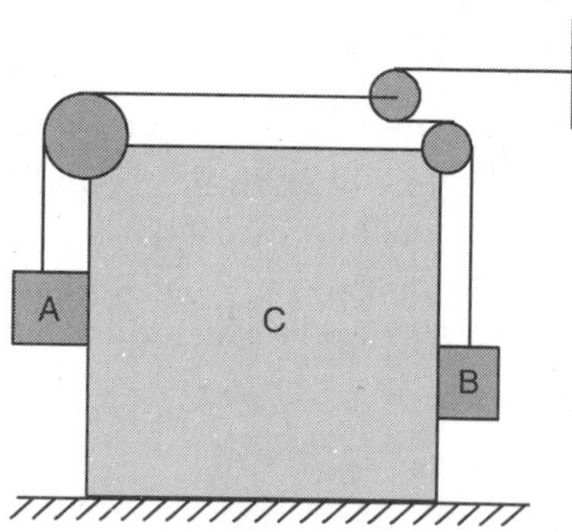

Fig. E10

3 Work, Energy and Momentum

Summary and Concept

Work

$$dw = \vec{F} \cdot d\vec{r}$$

The total work done in translating a body from A to B along a path is

$$W_{AB} = \int_A^B \vec{F} \cdot d\vec{r}$$

Power

$$P = \frac{dW}{dt} = \vec{F} \cdot \vec{v}$$

Conservative force if $\vec{F}$ is a conservative force. Then

(*a*) $\oint_C \vec{F} \cdot d\vec{r} = 0$ for any closed point path C.

(*b*) $\nabla \times \vec{F} = 0$

(*c*) There exists a function U such that

$$\vec{F} = -\nabla U$$

Work energy theorem

$$\sum W = \Delta T$$

ΔT = change in the kinetic energy

$$\sum W = \sum_j \left(-\Delta u_j\right) + \sum_i \left(\int \vec{F}_i \cdot d\vec{r}_i\right)$$

$-\Delta u_j$ is the work done by conservative forces

$\int \vec{F}_i \cdot d\vec{r}_i$ is the work done by non-conservative forces

Potential energy

The potential energy of a body at a point $\vec{r}$, when it moves from a point $\vec{A}$ to $\vec{r}$ under the action of a force $\vec{F}$

$$U(\vec{r}) = U(\vec{A}) - \int_{\vec{A}}^{\vec{r}} \vec{F} \cdot d\vec{r}$$

When $\vec{A} \to \infty$

$$U(\vec{r}) = \int_{r}^{\infty} \vec{F} \cdot d\vec{r}$$

Newton's second law for linear momentum

$$\frac{d\vec{P}}{dt} = \vec{F}$$

we can write it as in projection form

$$\frac{dP_x}{dt} = F_x$$

in the integral form

$$\Delta\vec{P} = \int \vec{F}\, dt$$

Conservation of momentum

$$\vec{P}_i = \vec{P}_f$$

when there are no external forces acting on the system.

An elastic collision is a collision between two particles in which the kinetic energy is conserved.

A plastic collision is a collision in which the kinetic energy is not conserved.

Coefficient of restitution

$$E = \frac{V'_{2n} - V'_{1n}}{V_{1n} - V_{2n}}$$

Where

V'_{1n} = velocity of particle 1 along the line of impact (common normal to the contact) before collision

V_{2n} = Velocity of particle 1 along the line of impact after collision

V_{2n} = Velocity of particle 2 along the line of impact before collision

V'_{2n} = Velocity of particle 2 along the line of impact after collision

Center of mass

$$\vec{R}_{cm} = \frac{\sum_{i=1}^{n} m_i \vec{r}_i}{\sum_{i=1}^{n} m_i}$$

in projection form

$$x_{\text{cm}} = \frac{\sum m_i x_i}{\sum m_i}$$

or

$$\Delta x_{\text{cm}} = \frac{\sum m_i \Delta x_i}{\sum m_i}$$

here $\Delta x_{cm} = 0$ when there is no horizontal external force acting on the system.

$$\sum m_i \Delta x_i = 0.$$

Level 1

Example 1

The small collar of mass m is released from rest at A and slides down the curved rod in the vertical plane with negligible friction. Express the velocity V of the collar as it strikes the base at B in terms of the given condition.

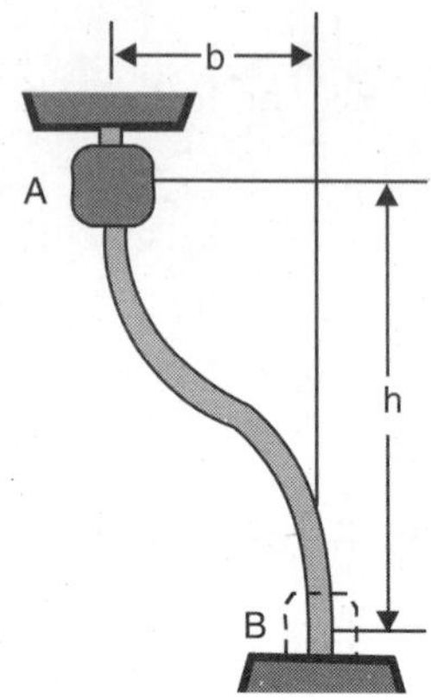

Fig. E1

Example 2

A force $F = (y^2 - x^2 + z^2)\hat{x} + (3xy - 5z)\ \hat{y} + 4z\hat{z}$ is applied on a particle. Find the work done by the force when the particle moves from the point (0,0,0) to the point (2,4,0) in the paths shown.

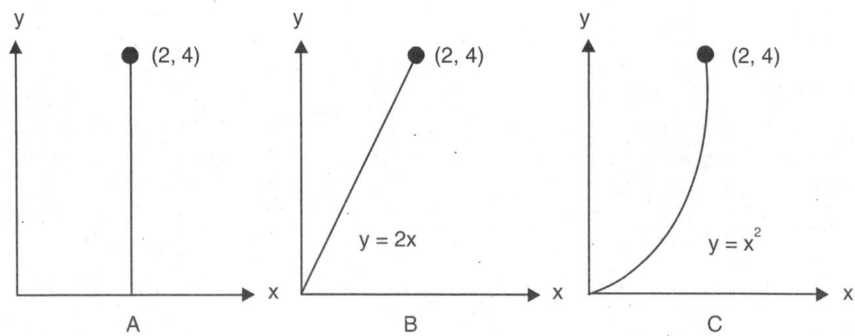

Fig. E2

Example 3

A bullet is fired at a speed of 100 m/s at an angle 37° with the horizontal. At the highest point, it breaks into two parts of mass ratio 1: 3 the smaller coming to rest. Find the distance from the launching point to the point where the heavier piece lands.

Example 4

The spring of constant k is unstretched when the slider of mass m passes position B. If the slider is released from rest in position A, determine its speed as it passes points B and C. What is the normal force exerted by the guide on the slider at position C? Neglect friction between the mass and the circular guide, which lies in a vertical plane.

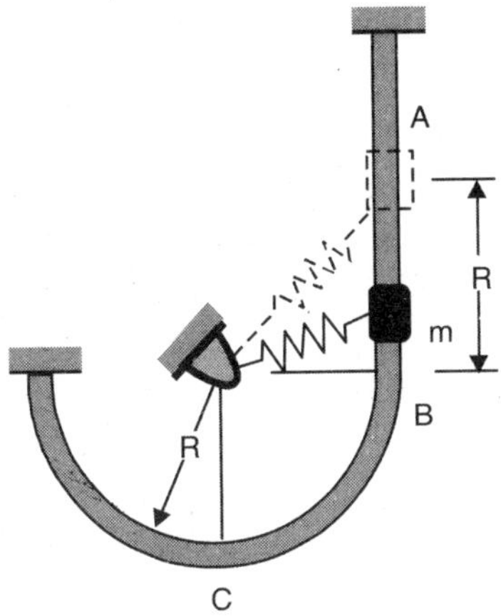

Fig. E4

Example 5

A 10 kg block is resting on the horizontal surface when the force T is applied to it for seconds. The variation of T with time is shown. Calculate the maximum velocity reached by the block and the total time Δt during which the block is in motion. The coefficients of static and kinetic friction are both 0.50.

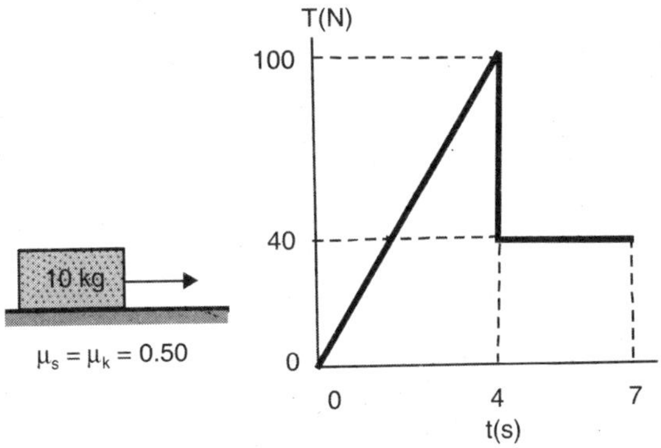

Fig. E5

Example 6

A small ball of mass m is tied by a string of length L to a cylindrical rod of radius a ($L >> a$). The ball moves in a plane perpendicular to the rod, and as a result of its movement, the string wraps around the rod. Since the string gets shorter all the time, the ball moves in a spiral path. Provided that the ball is released with a tangential velocity v_0 at $t = 0$.

1. Find the torque around an axis passing through the center of the rod.
2. Find the angular frequency of the ball after it circles the rod ten times.

Example 7

A small slider of mass m is released from rest while in position A and then slides along the vertical-plane track. The track is smooth from A to D and rough from point D. Determine (a) the normal N_B exerted by the track on the slider just after it passes point B, (b) the normal force N_C exerted by the track on the slider as it passes the bottom point C, and (c) the distance s traveled along the inclined point D before the slider stops.

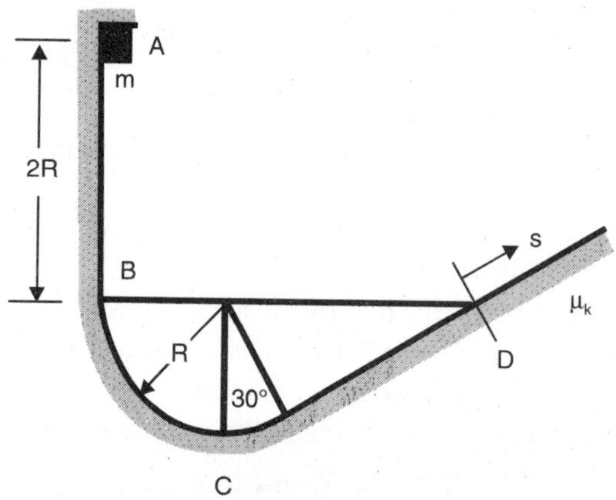

Fig. E7

Example 8

A 0.6 kg block slides on the circular path of radius 3 m in the vertical plane as shown. If the speed of the block at A is 5 m/s, determine the normal force exerted on the block by the surface at A and B.

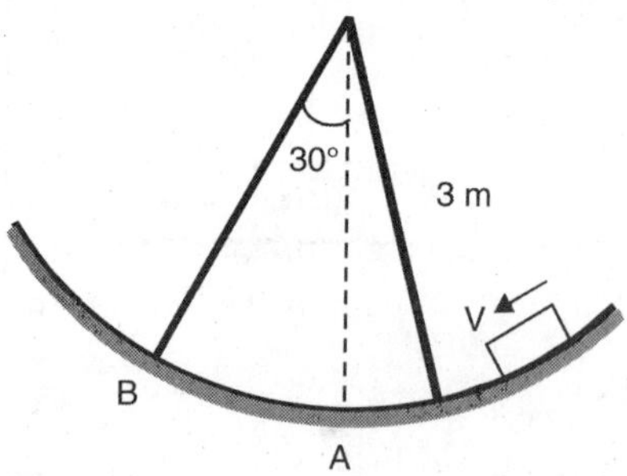

Fig. E8

Example 9

Identical packages are transferred from one horizontal conveyor belt to another using a ramp AB for which $\mu_s = 0.15$. A and B are respectively 2.5 m and 6 m apart vertically and horizontally. The top conveyor moves at 2 m/s and the packages are spaced 1 m apart. Find the required speed of the bottom conveyor so that no sliding occurs when the package come horizontally in contact with it. What is the spacing S between the packages on the bottom conveyor.

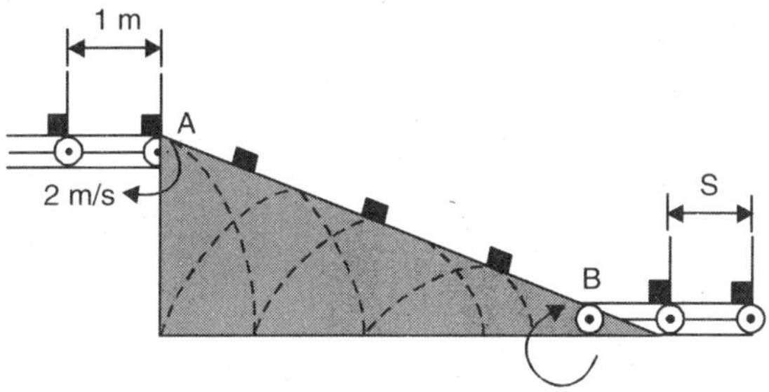

Fig. E9

Example 10

A block of mass m is suspended through a spring of spring constant k and is in equilibrium. A sharp instant jerk gives the block an initial downward velocity V. How far below the equilibrium position; the block comes to an instantaneous rest?

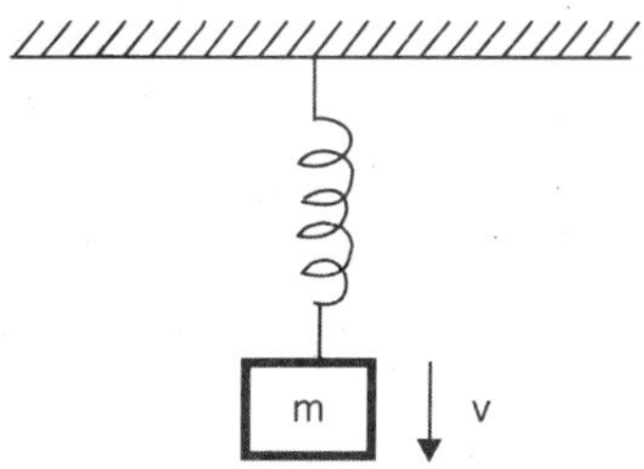

Fig. E10

Example 11

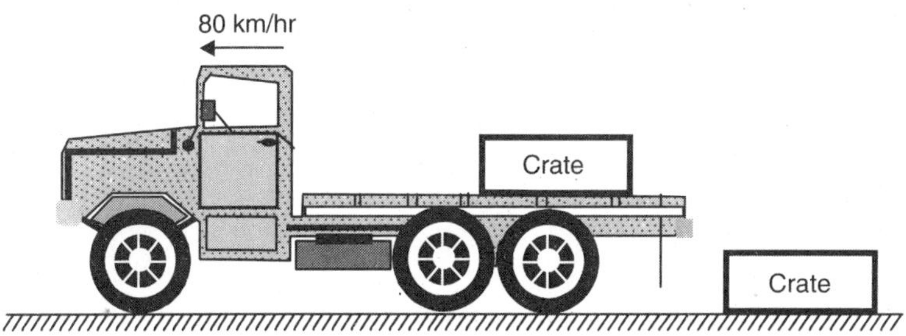

Fig. E11

A crate having mass of 60 kg falls horizontally off the back of a truck which is traveling at 80 km/h. Determine the coefficient of kinetic friction between the road and the crate if the crate slides 45 m to the ground with no tumbling along the road before coming to rest.

Level 2

Example 1

The 0.5 kg collar C starts from rest at A and slides with negligible friction on the fixed rod in the vertical plane. Determine the velocity with which the collar strikes end B when acted upon by the 5 N force, which is constant in direction. Neglect the small dimensions of the collar.

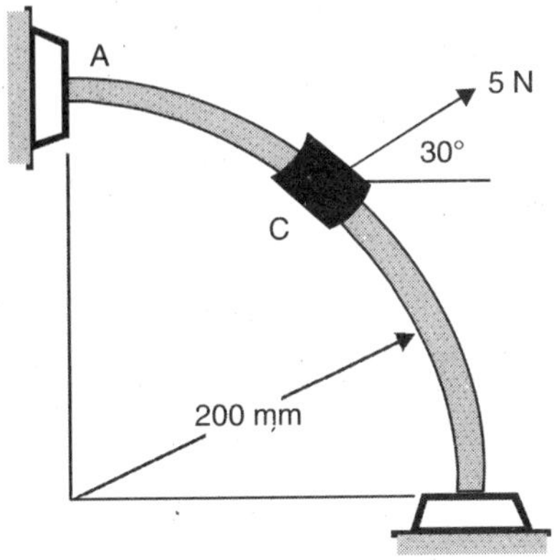

Fig. E1

Example 2

A uniform chain of length L and mass m overhangs a smooth table with its two-thirds part lying on the table. Find the kinetic energy of the chain as it completely slips off the table.

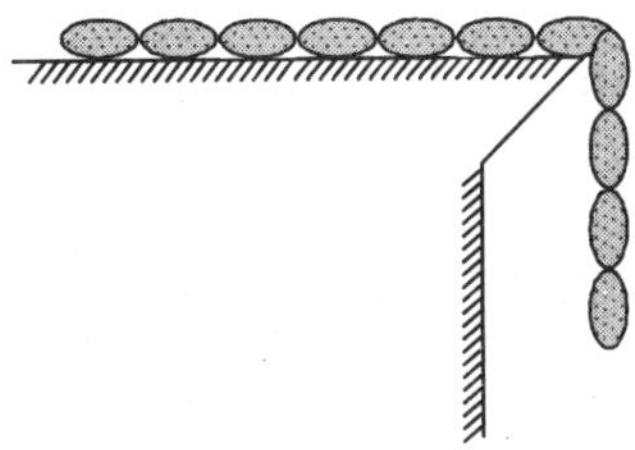

Fig. E2

Example 3

A homogeneous sphere of mass 10 kg and diameter 300 mm is released simultaneously along with a homogeneous cylinder of mass 30 kg and diameter 400 mm. Both are released from the same height 'h' meters, to roll down, bodies roll without slipping. Compute the velocity with which sphere and cylinder will reach the bottom of incline. Which body will reach the bottom first?

Example 4

The two springs, each of stiffness k = 1.2 kN/m are of equal length and undeformed when $\theta = 0$. If the mechanism is released from rest in the position $\theta = 20°$, determine its angular velocity when $\theta = 0$. The mass m of each sphere is 3 kg. Treat the spheres as particles and neglect the masses of the light rods and springs.

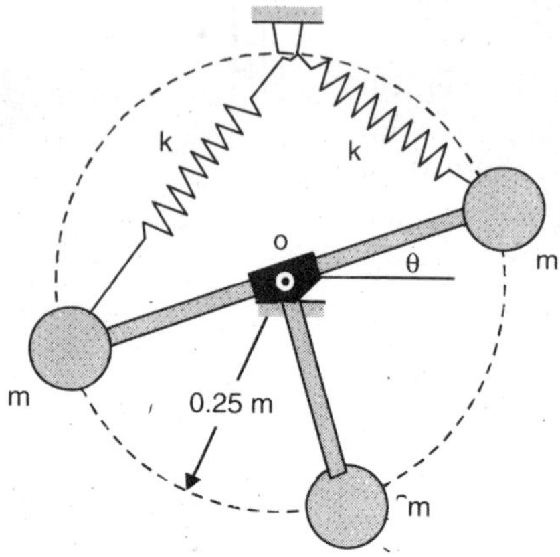

Fig. E4

Example 5

A father and his son are standing together at the right edge of a boat of length L, which is in quiet waters. They begin walking towards the other edge. The child reaches the middle of the boat and stops, while the father continues until he reaches the other edge.

The mass of the boat is M, the mass of the father is m_1 and the mass of his son is m_2. What distance does the boat pass? Assume that the water resistance is negligible.

Example 6

Three identical balls each of mass m = 5 kg are connected with each other as shown and rest over a smooth horizontal table. At moment $t = 0$, ball B is imparted a velocity $v_0 = 9$ ms^{-1}. Calculate velocity of A when it collides with ball C.

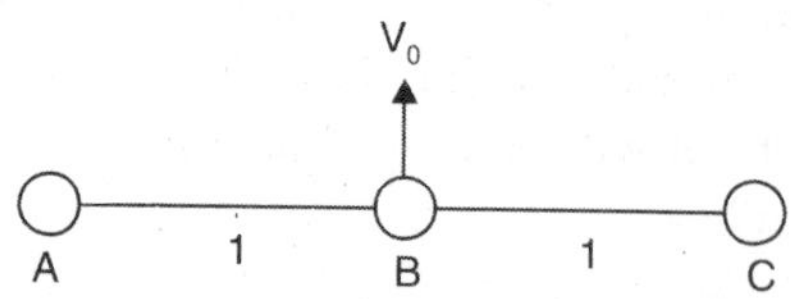

Fig. E6

Example 7

A man swings a stone tied to a string of length l in a vertical plane. The string remains stretched throughout the time of rotation.

The man is in an elevator going up with an acceleration a. What is the minimal velocity v_a of the stone, relative to the man, necessary at the lowest point of its path, to keep the string stretched throughout the time of rotation?

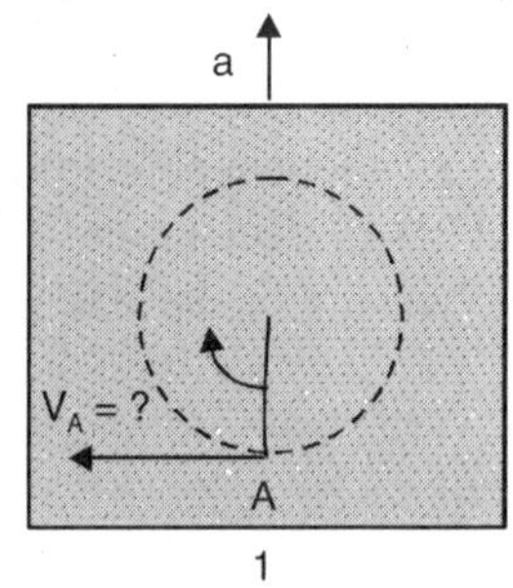

Fig. E7

Example 8

The two 1.5 kg spheres are released from rest and gently nudged outward from the position θ = 0 and then rotate in a vertical plane about the fixed centers of their attached gears, and maintaining centers of their attached gears, and maintaining the same angle θ for both rods. Determine the velocity v of each sphere as the rods pass the position θ = 30°. The spring is unstretched when θ = 0, and the masses of the two identical rods and the two gear wheels may be neglected.

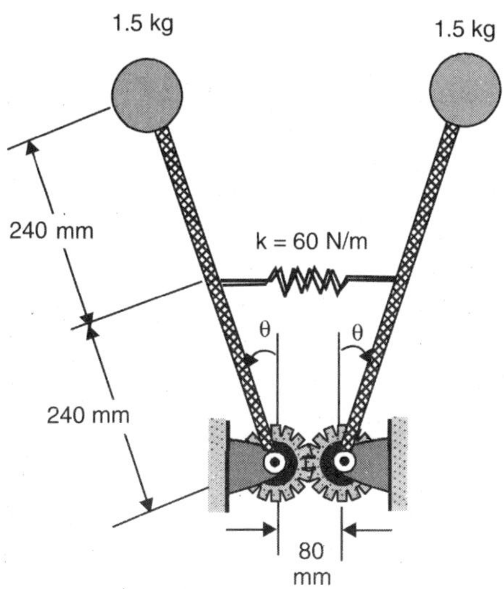

Fig. E8

Example 9

The 0.5 kg collar slides with negligible friction along the fixed spiral rod, which lies in the vertical plane. The rod has the shape of the spiral $r = 0.3\theta$, where r is in meters and θ is in radians. The collar is released from rest at A and slides to B under the action of a constant radial force $T = 10$ N. Calculate the velocity V of the slider as it reaches B.

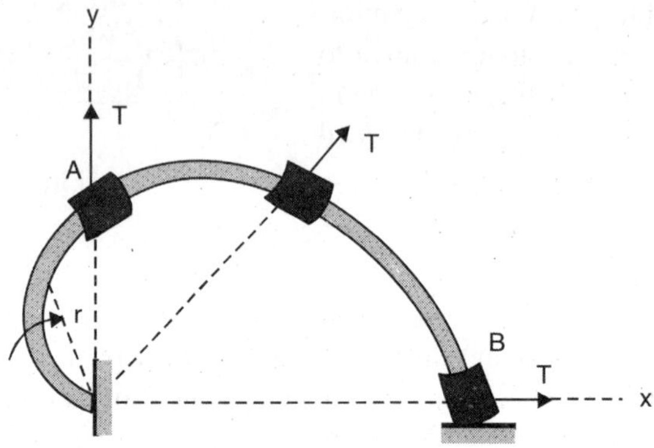

Fig. E9

Example 10

A heavy particle is suspended by a string of length L. The particle is given a horizontal velocity u. The string becomes slack at some angle and the particle proceeds on a parabola. Find the value of u if the particle passes through the point of suspension.

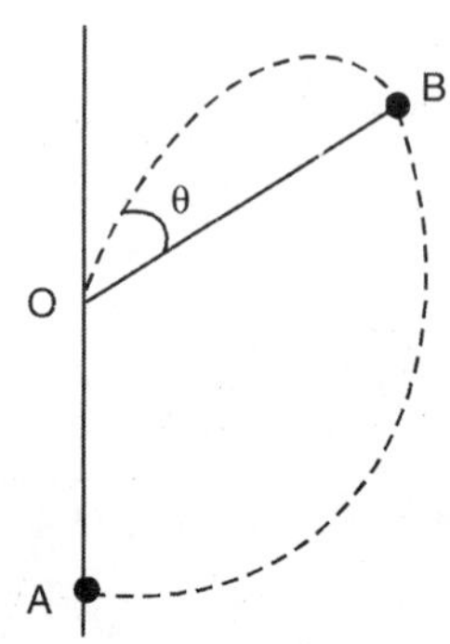

Fig. E10

Example 11

A bullet of mass 700 gm is fired from ground with a velocity 40 ms^{-1}. At highest point of its trajectory, it collides inelastically with a ball of mass M = 1.3 kg, suspended by a flexible thread of length 1.40 m. If thread deviates through an angle of 120°, calculate

(*i*) Maximum height of combined body from ground and

(*ii*) Distance between point of suspension of ball and point of projection of shell.

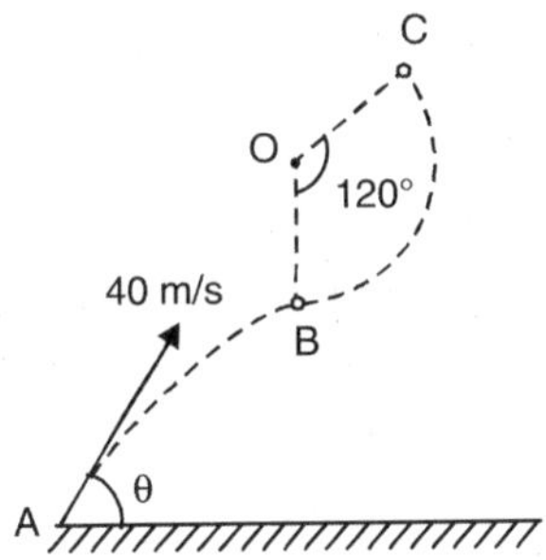

Fig. E11

Example 12

Two particles of masses m and m', are moving in parallel straight lines at a distance a apart with velocities v and v' ($v > v'$). The particles are connected by a string of length l which was loose in the beginning. Calculate the impulse of tension of the string when it becomes taut.

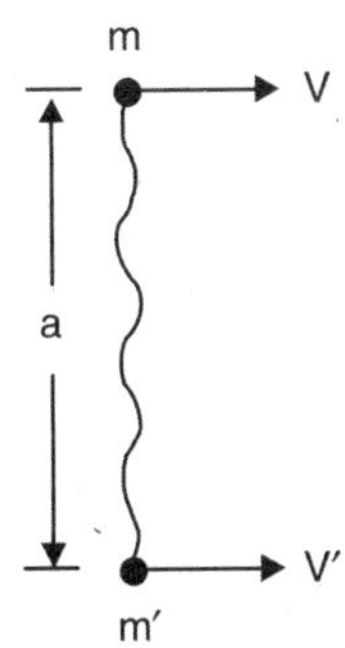

Fig. E12

Example 13

Two masses m_1 and m_2 are connected by a fine inelastic string, which passes over a smooth fixed pulley. The larger mass m_2 rests on a horizontal table; m_1 is then raised by the length h vertically above its hanging position of rest and let fall. Find the time interval between m_2 leaving the table and returning to it again.

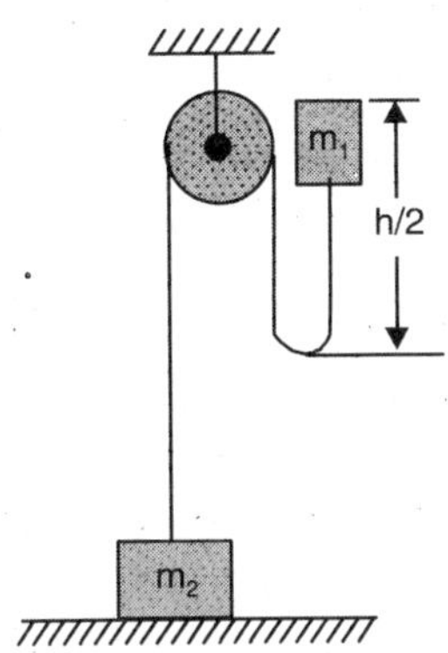

Fig. E13

Example 14

Three equal particles *A*, *B*, *C* of mass *m* are placed on a smooth horizontal plane. *A* is joined to *B* and *C* by light threads *AB*, *AC* and the angle *BAC* is 60°. An impulse *I* is applied to *A* in the direction *BA*. Find the initial velocities of the particles and show that *A* begins to move in a direction making an angle $\tan^{-1}\sqrt{3}/7$ with BA.

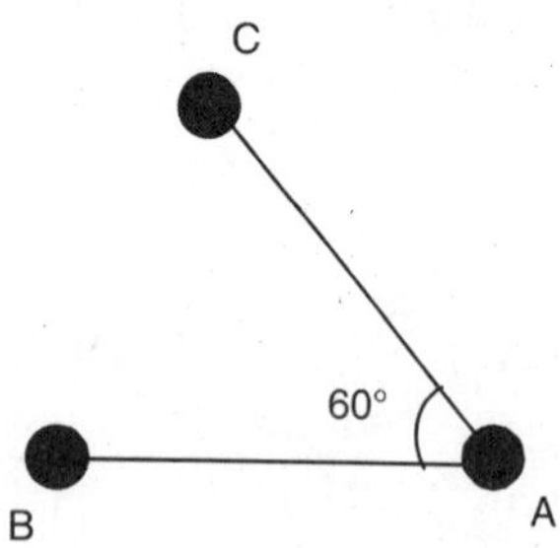

Fig. E14

Example 15

A string with one end fixed on a rigid wall, passing over a fixed frictionless pulley at a distance of 2 m from the wall, has a point mass *M* of 2 kg attached to it at a distance of 1 m from the wall. A mass *m* of 0.5 kg is attached to the free end. The system is initially held at rest so that the string is horizontal between wall and pulley and vertical beyond the pulley as shown.

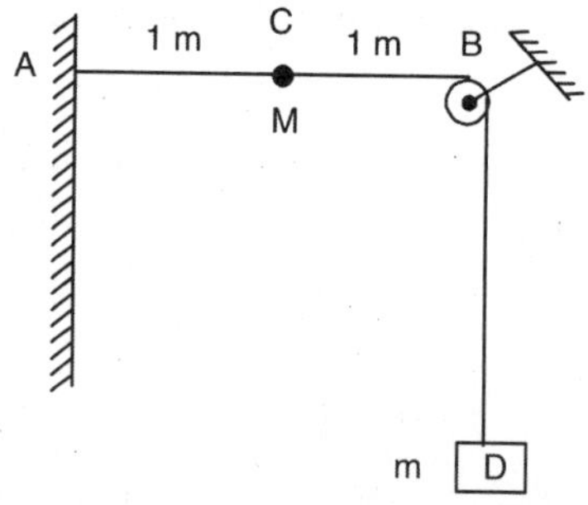

Fig. E15

What will be the speed with which the point mass *M* will hit the wall when the system is released? ($g = 10$ ms^{-2})

Level 3

Example 1

A man is traveling inside a cart that accelerates to the right with acceleration a. What is the minimal velocity V_A at the point A needed to keep the string stretched throughout the circular path?

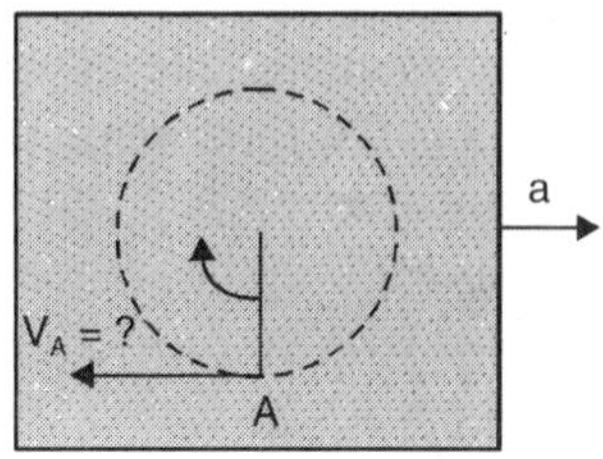

Fig. E1

Example 2

A ring of mass M can slide on a smooth straight horizontal wire and a particle of mass m is attached to the ring by a light string of length l. The particle is held in contact with the wire with the string taut, and is then let fall. Determine the angular velocity of string.

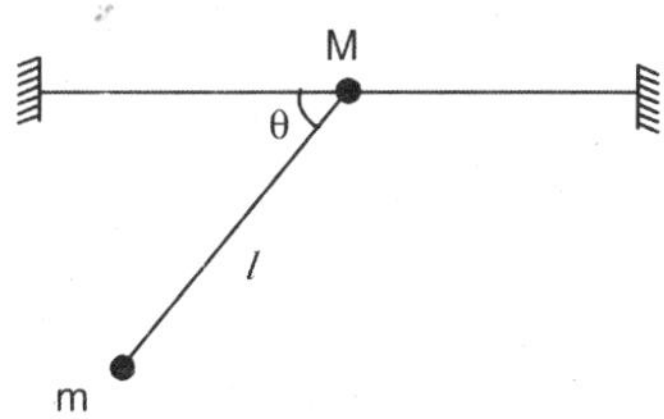

Fig. E2

Example 3

A small ball of mass m and a smooth hemisphere M are released simultaneously. Initially the small ball makes an angle α from the vertical.

(*i*) Find the angular velocity of the small ball, when it makes an angle β (β > α) from the vertical.

(*ii*) The angle from the vertical at the moment of break off from the hemispherical surface. (Assume all the contacts to be frictionless.)

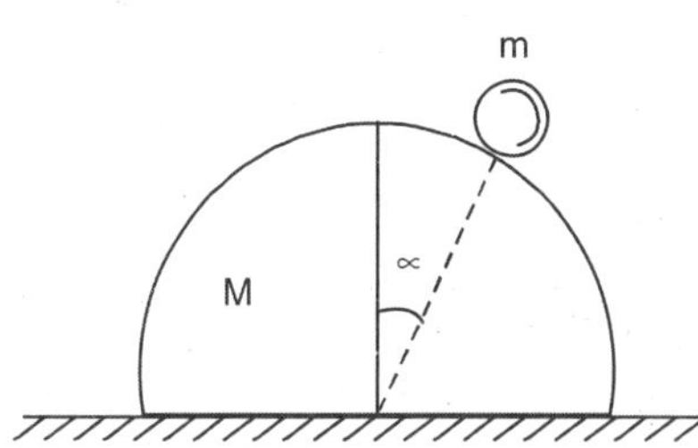

Fig. E3

Example 4

The inclined surfaces of two movable wedges of the same mass M are smoothly conjugated with the horizontal plane. A block of mass m slides down the left wedge from a height h. To what maximum height will the block rise along the right wedge? Friction should be neglected.

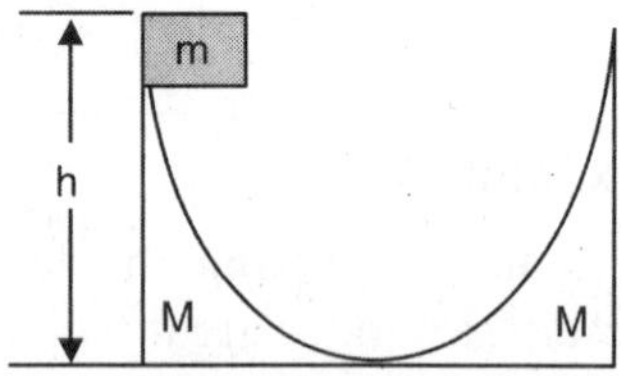

Fig. E4

Example 5

A ball of mass m is thrown at an angle of 45° to the horizontal from top of a 65 m high tower AB as shown. Another identical ball is thrown with velocity 20 ms^{-1} horizontally towards AB from top of a 30 m high tower CD one second after the projection of first ball. Both the balls move in same vertical plane. If they collide in mid air

(*i*) Calculate distance AC.

(*ii*) During collision the two balls get stuck together, calculate the distance between A and the point on the ground, at which the combined ball strikes. Given, $g = 10$ ms^{-2}.

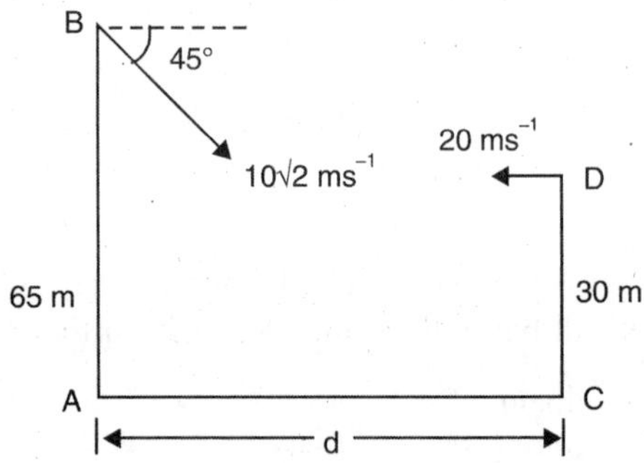

Fig. E5

Example 6

A proton with velocity V collides with an electron initially at rest. As a result, the proton diverges with an angle β from its original direction. What is the maximum value of β? Consider this is an elastic collision. Assume $V << c$, or there are not relativistic dynamics.

Example 7

Two men, A and B each of mass m, sit on two light platforms at the ends of a light flexible rope passing over a smooth pulley, A being higher than B by h.

In B's hands is placed a ball of mass m/10, he instantly throws to A, so that it just reaches him. Calculate the total distance by the time he ceases to ascend.

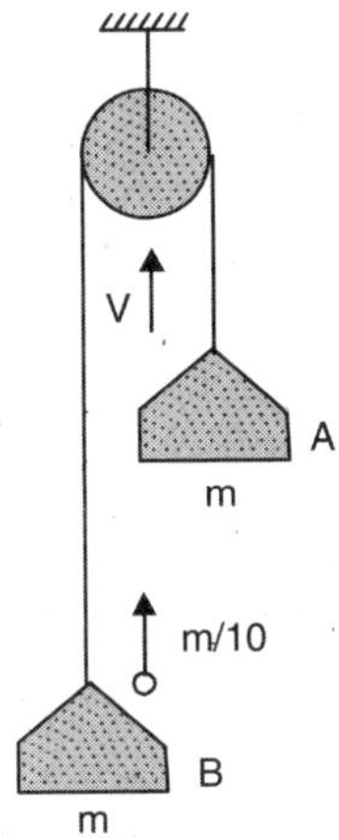

Fig. E7

Example 8

Two equal spheres of mass m' are suspended by vertical strings so that they are in contact with their centers at the same level. A third equal sphere of mass m falls vertically and strikes the other two simultaneously so that their centers at the instant of impact form an equilateral triangle in a vertical plane. If u is the velocity of m just before impact, find the velocities just after impact and the impulsive tension of the strings.

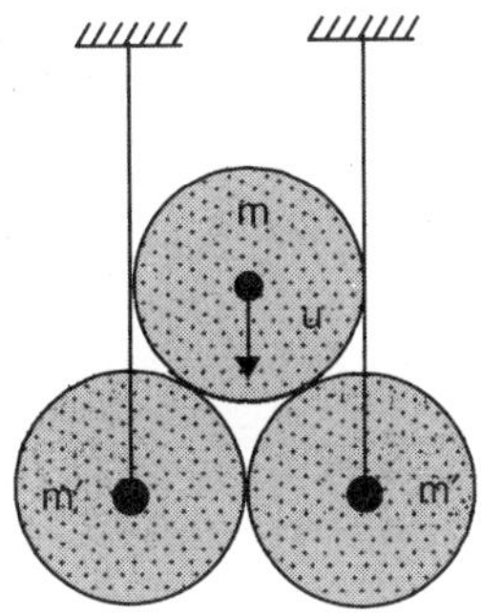

Fig. E8

Example 9

Two balls of mass m_1 = 100 gm, m_2 = 300 gm are suspended from point A by two equal inextensible threads, each of length l = 32/35 m. Ball of mass m_1 is drawn aside and held at the same level as A but at a distance $\frac{\sqrt{3}}{2}l$ from A. When ball m_1 is released, it collides elastically with the stationary ball of mass m_2. Calculate maximum rise of center of mass of the ball of mass m_2. (g = 10 ms^{-2})

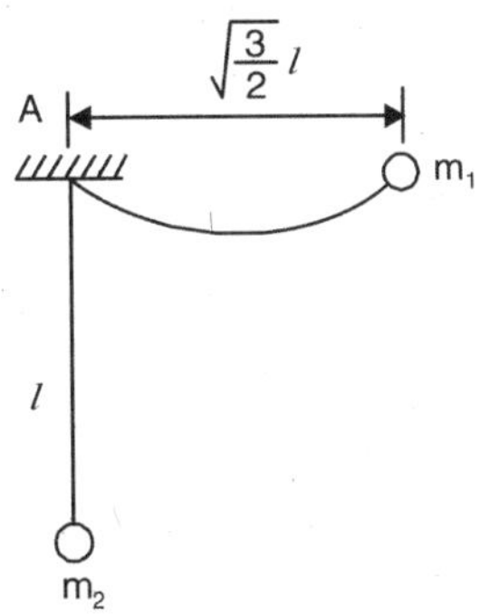

Fig. E9

Example 10

A particle of mass m moves with a uniform speed v along the upper horizontal segment of a hill. At the end of the horizontal segment, the shape of the hill is circular with a radius R. At what height will the particle become disengaged from the circular part? What is the distance d between the lower edge of the hill and the point where the particle lands on the ground?

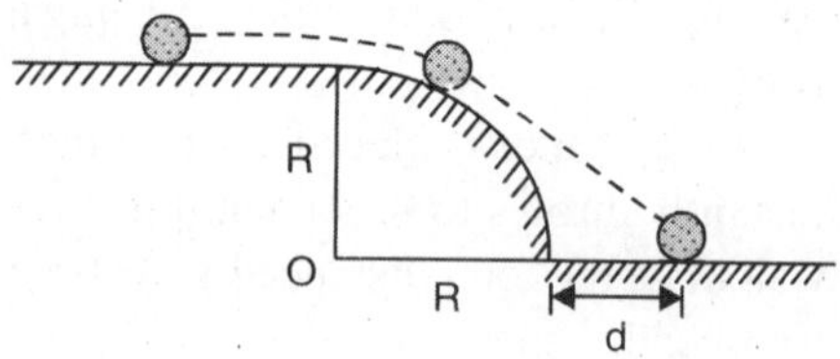

Fig. E10

Example 11

Two identical blocks A and B each of mass 2 kg are hanging stationary by a light inextensible flexible string, passing over a light and frictionless pulley, as shown. A shell C, of mass 1 kg moving vertically upwards with velocity 9 ms^{-1} collides with block B and gets stuck to it. Calculate maximum height reached by B.

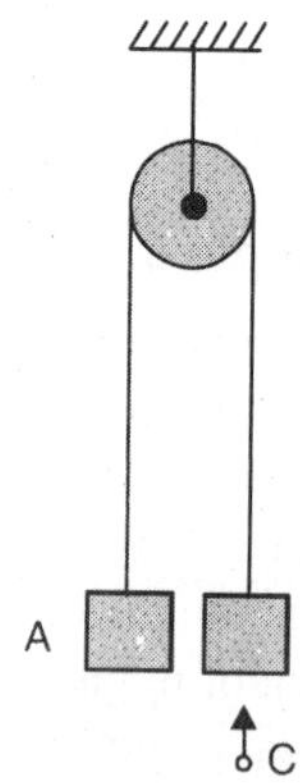

Fig. E11

4 Rotation and Simple Harmonic Motion

Summary and Concept

Angular momentum

For a particle

$$\vec{L} = \vec{r} \times \vec{P}$$

for N particles

$$\vec{L} = \sum_{i=1}^{N} \vec{r}_i \times \vec{P}_i$$

Angular impulse momentum theorem

$$\vec{\tau}_0 = \frac{d\vec{L}_0}{dt}$$

in the integral form

$$\Delta L = \int \vec{\tau}\, dt$$

Angular momentum of system relative to a point 0

$$\vec{L}_0 = \tilde{L} + \vec{r}_C \times \vec{P}$$

where $\tilde{L}$ is the interangular momentum or the angular momentum of the system in the frame of center of mass.

The kinetic energy

The kinetic energy of rotational motion is

$$E_k = \frac{1}{2} I \omega^2$$

I = Moment of interia
ω = Angular velocity of the body

Moment of interia of a rigid body about an axis

$$I = \int dm r^2$$

Steiner's theorem (parallel axis theorem)

$$I = I_C + md^2$$

Here d = distance between given axis and center of mass axis.

Angular momentum of rigid body rotating about a fixed axis.

$$L_z = I\omega_z$$

In vector form

$$\vec{L} = I\vec{\omega}$$

kinetic energy of a solid body in plane motion.

$$T = \frac{1}{2}I_C\omega^2 + \frac{1}{2}mV_C^2$$

The equation of harmonic motion in linear form

$$F_x = -kx$$

$$\ddot{x} = -\omega^2 x$$

The simple pendulum

The equation of motion is

$$\ddot{\theta} + \left(\frac{g}{L}\right)\theta = 0$$

in this case $\omega = \sqrt{\frac{g}{L}}$

Method to solve the SHM problems

(i) Force method

According to the problem, write the translational and rotational equations. Newton's second law of motion is

$$\vec{F} = m\vec{a}$$

$\Rightarrow$

$$\vec{\tau} = 1\vec{\alpha}$$

finally

$$\ddot{x} = -\omega^2 x$$

sought time period

$$T = \frac{2\pi}{\omega}$$

(ii) Energy method

The mechanical energy of the system is conserved forces only,

$$\Delta T + \Delta U = 0$$

finally

$$\ddot{x} = -\omega^2 x$$

sought time period

$$T = \frac{2\pi}{\omega}$$

(iii) The conversation of mechanical energy of oscillation of the system also means

$$E_{\text{mean}} = E_{\text{extreme}}$$

This method is used when it is given that system is in SHM.

Level 1

Example 1

The greatest and least velocities of a certain planet in its orbit round the sun are 30 and 29.2 kilometers per second. Find the eccentricity of the orbit.

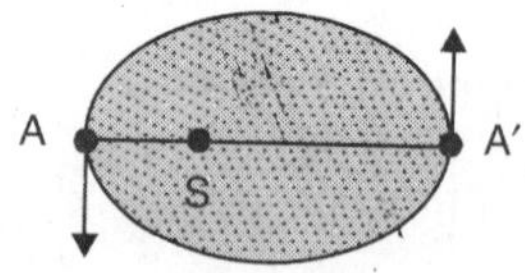

Fig. E1

Example 2

A square lamina *ABCD* of edge length $2l$ rests on a smooth horizontal plane. If the corner *A* is made to move with velocity along the line *BA* produced, determine the initial angular velocity of the lamina.

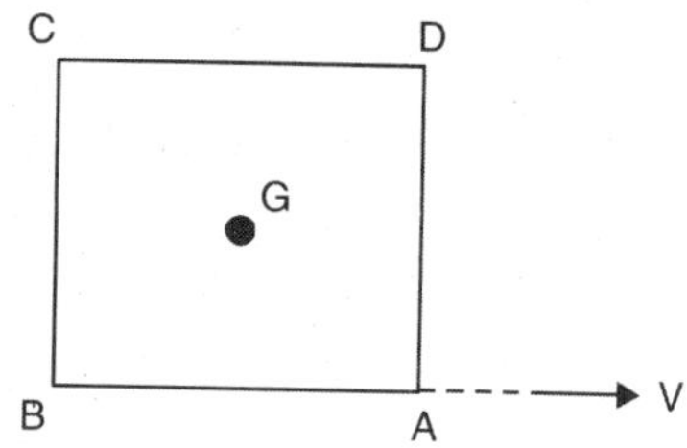

Fig. E2

Example 3

A circular disc is revolving in its planes about its center with angular velocity ω. What is the new angular velocity if a point on the rim is suddenly fixed?

Example 4

On a rough inclined plane of inclination 45° a block of mass m and a cylinder of mass 2 m are released. Coefficient of friction between all the surfaces of contact is 0.5. Find the acceleration of the block and the cylinder when (*a*) they move separately, (*b*) they move together.

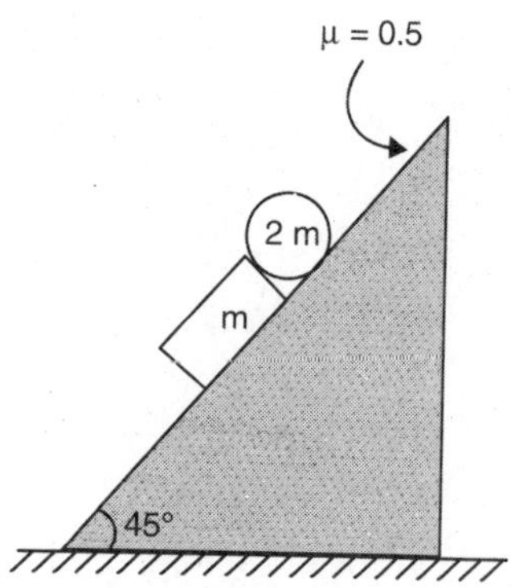

Fig. E4

Example 5

A small mass particle is given an initial velocity v_0 tangent to the horizontal rim of a smooth hemispherical bowl at a radius r_0 from the vertical centerline as shown at point *A*. As the particle slides past point *B*, a distance *h* below *A* and a distance *r* from the vertical centerline, it velocity *v* makes an angle θ with the horizontal tangent to the bowl through *B*. Determine θ.

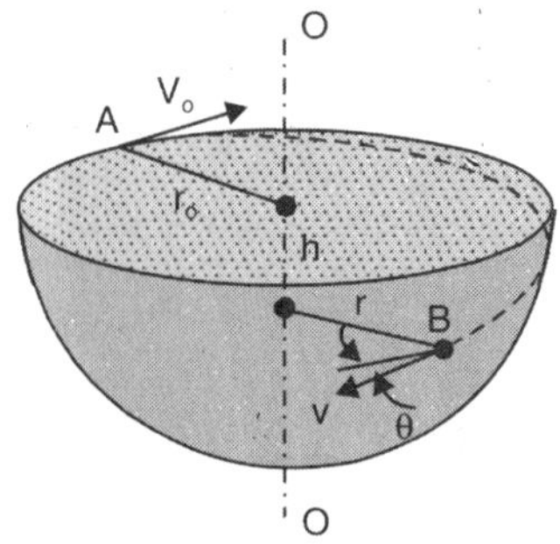

Fig. E5

Example 6

A uniform rectangular block with the given dimensions is dropped from rest from the position shown. Corner *A* strikes the edge at *B* and becomes latched to it. Determine the angular velocity ω of the block immediately after it becomes attached to *B*.

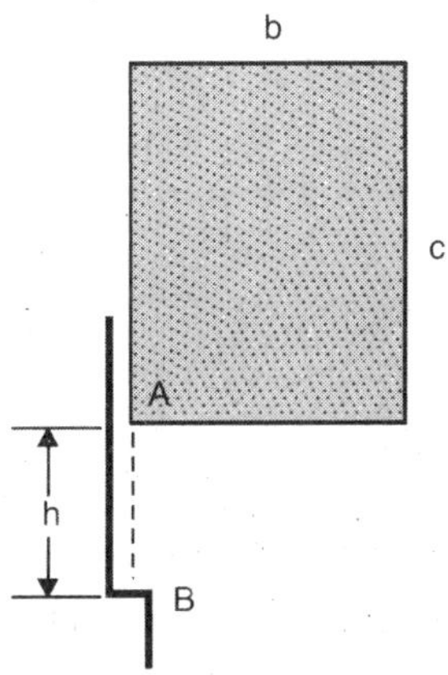

Fig. E6

Example 7

A uniform circular disc of mass *M* and radius *a* is pivoted at point *O* on its circumference so that it can rotate about the tangent at *O*. The center of the disc describing a vertical circle of center *O* in a plane perpendicular to the tangent. The point diametrically opposite to *O* is *A* and the disc is just displaced from rest. Find the angular velocity of the disc when *A* is vertically *O*.

Example 8

A sphere of mass *m* is placed on a rough plank of mass 2 *m*, which in turn is placed on a rough inclined plane. Coefficient of friction between the plank and sphere is μ_2 and that between plank and plane is μ_1. Determine the maximum value of μ_1/μ_2 to ensure pure rolling at plank and sphere surface.

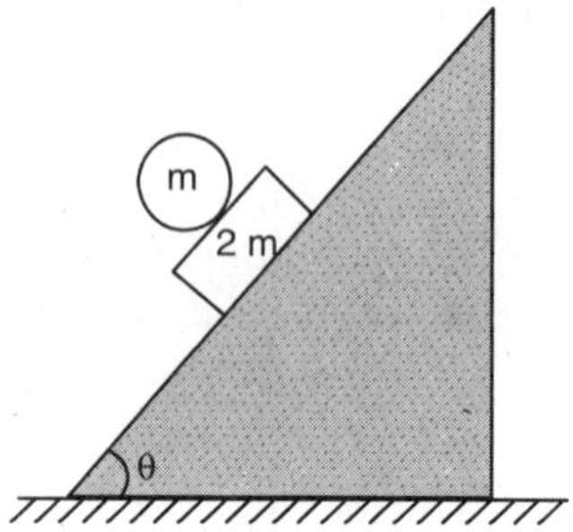

Fig. E8

Example 9

A particle is projected at time $t = 0$ from a point P with a speed v_0 at an angle of 45° to the horizontal. Find the magnitude and the direction of the angular momentum of the particle about the point P at time $t = v_0/g$.

Example 10

A uniform rod of mass m and length L is initially at rest in the position shown in figure on the circular surface of radius R. Find the angular velocity of the rod when it is released from rest at particular angle θ, it makes with the horizontal.

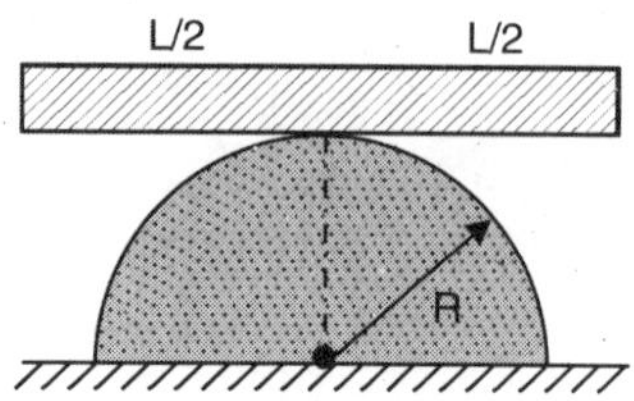

Fig. E10

Example 11

Two identical balls A and B, each of mass m are attached to the two identical massless springs. The spring mass system is constrained to move inside a rigid smooth pipe bent in the form of circle. The pipe is fixed in horizontal plane. The spring constant is K and πR is the natural length of the spring. Both balls are released at particular angle θ as shown in figure. Determine frequency of small oscillation.

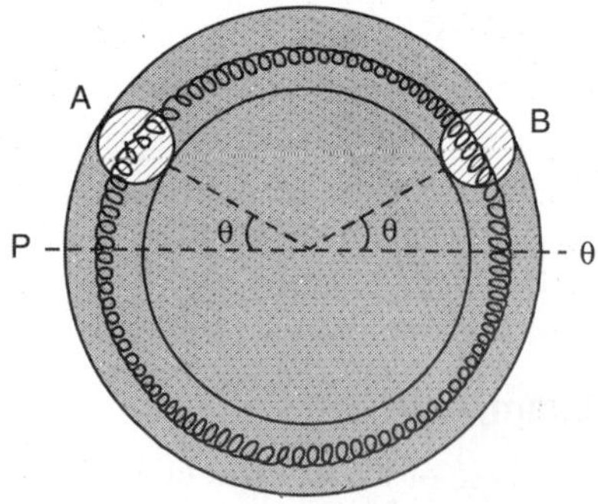

Fig. E11

Example 12

What should be the minimum coefficient of static friction to support pure rolling? Sphere rolls without slipping on inclined plane.

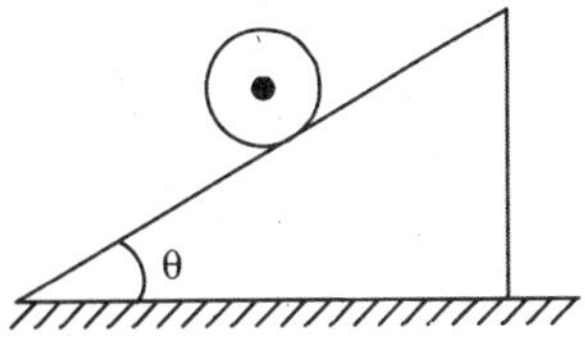

Fig. E12

Example 13

A rod of length L and mass m is pivoted at one end to constitute a pendulum. Determine its period of oscillation.

Example 14

A disc is thrown with a linear speed v_0 in the forward direction and an angular speed v/R in the anticlockwise direction as shown in figure. Find the linear speed of the disc when it starts pure rolling.

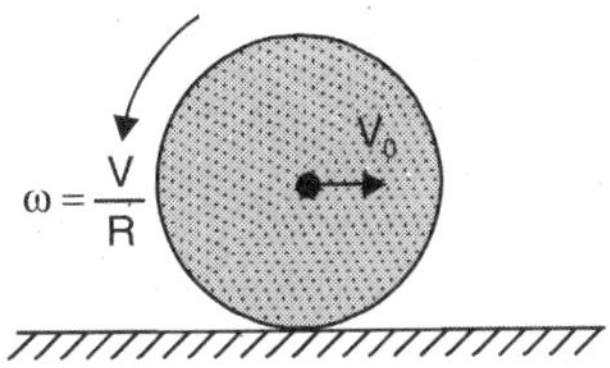

Fig. E14

Example 15

A disc of radius r is given an initial angular velocity ω_0 and is placed on the inclined plane as shown in figure. Find the distance the disc moves before sliding stops. Assume $\mu > \tan \alpha$.

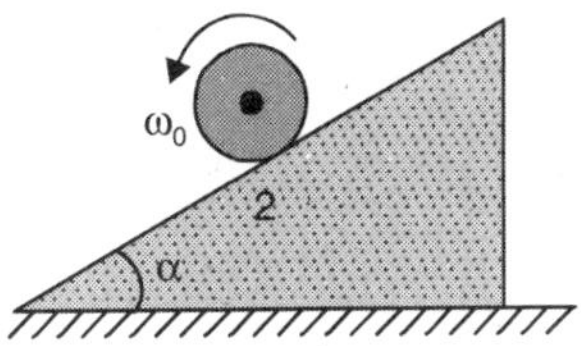

Fig. E15

Example 16

A rectangular plate (2 m × 3 m) of mass 100 mg/s^2 is supported by hinges at A and B as shown in figure. If the support A is removed, determine the reaction at B and acceleration of the center of mass.

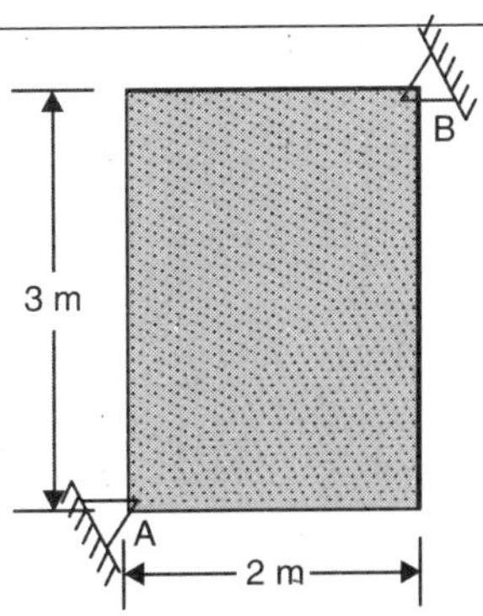

Fig. E16

Example 17

Find the moment of inertia of a thin disc of mass m and radius R with a hole of radius $R/2$ around the z-axis as shown.

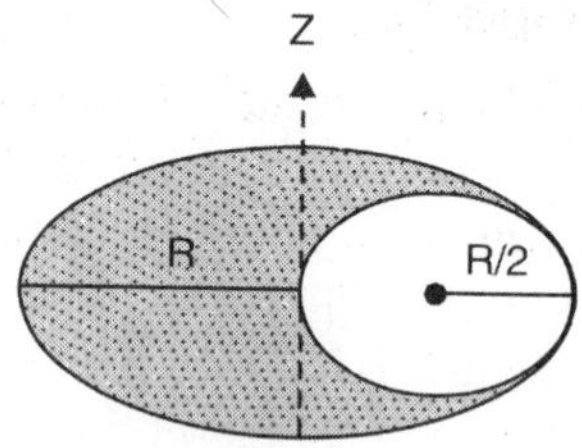

Fig. E17

Level 2

Example 1

A round board of mass M and radius R is placed on a fixed smooth horizontal plane, and is free to rotate about a smooth axis, which passes through its center. A man is standing on the circumference of the board. At first, the board and the man are at rest. Then, the man starts moving along the rim of the board at a constant velocity of v_0.

What will be the angle of the board's rotation when the man passes his starting point first time?

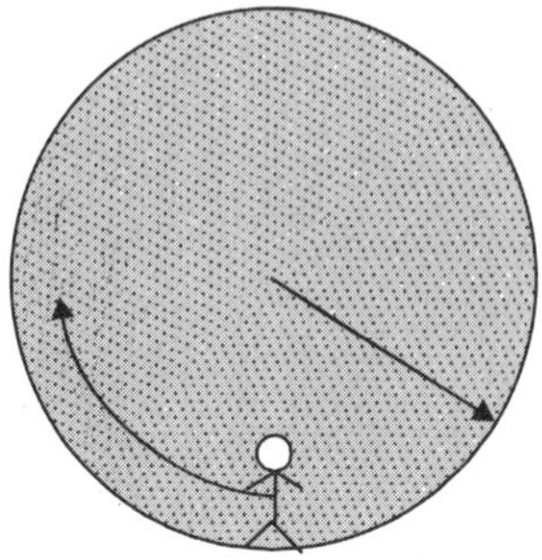

Fig. E1

Example 2

A uniform hemisphere of mass m and radius R is released from rest in the position as shown. Assuming that the hemisphere rolls without sliding, determine its angular velocity after it has rolled through 90°.

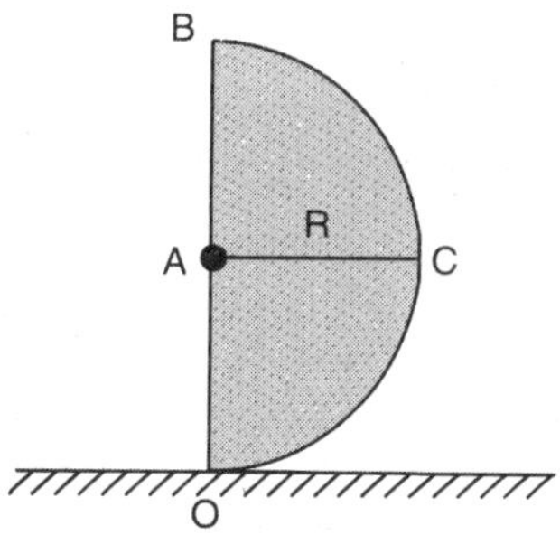

Fig. E2

Example 3

A uniform rectangular block is released from rest with θ essentially zero and pivots in the vertical plane about the center A of its lower face on the fixed corner. (a) If the block is observed to slip when $\theta = 30°$, find the coefficient of static friction between the block and the corner. (b) If the bottom face of the block is notched so that it cannot slip, find the angle θ at which contact between the block and the corner ceases.

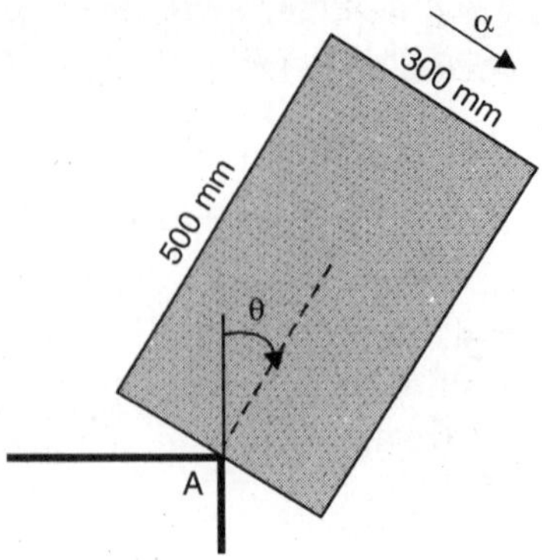

Fig. E3

Example 4

A solid sphere of radius R rolls without slipping in a cylindrical shell of radius $5R$ as shown in figure. Find the time period of small oscillations.

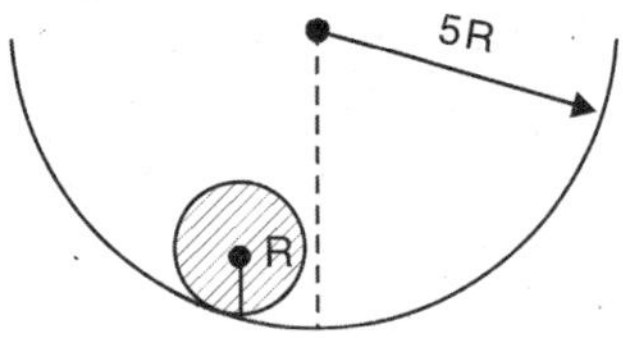

Fig. E4

Example 5

For an arrangement shown in the figure, radius of the pulley is R, its moment of inertia relative to the rotation axis is I, the mass of the body is m and the spring constant is K. Find the time period of small oscillation.

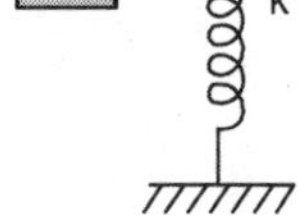

Fig. E5

Example 6

A particle of mass m is attached with four identical springs, each of length l. Initial tension in each spring is I. Neglecting gravity, calculate period of small oscillations.

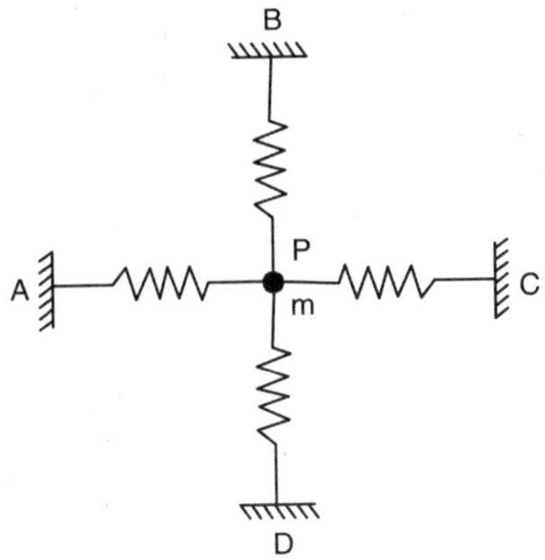

Fig. E6

Example 7

A uniform straight rod is held at an inclination α to the vertical with its lower end in contact with a smooth horizontal plane and let go. Find the angular velocity when the rod becomes horizontal.

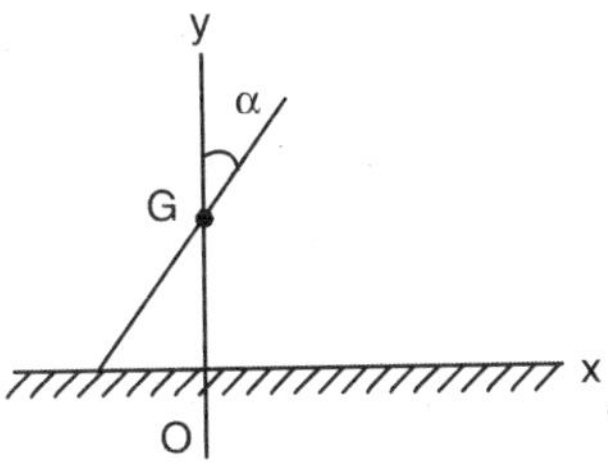

Fig. E7

Example 8

A square package of side a and mass m slides down an inclined smooth plane A. At the end of the incline, corner of the package strikes a rigid support at B with linear velocity v_0. Assuming that the impact at B is perfectly plastic, find the minimum value of v_0 for which the package will rotate about B and reach the horizontal floor.

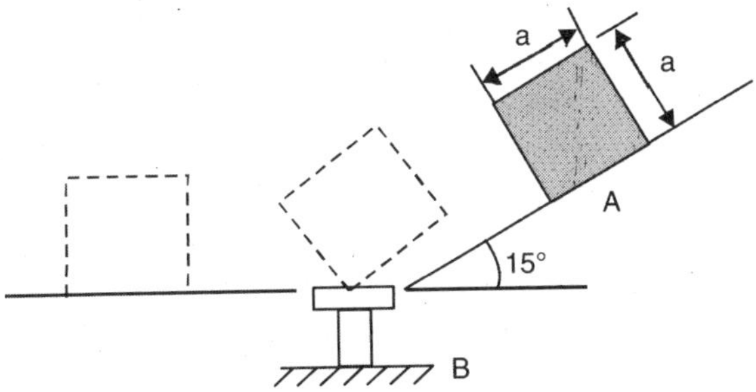

Fig. E8

Example 9

A sphere of radius a, and of radius of gyration k about any axis through its center, rolls with linear velocity v on a horizontal plane, the direction of motion being perpendicular to a vertical face of a fixed rectangular block of height h, where $h < a$. The sphere strikes the block; the sphere and the block are perfectly rough and perfectly inelastic. Show that the sphere will surmount the block if $(a^2 - ah + k^2)\, v^2 > 2gha^2 (a^2 + k^2)$.

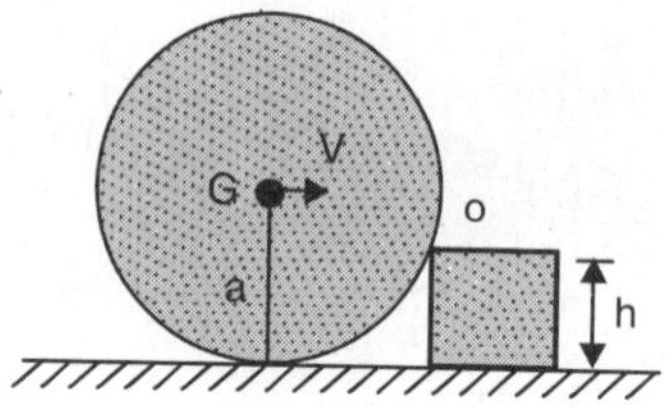

Fig. E9

Example 10

A solid hemisphere of mass m is held with its base against a smooth vertical wall and its lowest point in a smooth floor. The hemisphere is released. Find the initial pressures on the wall and the floor.

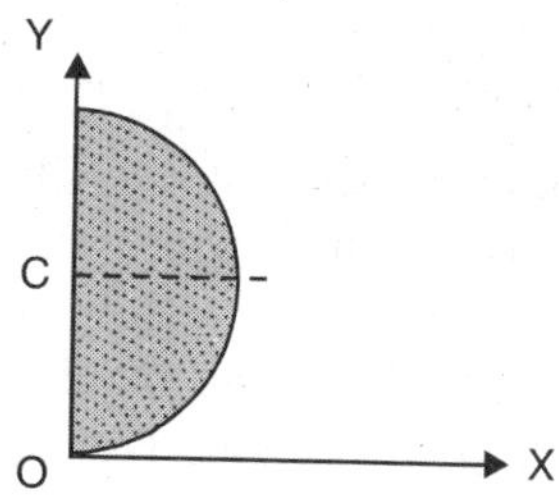

Fig. E10

Example 11

A solid ball of diameter $d = 11$ cm is rotating about its one of the horizontal diameters with angular velocity $\omega_0 = 120$ rad/sec. It is released from a height so that it falls $h = 1.8$ m freely and then collides with the horizontal floor. Coefficient of restitution is $e = 5/6$ and coefficient of friction between the ball and the ground is $\mu = 0.2$. Calculate distance between the points where the ball strikes the floor for the first and second time.

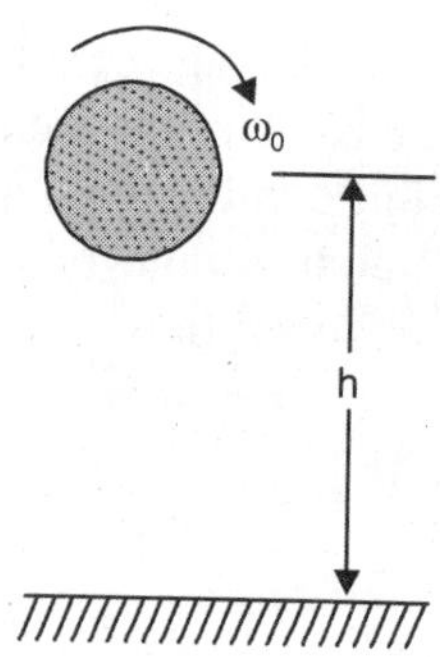

Fig. E11

Example12

A triangular prism of mass = 1.12 kg having base angle 37° is placed on a smooth horizontal floor. A solid cylinder of radius $R = 20$ cm and mass

$m = 4$ kg is placed over the inclined surface of the prism. If sufficient friction exists between the cylinder surface and the prism, so that cylinder does not slip, calculate acceleration of prism when the system is released. Calculate also force of interaction between cylinder and the prism.

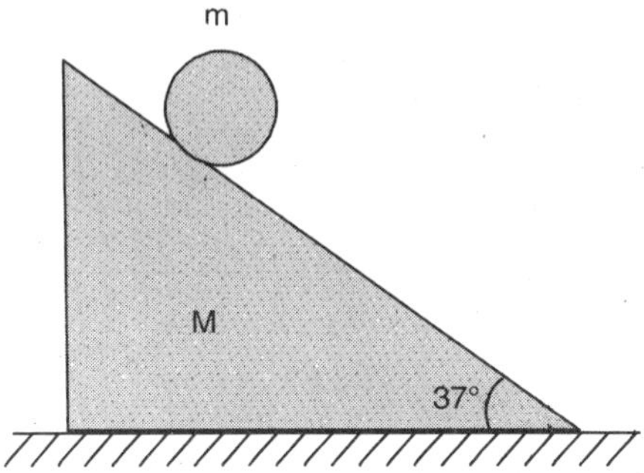

Fig. E12

Example 13

A uniform circular disc of mass M and radius R is free to rotate about a vertical axis O passing through its rim. An insect of mass m is at point A such that line OA is the diameter of the disc as shown in Fig. 13. The insect describes a complete circle relative to disc and returns to the starting point A. Calculate the angle moved by the disc relative to the ground.

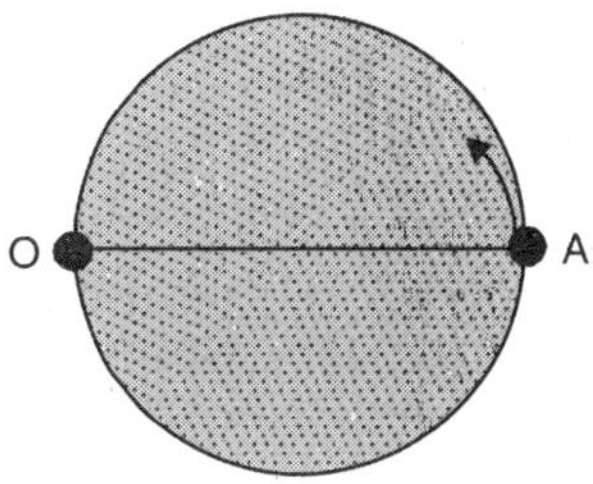

Fig. E13

Example 14

A full cylinder of mass M and radius R is to be placed on a rough surface. A thin rod of same mass M and same length as the cylinder, is fixed parallel to the cylinder's axis, at a distance $r = \frac{R}{3}$ from it. The system is released from rest at the state described in Fig. 14. The initial angle between $\hat{r}$ (which is a unit position vector of the rod relative to the cylinder axis) and $\hat{g}$ (which is the constant direction of gravity acceleration) is θ_0. It is also known that the cylinder does not slide all through its motion.

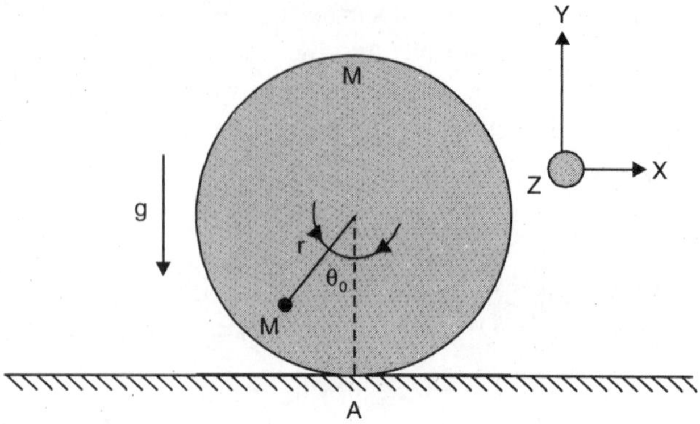

Fig. E14

What is the cylinder's axis velocity when the rod is at its lowest position (i.e. closest to the plane)?

Example 15

A uniform circular disc of mass M and radius a is pivoted at a point O on its circumference so that it can rotate about the tangent at O, which is horizontal. The center of the disc describing a vertical circle of center O in a plane perpendicular to the tangent. The point diametrically opposite to O is A and the disc is just displaced from rest. When A is vertically above O, find the angular velocity of the disc when A is vertically O.

At this instant a particle of mass M traveling with velocity u in the opposite direction of motion of the center of disc hits the disc at its center, and adheres it. Find the angular velocity of the system immediately after the impact. If the disc just reaches its initial position show that

$$u = \left(3\sqrt{2} + \sqrt{3}\right)\sqrt{ag}$$

Example 16

A 4 kg cylinder B and 3 kg wedge A are held at rest in the position as shown in Fig. E16. Suddenly cord has been cut, determine acceleration of the cylinder B.

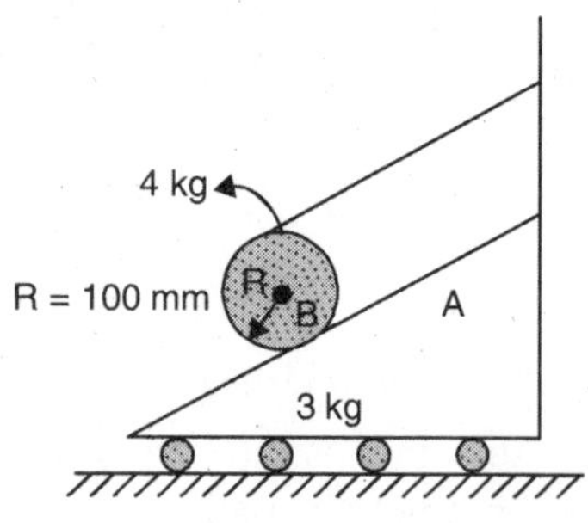

Fig. E16

Example 17

A spherical ball of radius r and mass m collides with a plank of mass M kept on a smooth horizontal surface. Before impact, the center of the ball has a velocity V_0 and angular velocity w_0 as shown. The normal velocity is reversed with some magnitude and the ball stops rotating after the impact. Find the distance on the plank between first two impacts of the ball. The coefficient of friction between the ball and the plank is μ. Assume that plank is large enough and elastic.

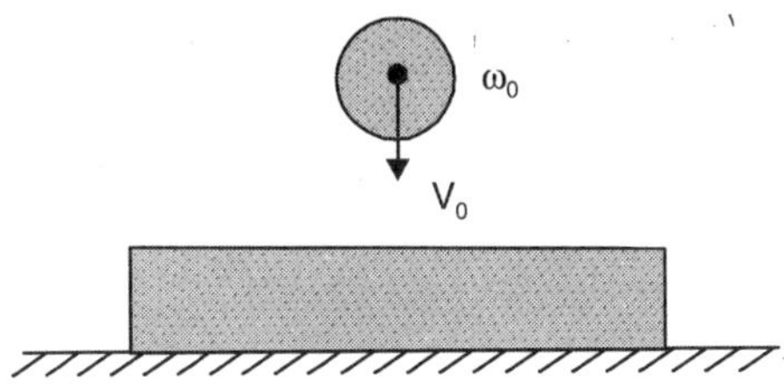

Fig. E17

Example 18

A pendulum clock is kept in an elevator which starts going up with a constant acceleration a ($< g$). At a height h the acceleration of the elavator reverses, its magnitude remaining the same. How soon after the start of the motion will the clock show the right time again?

Example 19

At the instant shown in Fig. 19. $\theta = 60°$ and rod AB is subjected to a decceleration of 16 m/s^2 when the velocity is 10 m/s. Determine the angular velocity and angular acceleration of the link CD at this instant.

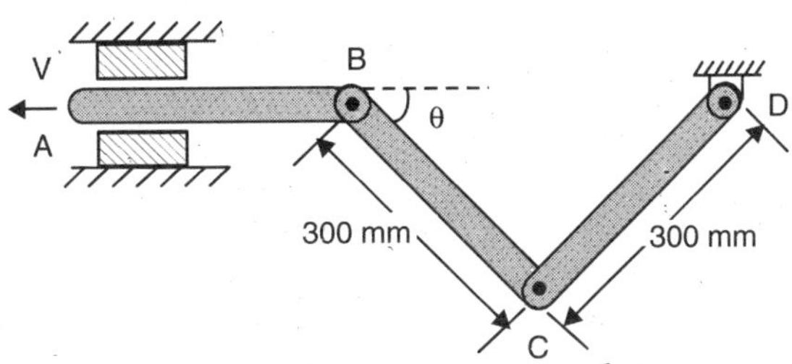

Fig. E19

Example 20

A uniform horizontal plank is resting symmetrically on two counter rotating drums. The distance between the drums is $2l$ and μ is the coefficient of friction between the plank and drums. For the small displacement from the equilibrium position, calculate the angular frequency of the plank.

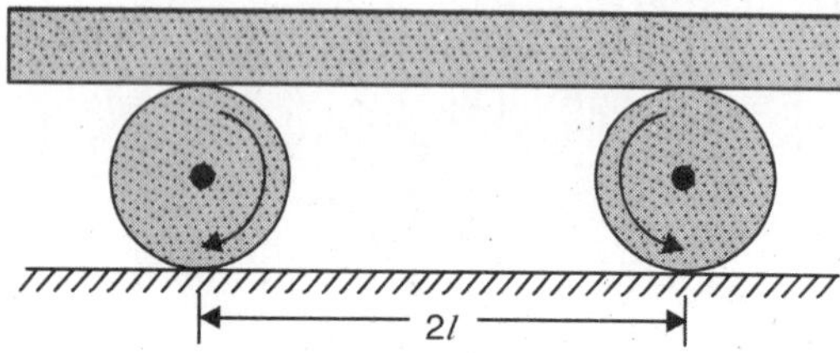

Fig. E20

Example 21

For the arrangement shown in figure, K_1, K_2, are the spring constant and m is the mass of block. Neglect the mass of pulley. Calculate the angular frequency for small oscillation.

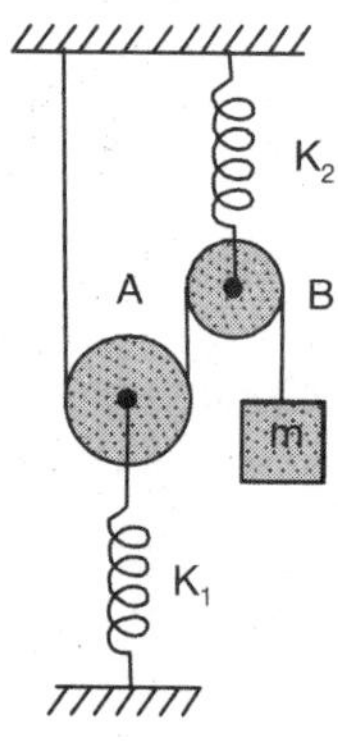

Fig. E21

Example 22

Find the natural frequency of the system as shown in figure. The pulleys are smooth and massless.

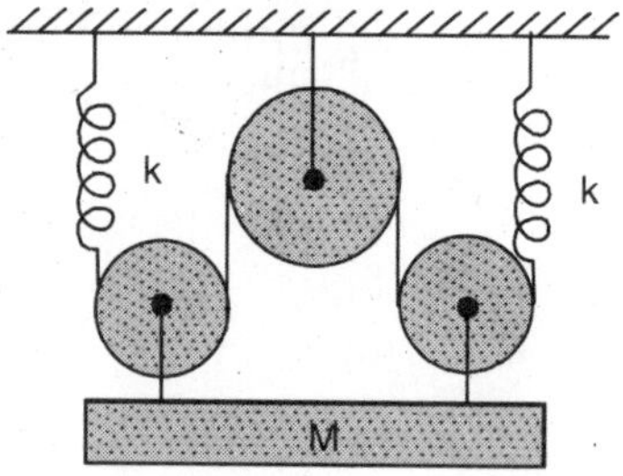

Fig. E22

Example 23

A uniform straight bar 1.2 m long having mass 14 kg hangs vertically from a thin cord as shown. A bullet of mass 30 gm moving horizontally with velocity 600 mm/sec. pierces through the bar at *C* and emerges with a velocity 240 mm/sec. Determine the velocity of the end *A* of the bar just after impact.

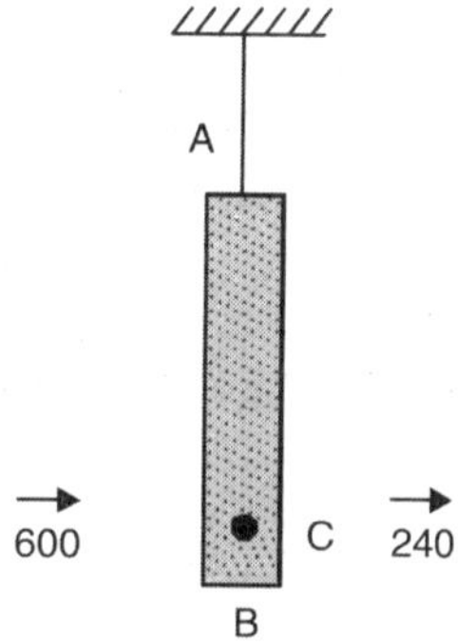

Fig. E23

Example 24

A 2 m long rod *AB* moves with its ends in contact with the floor and the inclined surface. At the instant shown the velocity and acceleration of *A* are 6 m/s leftward and 6.3 m/s^2 rightward respectively. Determine the acceleration of end *B*.

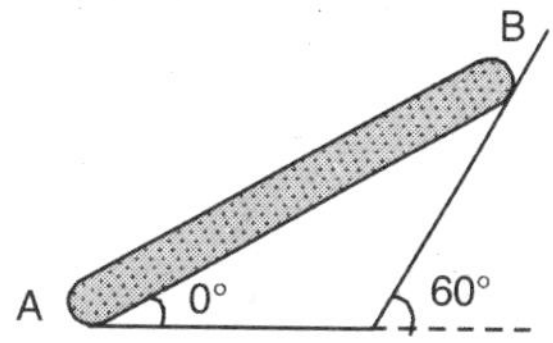

Fig. E24

Example 25

In the system shown in Fig. E25 blocks *A* and *B* have mass $m_1 = 2$ kg and $m_2 = \frac{26}{7}$ kg respectively. Pulley having moment of inertia $I = 0.11$ kgm^2 can rotate without friction about a fixed axis. Inner and outer radii of pulley are 10 cm and 15 cm respectively.

Calculate

(i) tension in each thread

(*ii*) acceleration of each block

($g = 10$ m/s^2)

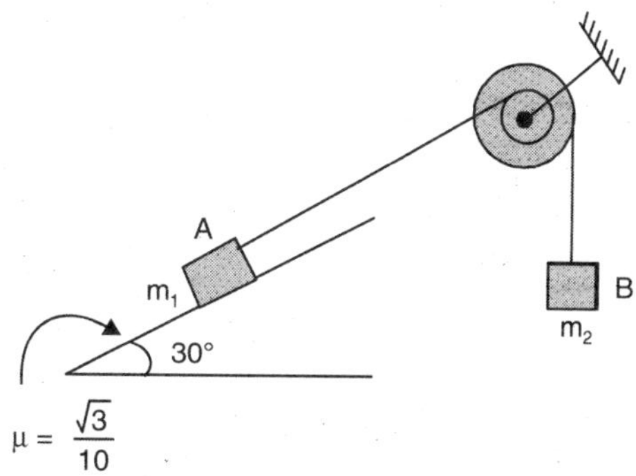

Fig. E25

Example 26

A homogeneous rod *AB* of length L and mass M is hinged at the center *O* in such a way that it can rotate freely in the vertical plane. The rod is initially in horizontal position. An insect S of the same mass M falls vertically with speed V on point C, midway between the points O and B. Immediately after falling, the insect starts to move towards B such that the rod rotates with a constant angular velocity ω.

(*i*) Calculate angular velocity ω in terms of *V* and *L*.

(*ii*) If insect reaches the end *B* when the rod has turned through angle of 90° calculate *V* in terms of *L*.

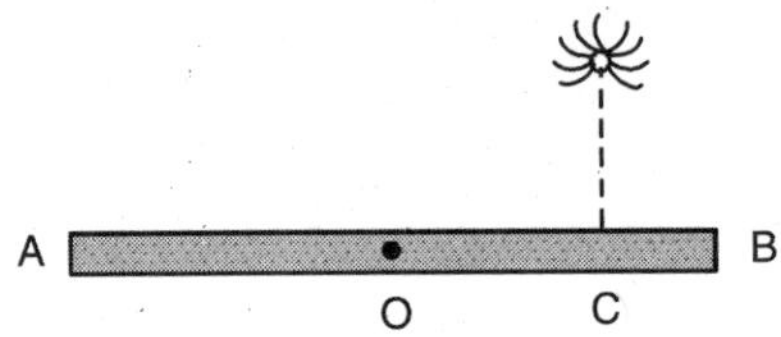

Fig. E26

Example 27

A uniform straight rod of mass *m* is suspended from a point by two strings of the same length at the rod attached to its ends. If one string is cut, prove that the initial angular acceleration of the rod is nine times that of the remaining string. Find the tension in the string.

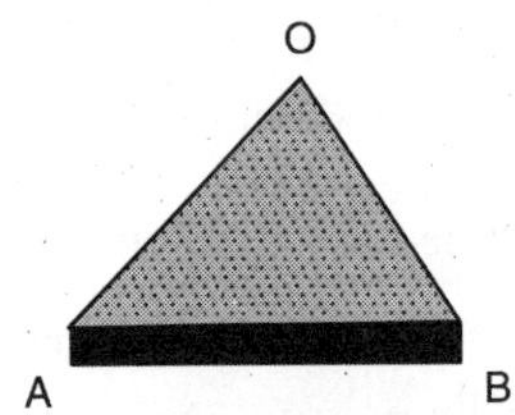

Fig. E27

Level 3

Example 1

A heavy particle is supported in equilibrium by two equal elastic strings with their ends attached to two points in a horizontal plane and each inclined at an angle of 30° to the horizontal. The modulus of elasticity is such that when the particle is suspended from any portion of the string its extension is equal to its natural length. The particle is displaced vertically a small distance and then release. Determine period of its small oscillations.

Example 2

In the arrangement shown in figure, pulleys are small and light and springs are ideal. The springs constant of the springs are K_1, K_2, K_3 and K_4. Calculate period of small oscillations of block of mass m.

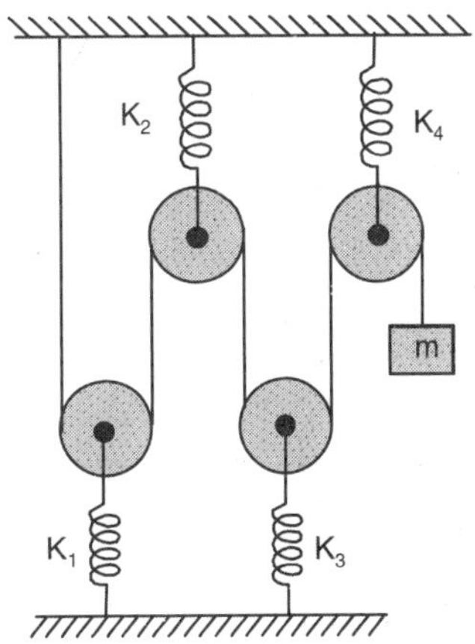

Fig. E2

Example 3

A helicopter is standing on the ground. Three bladed rotor (fan) of the helicopter consists of a 0.2 m diameter shaft 1 m long and 1.5 m long radial blades. The shaft is made of steel. Specific gravity 7 and the blades have a mass of 20 kg/m. Two birds flying parallel to *OA* collides *OB* and *OC* at a distance 1.2 m and 0.9 m from center *O* with velocity 20 m/s and 30 m/s respectively.

Determine the number of revolutions of the shaft blades system after the collision.

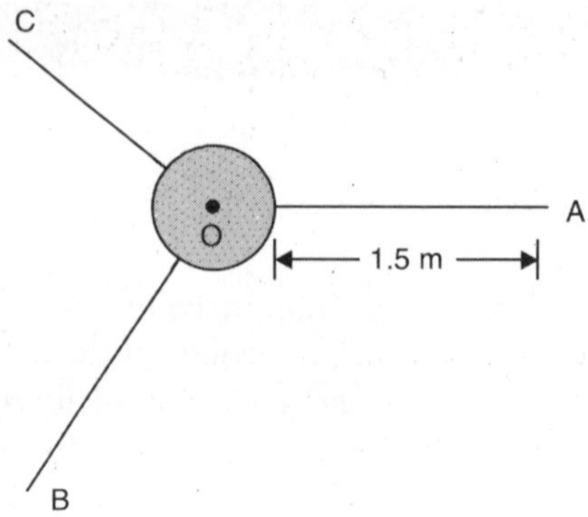

Fig. E3

Example 4

A small body of mass m is attached to the inside of a ring of radius R and mass m. The ring starts rolling down on a rough incline from the initial position as shown in figure.

(*a*) Find the velocity of center of mass of the ring when it has moved down a distance S_0 down the plane

(*b*) Find the minimum value of the angle of incline at which the ring leaves contact with the surface.

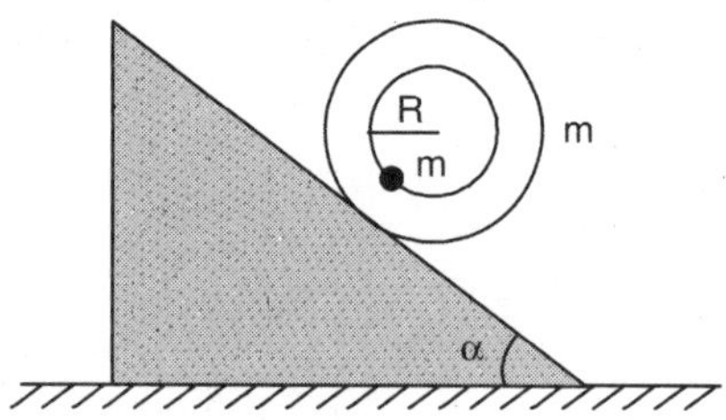

Fig. E4

Example 5

A thin rod of length L and mass m moves on a smooth horizontal surface at a velocity $\vec{v}_0$ in the direction of its length. The end of the rod hits the end of a second rod of the same length and mass, which is initially at rest, perpendicular to the first rod. At the moment they touch, the two poles stick together, while the angle between them remains $\frac{\pi}{2}$.

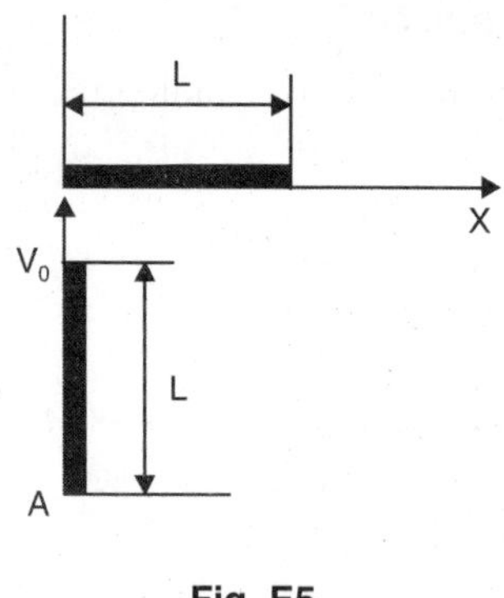

Fig. E5

What is the angular velocity of the system about the center of mass?

What is the maximal velocity of the pole's end, denoted by A, after the collision?

When does the point A reach this velocity for the first time?

Example 6

In the arrangement shown in Fig. E6, mass of blocks A and B is $m_1 = 0.5$ kg and $m_2 = 10$ kg, respectively and mass of spool is $M = 8$ kg. Inner and outer radii of the spool are $a = 10$ cm and $b = 15$ cm respectively. Its moment of inertia about its own axis is $I_0 = 0.10$ kg m^2. If friction is sufficient to prevent sliding, calculate acceleration of blocks A and B. Take $g = 10$ m/s^2.

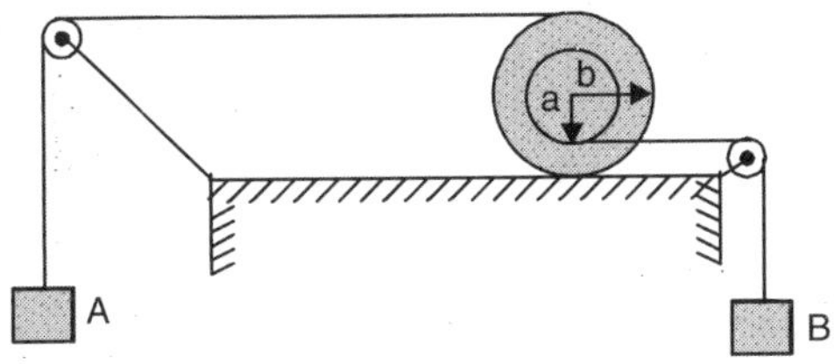

Fig. E6

Example 7

A body of mass m is placed on a perfectly smooth table and tied to a massless string which passes through a hole in the table. The other end of the string (the one below the table) is tied to another mass M. Initially, M is held so that the mass m is at a distance R from the hole, and moves in a circle at a tangential velocity $v_0\hat{y}$. M is then released.

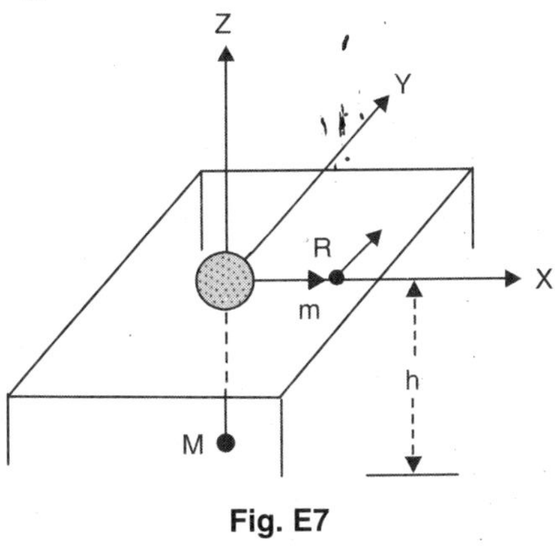

Fig. E7

Find the final radius, r_1, which the mass m reaches when M stops (only).

Example 8

A small rubber ball of radius r is thrown against a rough floor with a velocity V_0 and a backspin w_0. It is observed that the ball bounces from A to B, then from B to A, then from A to B, etc. Assuming perfectly elastic impact, determine the required magnitude w_o of the back spin in terms of V_0 and r.

Example 9

Assuming all the contacts to be frictionless, find the angular frequency of wedge and block system.

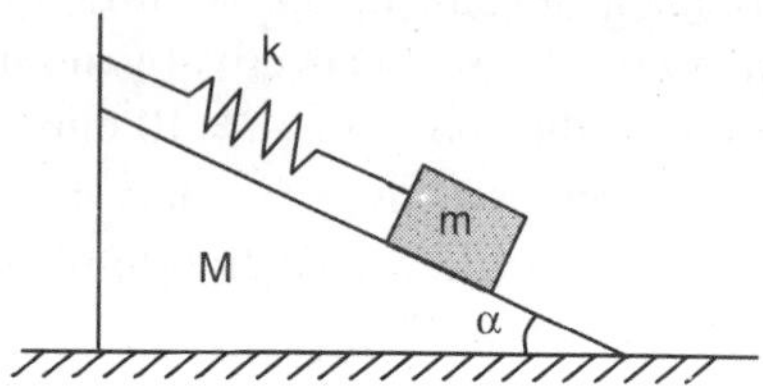

Fig. E9

Example 10

A semi cylindrical shell of radius r is set into small oscillation on the horizontal surface. If no slipping occurs, determine the time period of small oscillation.

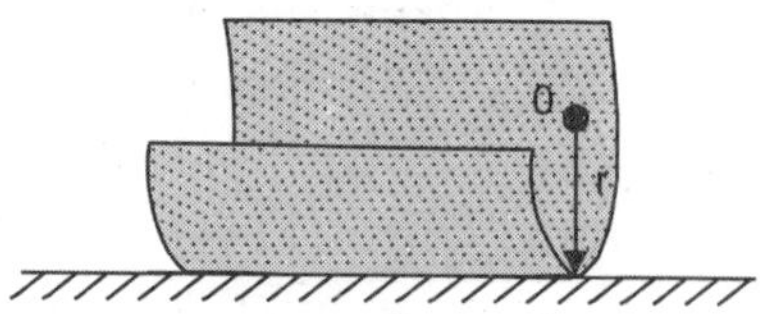

Fig. E10

Example 11

In an acrobatic stunt, man A of mass m_A drops from a raised platform onto the end of the light but strong beam with a velocity v_0. The body of mass m_B is propelled upward with a velocity v_B. For a given ratio $n = m_B/m_A$, determine b in terms of L to maximize the upward velocity of the boy. Assume that both man and boy act as rigid bodies.

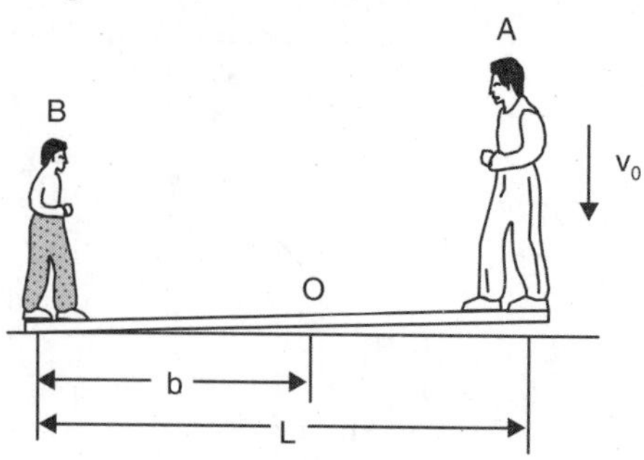

Fig. E11

Example 12

ABC is a straight light and rigid rod of length 90 cm. End *A* is pivoted so that the rod can rotate freely about it, in vertical plane. A pulley, having internal and external radius $R = 7.5$ cm and $r = 5$ cm is fixed to a shaft of radius 5 cm. The pulley-shi system can rotate about a fixed horizontal axis *O*. *B* is point of contact of the pulley and the rod. From free end *C* of the rod a mass $m_2 = 2$ kg is suspended by a thread. Another thread is wound over the shaft and a block of mass $m_1 = 4$ kg is suspended from it. If coefficient of friction between the rod and pulley surface is $\mu = 0.4$ and moment of inertia of pulley-shaft system about axis *O* is $I = 0.045$ kg-m^2, calculate acceleration of block m_1; when the system is released. (Take $g = 10$ m/s^2)

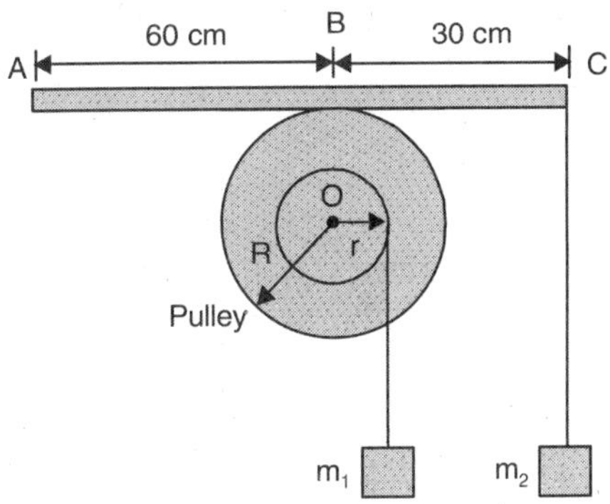

Fig. E12

Example 13

A disc of mass *M* and radius *R* rotates about its vertical axis of symmetry with an angular velocity ω_0. A bullet of mass $\left(m = \frac{1}{2}M\right)$ is fired at a velocity of v_0 towards the disc. The vertical distance of the mass from the center of the disc's rotation is *a* (*a* is an impact parameter). The bullet sticks to the disc:

The disc rotates about its axis with an angular velocity ω_0 while its axis move at $r\omega_0$.

What is the value of v_0 for which the system stops rotating after the collision?

Think Without Ink

1. A ball is projected with a speed V at an angle θ from a point on a horizontal surface. Find the time after which its velocity becomes perpendicular to the velocity of projection. Neglect air resistance.
2. Compare the velocity of point B, C, D, and E. Disc rolls on the surface without slipping.

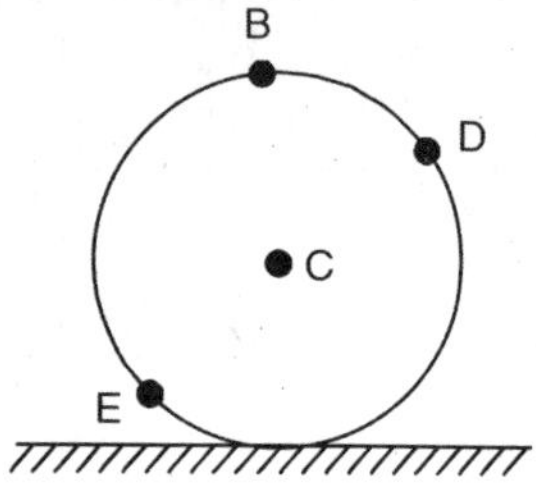

Fig. 2

3. The cylinder A is moving upward according to the function $y = t^2/8$ where t is in seconds. Calculate the downward acceleration of block B.

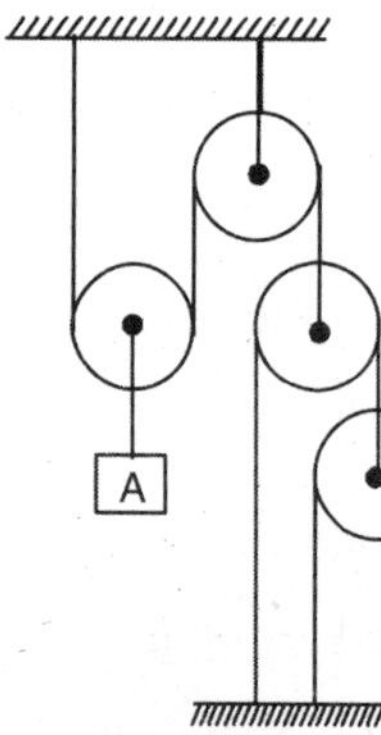

Fig. 3

4. Compare the angular velocities of the rod, square and sphere as shown in figure.

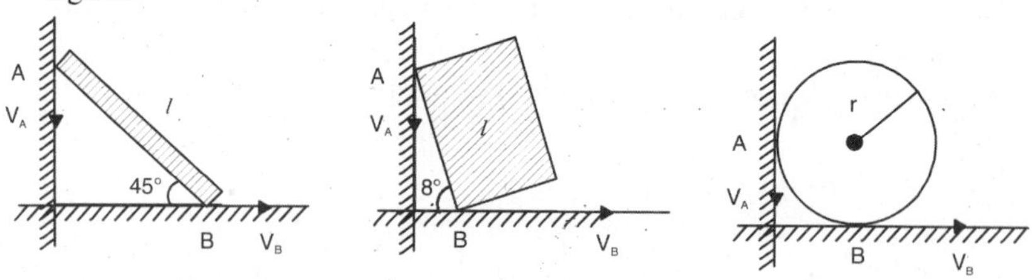

Fig. 4

5. Jack is moving in a circle of radius R_1 with uniform speed V. Sam and Mickey is rest on the diameter of circle. R_2 is the distance between the Sam and Jack. What is the angular velocity of Jack with respect to Sam and Mickey and center of circle C?

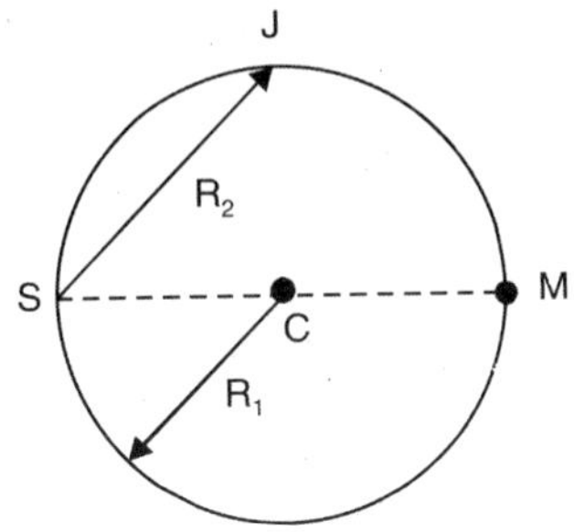

Fig. 5

6. Three identical balls, each of mass m is welded to the rigid equiangular frame of negligible mass. There are two arrangements of these balls as shown in figure. If a force F is suddenly applied on each frame as shown, determine the instantaneous acceleration of each frame.

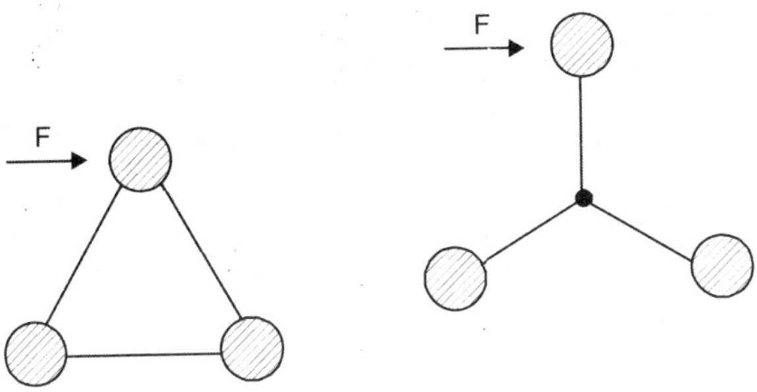

Fig. 6

7. Two blocks A and B are placed on an elevator and elevator is descending with an acceleration of a m/s^2. What force is exerted by the block A on the block B, the mass of the block A is m.

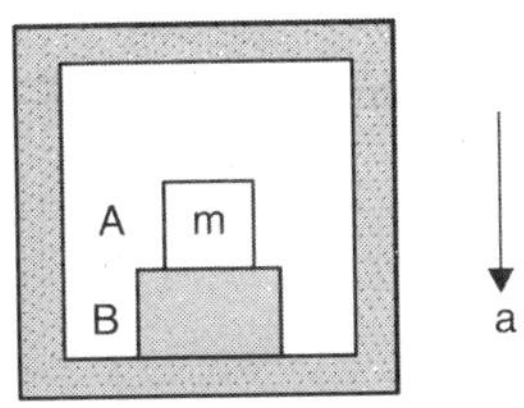

Fig. 7

8. A man has fallen into a well of width d and two of his friends are slowly pulling him out using light rope as shown in figure. Find the force when the man is at a depth h.

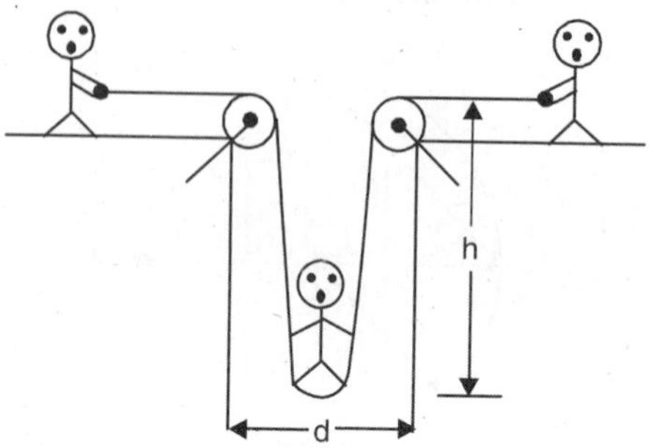

Fig. 8

9. A block of mass 5 kg is suspended through two light spring balances as shown in figure, compare the reading of both scale.

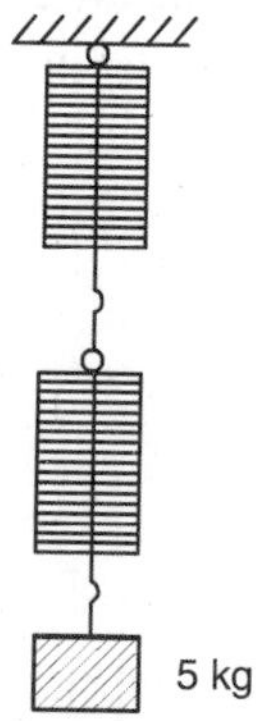

Fig. 9

10. A pipe of length l and altitude difference h connects two points A and B. A string passed through the pipe is fixed at point A. Neglecting friction, determine the initial acceleration a of the string when it is released.

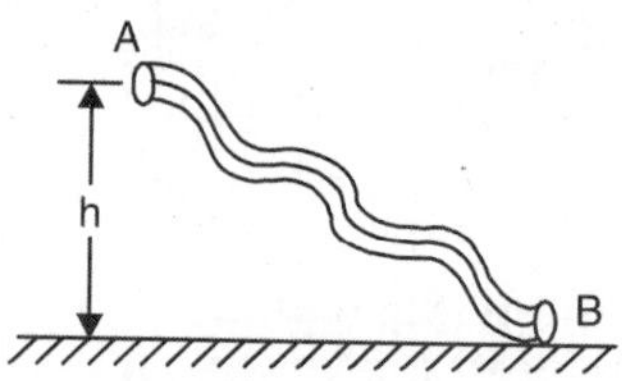

Fig. 10

11. A block of mass m is released from rest onto an ideal non-deformed spring of spring constant K from a negligible height. Neglecting air resistance, find the compression d of the spring.

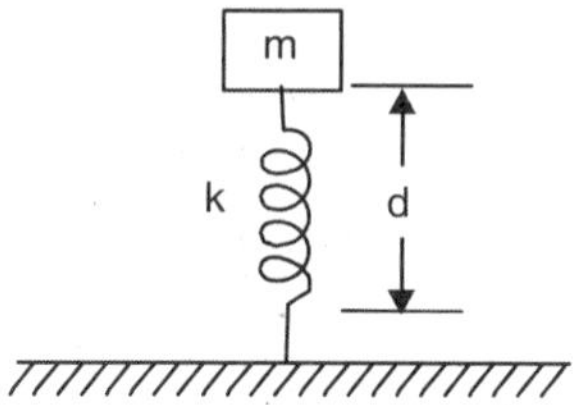

Fig. 11

12. Two cylindrical containers of equal cross sectional area A contain oil upto heights h_1 and h_2. The containers are interconnected so that the levels in them become equal. What will be the work done by the gravity for this process. The density of oil is r.

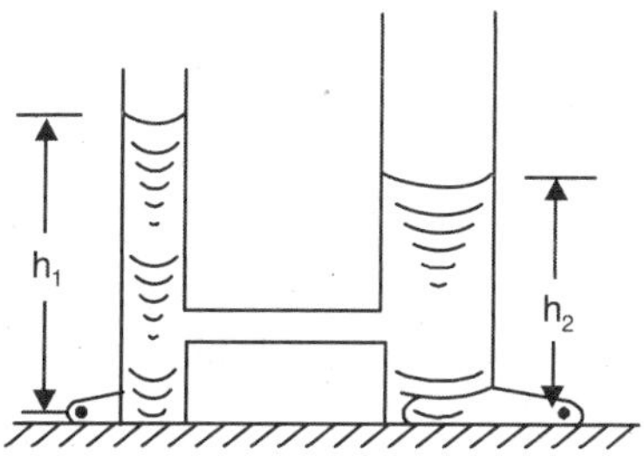

Fig. 12

13. Two identical balls A and B kept on the surface. Ball A is moving at a speed v towards right and collide with ball B which was in rest. Assume all collisions to be elastic. Discuss the speed of the each ball after all collisions.

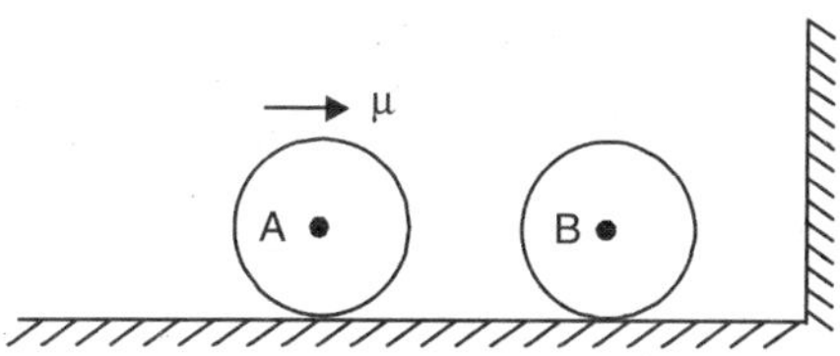

Fig. 13

14. A ring of mass M is placed on a smooth horizontal surface as shown in figure. An ant of mass m alights on the ring. Describe the trajectory of the ant when it begins to move along the ring.

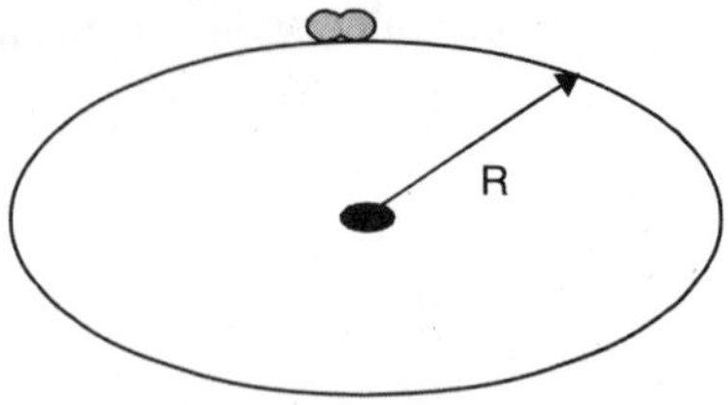

Fig. 14

15. A particle of mass m_1 is released from the top of smooth wedge of mass m_2, which rests on a smooth horizontal floor. Calculate the distance moved by the wedge till m_1 touches the floor.

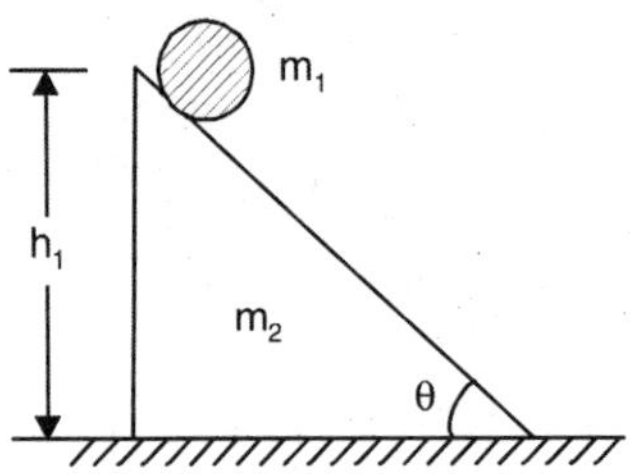

Fig. 15

16. A solid sphere is gently placed on a rough horizontal ground with an initial angular velocity ω_0 and no linear velocity. Determine angular velocity ω at the end of slipping

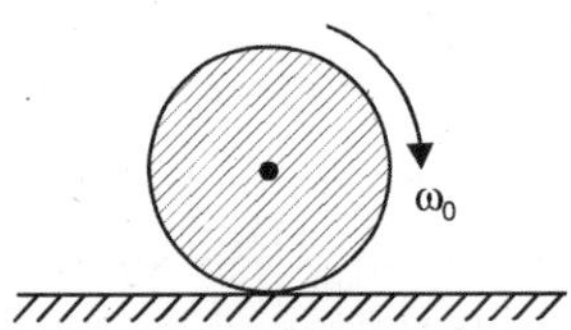

Fig. 16

17. A cube is resting on an inclined plane. If the angle of inclination is gradually increased, what must be the coefficient of friction between the cube and plane so that,

(*a*) Cube slides before toppling?
(*b*) Cube topples before sliding?

18. In the figure shown five identical yo-yos initially are at rest on a horizontal surface. There is sufficient friction for the yo-yo to roll without slipping. A string is pulled on different yo-yos in different directions. In what direction will each yo-yo rotate?

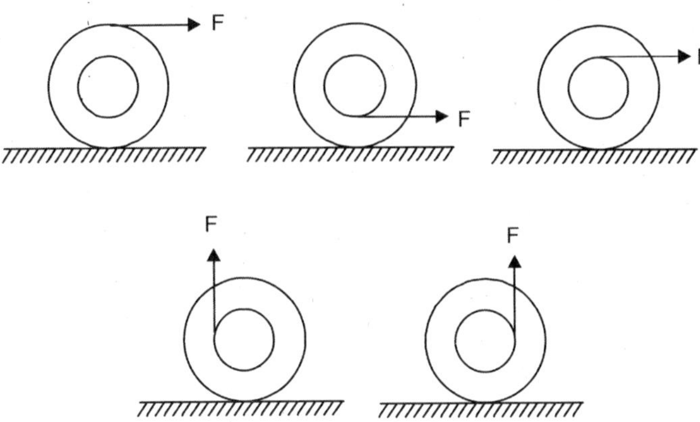

Fig. 18

19. Front disc of an assembly of two discs is given an angular velocity ω_0 and is placed on the rough horizontal surface as shown in figure. Determine the angular velocity when both the disc starts rolling.

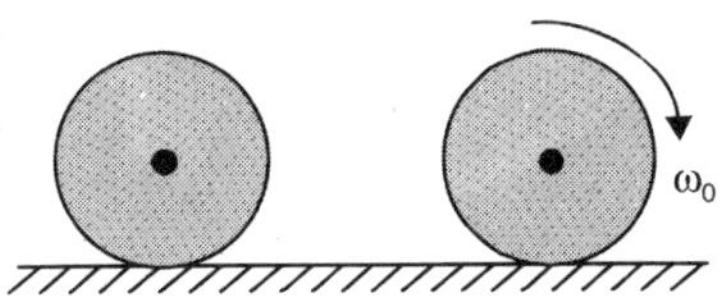

Fig. 19

20. A simple pendulum of length L is suspended in a cart that is moving with constant speed V around a circle of radius r. Find the period of oscillation.

21. A spring mass system is hanging from the ceiling of an elevator in equilibrium. At any instant elevator starts accelerating upwards with acceleration a, find the frequency of oscillation.

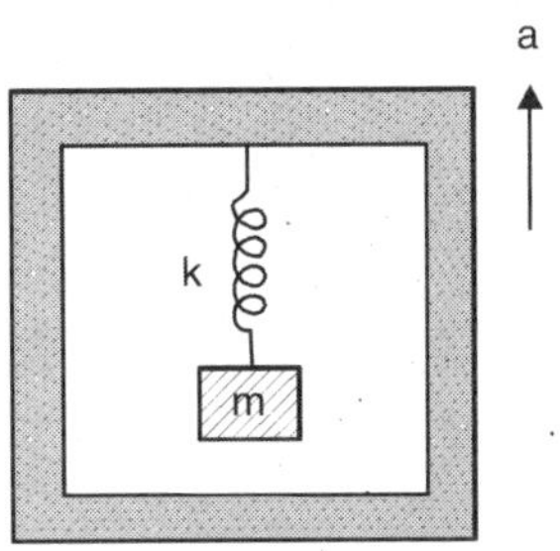

Fig. 21

Part III

Answers and Solutions

Answers

Equilibrium

LEVEL 1

1. 40 mm
2. 45, 2.25, $\theta = 53.14°$
3. $P = 115.47$ kN
   $Q = 115.47$ kN
   $R = 0$
4. $y = 210$ mm
5. $A_x = 100$ N $\rightarrow$
   $C_x = 100$ N $\leftarrow$
6. $W_C = 38.2$ N
7. $\bar{y} = 1034.2$ mm
8. $\theta = 56.69°$
9. $R \approx 231$ N
11. $\frac{w}{R_0}(2R - R_1 - R_2)$
12. 351 N
13. $P = \frac{mg\sqrt{2rhr^2}}{rh}$
14. $P = 654$ N
    $A = 327$ N
    $\theta = 22°$
15. $T_{BC} = 48$ N, $T_{AC} = 13.42$ N
16. $L = \sqrt{3}\,r$
17. $h = \sqrt{3}\,r$
18. $x = 41.2$ m
19. $G(\bar{x}, \bar{y}) = (5.08, 10.89)$
20. $T = 2\,mg\cos\theta$, $N = mg$

LEVEL 2

1. $\frac{3a}{4}$
2. $(\bar{x}, \bar{y}) = \left(\frac{2}{5}, 1\right)$
3. 99 N, 973 N
4. $R \le r\,(1 + 2\sqrt{7}\,)$
5. $x_c = (5/9)\,l$
6. $\theta = 23.2°$
7. $(\bar{x}, \bar{y}) = \left(\frac{256a}{315\pi}, \frac{256a}{315\pi}\right)$
8. $C = 60$ mm
9. $h = a\sqrt{3}$
10. $\alpha = 26.6°$
11. $\theta = 48.2°$
12. $y = 11.81$ cm
13. $\theta = \tan^{-1}\frac{\pi m_1}{2m_2}$
14. $M = \left(\frac{W}{4}\right)(ab/h)\sin\theta$
15. $N = 4380$ kN
16. $L\cos^2\theta = C - r\sin\theta$
17. $y = \frac{L}{2} - \frac{L}{2}\sqrt{R^2x^2}$
18. $R = \frac{2}{3}\,pgr^2$
    $\bar{y} = \frac{3\pi r}{16}$
19. (2/7, 4/5)

LEVEL 3

1. $\bar{x} = \left(\frac{\pi a}{4\sqrt{2}}, 0\right)$
2. $\bar{x}, \bar{y} = \left(\frac{3h}{4}, \frac{a}{\pi}\right)$
3. $R_{min} = \frac{mg}{2}$
5. $\theta = \sin^{-1}\left[\frac{r}{b}\left(1 + \frac{m}{m_0}\right)\sin\alpha\right]$
6. $\alpha = \tan^{-1}\left[\left(\frac{m_1 m_2}{m_1 + m_2}\right)\tan\beta\right]$
8. $A = 2\pi r^2 (\pi - 1)$

Friction

LEVEL 1

1. P = 2086.03 N
2. P = 70.285 N
3. P = 176.5 N
4. $\mu = b/a$
5. $\mu = \frac{p}{mg}$
6. P = 86.88 N
7. P_{min} = 247.12 N
8. $M = \frac{\mu Wr(1+\mu)}{(1+\mu^2)}$
9. P = 43.96 kN
10. 3.2 turns
11. $R_{Least} = W \sin(\alpha - \lambda)$
12. x = 2.94 m
13. F = 81.11 N
14. $\frac{P}{W} < \frac{\mu}{\cos\theta + \mu\sin\theta}$

    $\frac{P}{W} \le \frac{a}{h\cos\theta + 2a\sin\theta}$
15. 30 N, 60 N, 50 N, 50 N
16. M = 19.57 Nm
17. $\theta = 33.4°$
18. (a) P = 176.6 N

    (b) P = 147.3 N

    (c) P = 63.1 N

LEVEL 2

1. P = 2500N
2. $\theta = 65.5°$
3. $M = Wr \tan\left[\tan^1\left(\frac{L}{2\pi r}\right) + \phi\right]$
4. m = 135.25 kg
5. $\theta = 40.6°$
6. T_{min} = 0.33 W
7. $x = \frac{a}{2\mu s}$
8. $\frac{\mu_S h}{(1+\mu 2S)L} > \frac{1}{\sqrt{27}}$
9. F = 305 N
10. x = 8.28 m
11. 31.765 Nm
12. P = 1796.7 N
14. $\alpha/2$

LEVEL 3

1. $\mu W\, l\left(\sqrt{2}-1\right)$
3. $M = \dfrac{\mu W}{3}(AB + AC)$
4. $\dfrac{1}{\sqrt{3}}$
5. $\theta = \tan^{-1}$

$$\left[\frac{2\sin(\alpha-\lambda)\sin(\beta+\lambda)}{\sin(2\lambda+\beta\alpha)}\right]$$

6. $\theta = 45°$
7. (i) $Wa \sin \alpha$
   (ii) $Wa \sin \lambda \cos \lambda \sec \alpha$
8. $\sin \lambda > \dfrac{m}{M+m}$
9. $\theta = \tan^{-1}\left(\mu \dfrac{a+b}{a}\right)$

Analysis of Structures

LEVEL 1

1. $A_x = \left(2-\sqrt{3}\right)W$

$$A_y = \left(\frac{\sqrt{3}-1}{\sqrt{3}}\right)W \uparrow$$

$C_x = (\sqrt{3}-1)\,W$

$\therefore\ C_y = (w/\sqrt{3})$

3. $A_x = 24.83$ kN
   $D_y = 10$ kN $\uparrow$
   $R_A = 35.24$ kN
   $R_D = 10.07$ kN
4. $A_y = 5$ kN
   $R_B = 30$ kN
   $R_C = 5$ kN
5. $F_{CD} = 15.9$ kN (tensile)
   $F_{BC} = 16.8$ kN (comp)
   $F_{BD} = 13.5$ kN (comp)
6. $R_B = 257$ N
   $\phi = 63.9°$
7. $\theta = 39.8°$
8. $F_{AC} = 16.67$ kN (tensile)
   $F_{BC} = 23.33$ kN (Comp)
   $A_x = 36.67$ kN $\leftarrow$
   $A_y = 10$ kN $\uparrow$
   $B_x = 36.67$ kN $\rightarrow$
   $B_y = 10$ kN $\uparrow$
9. $w = 34.64$ kN
   $R_A = 73$ kN $\uparrow$
   $F_{AE} = 84.29$ kN (Comp)
   $F_{AC} = 42.15$ kN (Tensile)
10. $A_x = 24.25$ N $\leftarrow$
    $A_y = 30$ N $\uparrow$
    $R_F = 24.25$ N $\rightarrow$
    $C_X = 24.75$ N
    $C_y = 47.14$ N
    $D_x = 24.75$ N
    $D_y = 77.14$ N
    $B_x = 0.5$ N
    $B_y = 47.14$ N
11. $Q = P\dfrac{b}{a}$
12. $AC = \dfrac{L}{2}T$
13. $P = 3.73$ N

LEVEL 2

1. $A_x = B_x = 0$
   $B_y = 11.29$ kN
   $A_y = 6.74$ kN
2. $R_A = 500$ N ↑
   $R_B = 1500$ N ↑
3. $T_F = T_{max} = 12.29$ KN
4. $F_{DG} = 13.71$ kN (tensile)
   $F_{DF} = 40.62$ kN (comp)
   $F_{HF} = 42.86$ kN (comp)
   $F_{GH} = 41.14$ kN (tensile)
   $F_{FG} = 7.86$ kN (Comp.)
5. $$R_E = P\sqrt{1+\left(1\frac{2x}{a}\right)^2}$$
   $R_{E(max)} = P\sqrt{2}$
6. $X = 0.882$ m
7. $F_{FH} = -13.81$ kN
   $F_{GI} = 13.13$ kN
8. $H_y = 66$ kN ↓
   $A_y = 42$ lM ↑
   $B_y = 78$ kN ↑
   $B_x = 117$ kN →
9. $F_{BC} = 0.125$ kN
   $T_D = 1.3$ kN
10. $T = \dfrac{\sqrt{2}PL}{x}$
12. $A_y = 22.11$ kN ↑
13. $F_S(x) =$
$$\begin{cases} (515-40x)\text{N} & 0<x<12 \\ 35\text{N} & 12<x<18 \\ -365\text{ N} & 18<x<32 \end{cases}$$
$M_B(x) =$
$$\begin{cases} (515x-20x^2) & 0<x<12 \\ (2880+35x) & 12<x<18 \\ (11680-365x) & 18<x<32 \end{cases}$$
14. $F_S = \dfrac{2mg}{\pi}(\theta \sin\theta - \cos\theta)$
    $F_C = \dfrac{2mg}{\pi}(\theta \cos\theta + \sin\theta)$
    $M = \dfrac{2mgr}{\pi}\theta\cos\theta$
15. 2.7 kN
16. $R_C = 510$ N

LEVEL 3

2. $R_A = 40.06$ kN↓
   $R_B = 198.19$ kN
   $R_D = 172.02$ kN↑
   $R_E = 33.67$ kN
   $R_F = 14.67$ kN↑
   $W = 122.13$ kN ↓
3. $R_B = 1100$ N↑
   $R_A = 650$ N ↑
4. $\theta = 47.75°$
5. $E_D = 5.83$ m
   $E_B = 5.56$ m
7. $T = 53{,}850$ kg
8. 3° , 40 m
9. $F_4 = \dfrac{P}{4}$ (Tensile)
10. $R_A = 379.68$ N
    $T_{ACB} = 379.68$ N

11. (a) $F_S = \left(\frac{w_0 l}{\pi}\right)\left(1+\cos\frac{\pi x}{l}\right)$

(b) $M = \left(\frac{w_0 l^2}{\pi}\right)\left(\frac{x}{l} - 1 + \frac{1}{\pi}\sin\frac{\pi x}{l}\right)$

12. $AB = \frac{l}{3}$

$F_S = Wx - \frac{3}{4}Wl$

$M = \frac{1}{4}W(l-x)(l-2x)$

Virtual Work

LEVEL 1

1. $M = 5\,Pl\sin\theta$
2. $C_y = 26.67$ kN
   $A_y = 23.33$ kN and $A_x = 0$
3. $\tan\theta = \frac{P}{q}$
4. $B_y = \frac{Pa}{b}$
5. $\frac{\sin\alpha_1 + \mu_1\cos\alpha_1}{\sin\alpha_2 + \mu_2\cos\alpha_2} = \frac{m_2}{m_1}$
6. $P = \frac{Mx}{ha}$
7. $P = \frac{1}{2}W\tan\theta$
8. $\theta = \tan^{-1}\left(\frac{F_1}{2F_2}\right)$
9. $M = mgl$
10. $M = Pl(\sin\theta + \cos\theta)$
11. $\theta = 60°$
12. $P = \frac{Wr}{2L\sin^2\theta}$
13. $q = P\left(\frac{L\cos^2\theta}{a} - 1\right)$

LEVEL 2

1. $D_y = 15$ kN
   $R_B = 58.75$ kN
   $A_y = 13.75$ kN
   $A_x = 0$, $D_x = 0$
2. $\alpha = 33.6°$
   $\theta = 22.7°$
3. $M = 106.48$ Nm
4. $\theta = 60°$
5. $Cx = \frac{5}{12}P\cot\theta$
6. $\frac{2q}{3P} = \tan\theta$
7. $C_y = 56.66$ N
   $B_y = 16.66$ N
8. 65.1°
9. $F = P\sqrt{\left(\frac{2b^2}{h}\right) - 1}$
10. $P = 10.7$ kN
11. $R_A = 10$ kN
    $R_C = 58.3$ kN
    $R_D = 26.7$ kN

12. 450 kN-m

$R_A = 200\uparrow$

$R_G = 100\uparrow$

$R_D = 1050 \text{ N}\uparrow$

14. $\theta = 40.2°$

LEVEL 3

1. $\theta = \tan^{-1}\dfrac{\sin(\beta+\alpha)}{2\sin\alpha\sin\beta}$

2. $M = FR\sin\theta$

$$\left[1+\frac{R\cos\theta}{l\sqrt{1-\left(\frac{R}{l}\right)^2\sin^2\theta}}\right]$$

3. $\theta_2 = \tan^{-1}\left(\dfrac{W}{2P}\right)$

$\theta_1 = \tan^{-1}\left(\dfrac{2W}{P}\right)$

4. $\theta = \cos^{-1}\left(\dfrac{W}{4kl}\right)$

5. $\phi = \tan^{-1}\left(\dfrac{3}{8}+\tan\theta\right)$

6. $T = W' + (W + W')$

$$\left(\frac{\tan\theta}{\tan\theta+\cot\theta}\right)$$

7. $P = \dfrac{W\cos\alpha\cos\beta}{2\sin(\alpha+\beta)}$

9. $\dfrac{h^2}{2\left(h+\dfrac{d}{4}\right)} < r$

Space Forces

LEVEL 1

1. $T_C = 1.4$ kN

$T_B = 0.8$ kN

$T_A = 1.4$ kN

2. $T_A = 0.627$ kN

$T_B = 0.36$ Kn

$T_C = 0.514$ kN

3. $(\bar{x}, \bar{y}) = (0.28 \text{ m}, 0.13 \text{ m})$

4. 3.5, 0,3

5. $\dfrac{Pa}{\sqrt{6}}$

7. $T_A = T_B = 5.41$ kN, $T_C = 9.8$ kN

LEVEL 2

1. $F_2 = 1.032$ W (comp)

$F_1 = 0.108$ W (tensile)

$F_3 = 0.741$ W (tensile)

2. $R_A = 1425.36$ N

3. $T_1 = 347$ N, $T_2 = 431$ N

4. $T = 277$ N, $R = 169.9$ N

6. $N_A = 0.533$ L, $N_B = N_C$

$= 0.233$L

Basics of Dynamics

LEVEL 1

1.

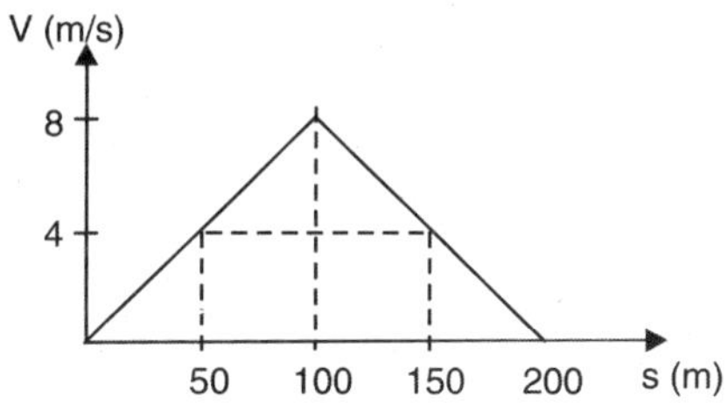

Fig. 1

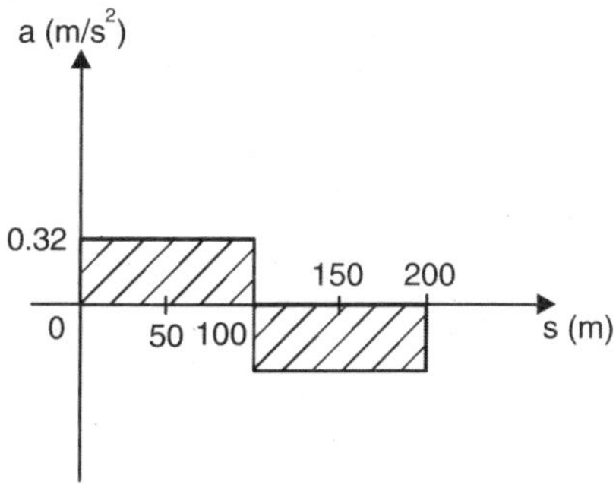

Fig. 2

2. 152 m, 41 m, 200 m

3. $\dfrac{3y}{2\sqrt{b^2+y^2}} \cdot V_A$

5. 29.22 m/s^2

6. $V_B = 1.2$ m/s ↑ $a_B = 3$ m/s^2

7. 0, 15 m/s^2, 1 rad/s, 3 rad/s^2, 2.5 m

8. 6.00 m

9. $r' = 56.6$ rounds/s
 $r' = 26.6$ rounds/s

10. 60 m/s, 1194 m 59.25 m/s

13. 4 rad/s, 9.6 m/s

15. $V_{B/A} = 0.5$ m/s, $a_{B/A} = 0.75$ m/s^2,
 $V_C = 1$ m/s

17. $\dot{r} = \dfrac{\frac{1}{2}at(2R\cos\alpha + at^2)}{\sqrt{R^2 + Rat^2\cos\alpha + \frac{1}{4}a^2t^4}}$

19. 1.5 m/s

LEVEL 2

1. 71.5 km/hr
2. 1833.33 km/hr^2
3. 0.06 rad/s
4. 4 sec, 36 m/s^2
5. 0.2 m/s ↑
6. $\Delta x_B = 400$ mm ↑
7. $t = 1.82$ sec.
8. 244.3 mm $< h <$ 386.2 mm
9. 325 m
10. $R_1 = 150$ mm, $R_2 = 100$ mm
11. 2 m
13. $\left(\dfrac{a}{b^2}\right)V_0^2$
14. 63.22 m
15. 41.8 m
16. $\dfrac{3\sqrt{(y+5)^2 - 25}}{y+5}$
17. 2 m/s^2
18. (*i*) $t_0 = \dfrac{2\pi(a+b)}{b(\omega - \omega)}$

 (*ii*) $t_0 = \dfrac{2\pi(a+b)}{a\omega + b\omega'}$

 $a_p = -\,b\omega'^2 + \dfrac{(a\omega + b\omega')^2}{a+b}$

 $a_{max} = b\omega'^2 + \dfrac{(a\omega + b\omega')^2}{a+b}$

19. $\theta_{min} = 30°$
 $u = 20$ m/s

20. $R = \dfrac{v_0^2}{g(\sin g\alpha + 1)}$

 $\theta_{max} = \tan^{-1}\left(\dfrac{1}{\tan\alpha + \sec\alpha}\right)$

23. $V_B = \dfrac{xV_A}{2\sqrt{h^2 + x^2}}$

24. $V_x = 5.77$ cm/sec ←
 $a_x = 15.39$ cm/s^2 ←

26. $\dfrac{d}{2}\left(\dfrac{1}{t_1} + \dfrac{1}{t_2}\right), \dfrac{d}{2}\left(\dfrac{1}{t_1} - \dfrac{1}{t_2}\right)$

27. $a_A = \sqrt{r\alpha\cos\theta + r0w^2}^{\,2}$

 $\sqrt{r\alpha\cos\theta + r0w^2}^{\,2}$

 $a_c = rw^2$

28. $\cos\theta = \dfrac{\sqrt{R_1}\sqrt{R_2}}{(R_1\sqrt{R_1R_2} + R_2)}$

29. $v = \dfrac{8}{1 + 6t}$

 $s = \dfrac{4}{3}\ln(1 + 6t)$

30. $a_x = -\dfrac{L^2V_A^2}{(L^2 - y^2)^{3/2}}$

31. $V_\theta = 4.24$ m/s

32. $\dfrac{2l}{3V_0}$

34. $V_B = V_A\sqrt{\dfrac{2x^2 + h^2}{x^2 + h^2}}$

LEVEL 3

1. $V_A = 5.94$ m/s 9.47 m/s
 $V_B = 3.96$ m/s, 6.31 m/s
2. $t_2 = (\sqrt{5} - 1)\, t_1$
3. $T = \dfrac{Vl}{V^2 - u^2}$
4. $a_1 = 2a \sin(\theta/2)$

5 $\dfrac{r_2}{\rho_1\rho_3}$V

6. $d_{min} =$

 $\dfrac{l(v_2\sin\beta v_1\sin\alpha)}{\sqrt{v_1^2 + v_2^2 + 2v_1v_2}\cos(\alpha + \beta)}$

7. $T = 2.10\,(10^{-3})$ s
8. 233 rad/s^2
10. 32 rad/s^2
 89.1 cm/s^2
11. 4.259 m
12. $\omega = \dfrac{2V\sin^2(\alpha/2)}{(R_1 + R_2)\cos(\alpha/2)}$

Newton's Laws

LEVEL 1

1. $\dfrac{mg}{2k}$

2. $T = \sqrt{\dfrac{2L}{g\sin\alpha + a\cos\alpha}}$

3. $\dfrac{g}{1+\dfrac{m_1}{4}\left(\dfrac{1}{m_2}+\dfrac{1}{m_3}\right)}$

4. $a = 9.816$ m/s^2

5. 6.15 kg

6. $a_A = 0$

   $a_B = 4.114$ m/s^2

   $a_C = 0.162$ m/s^2

7. $\tan^{-1}(0.23)$

8. 0, 526.41 N

9. $R_B = 163$ m

   $N_A = 1264.655$ N

   $N_B = 3640.25$ N

10. $a = \dfrac{F}{5m}$

11. 4 sec

12. $t = 19.8$ sec

13. $h/d = 4$

14. 88 km/h

15. 0.696 m, 0.376 m

16. 317.682 N

17. $\dfrac{1-\mu}{1+\mu}(M + 2m) \leq F \leq \dfrac{1+\mu}{1\mu}$

    $(M + 2m)$

18. $\dfrac{(M+m)g\sin\theta}{(M+m\sin^2\theta)}$

19. $T = 20$ N

    $T = 50$ N

20. $\omega = \dfrac{\sqrt{2gR}}{r}$, $N = 3\ mg\sin\theta$

22. $\left[\dfrac{2L}{(g+a)\sin\theta\cos\theta}\right]^{1/2}$

23. $a = \left(\dfrac{2}{Mm}\right)F - g$

    $T = F(M + m) - 2\ m\ Mg$

24. $\theta = 43.3°$

LEVEL 2

1. 2 cm

2. $a_1 = 19.52$ m/s^2

   $a_2 = 4.238$ m/s^2

3. 0.102, 0.32

4. $a_A = 3.75$ m/s^2

   $a_B = 1.25$ m/s^2

5. 11.66 rad/s

6. $a = 4.9$ m/s^2 < 30°

   $T_{AE} = 31.03$ N

   $T_{BD} = 11.42$ N

7. 2.75 N

   3.75 m/s^2

8. 9.689 rad/s^2

   17.44 rad/s^2

9. 49.05 N

10. $\alpha = \dfrac{2mg}{R(2m+M)}$

    $V = 2\sqrt{\dfrac{2mgR}{2m+M}}$

11. 30 kN

12. $\dfrac{3g}{2}$

13. $M\,[3g + 2\,a_0]$

14. $V = \sqrt{2ghln\left(\dfrac{l}{n}\right)}$

15. $5Mg$.

16. $x = 2\sqrt{6}$ m

18. $a_m = \dfrac{(M+m)g\sin^2\alpha}{(M+m\sin^2\alpha)}$

19. $v = \sqrt{\dfrac{2gx}{3}}$

20. $T = \left(\dfrac{m_1 m_2}{4m_2 + m_1}\right)g$

21. 37°, 315 N

LEVEL 3

1. $T' - 2T - m_1 g = m_1 a_2$

   $2T' - m_2 g = \dfrac{m_2}{2}(a_3 - a_2)$

   $T - m_3 g = m_3 (a_1 + a_2)$

   $T - m_4 g = m_4 (a_2 - a_1)$

   $m_5 g - T' = m_5 a_3$

2. $\sqrt{5}\,\text{m/s}^2$

3. $\left(\dfrac{2m_1 m_2}{m_1 + m_2}\right)\tan\alpha$

4. $T = \dfrac{\lambda V^2}{4}$

   $F = \dfrac{\lambda}{2}\left(\dfrac{\lambda V^2}{2} + gx\right)$

   $R = \dfrac{\lambda V^2}{4} + \lambda g\,(L - x/2)$

5. 7.58 m/s^2

6. 2.5 m/s^2, 0.25

7. $v = \sqrt{\dfrac{2p}{p}\mu_k g L}$

9. $T = \lambda\,(g + a) + \dfrac{3}{4}\lambda\,(g + a^2)\,t^2$

10. $\sqrt{2}\ \text{m/s}^2$, $\sqrt{10}\ \text{m/s}^2$, $1\ \text{m/s}^2$

Work, Energy and Momentum

LEVEL 1

1. $\sqrt{2gh}$

2. $\dfrac{280}{3}$, 40, $\dfrac{632}{15}$

3. 1120 m

5. $v_{max} = 5.19$ m/s, $\Delta t = 5.54$

7. 4mg, 7 mg, $\dfrac{4R}{1 + \mu k\sqrt{3}}$

10. $V\sqrt{m/k}$

LEVEL 2

1. $v = 2.32$ m/s

2. $\dfrac{4}{9}$ mg$l$

4. $\omega = 4.22$ rad/s

5. $l = \dfrac{\left(m_1 + \dfrac{m_2}{2}\right)L}{m_1 + m_2 + M}$

6. 6 m/s

8. 0.331 m/s

9. 5.3 m/s

10. $u = [gl\,(2 + \sqrt{3}\,)]^{1/2}$

11. $H_{max} = 62.36$ m
 $AO = 92.57$ m

12. $mm' \dfrac{(vv')}{(m+m')} \dfrac{\sqrt{l^2 a^2}}{l}$

14. $u = \dfrac{7I}{15m}$

 $V = \dfrac{\sqrt{3}I}{15m}$

15. $V = 5\sqrt{\dfrac{5-\sqrt{5}}{6}}$

LEVEL 3

1. $V_A = \sqrt{l\sqrt{a^2 + g^2}\,(2\sin\theta)}$

2. $\omega = \sqrt{\dfrac{2g(M+m)\sin\theta}{(M + m\cos^2\theta)l}}$

3. $\omega = \dfrac{V_{mM}}{R}$

 $= \sqrt{\dfrac{2(m+M)(\cos\alpha\cos\beta)}{(M + m\sin^2\beta)R^2}}\ .\ \theta_0$

 $= \cos^{-1}\left(\dfrac{2(m+M)\cos\alpha}{\left[3M + m(2+\sin^2\beta)\right]}\right)$

4. $h\left(\dfrac{M}{M+m}\right)^2$

5. 40 m, 15 m

6. $\beta = \tan^{-1}\left(\dfrac{m_{\cdot e}^2}{m_{\cdot p}^2 - m_{\cdot e}^2}\right)$

7. $\dfrac{59h}{369}$

8. $T = 3\ mm'\ u\ (1 + e)/(m + 6m')$

9. 0.20 m

10. $\dfrac{(V^2 + 2gR)^2}{(27g^3R^2)}$

 $\left(\sqrt{20g^2R^2gRV^2V^4} - \sqrt{2g^2R^2gRV^2V^4}\right) - R$

 $\left(1 - \sqrt{\dfrac{1}{3}\dfrac{V^2}{3gR}}\right)$

11. 0.81 m

Rotation and Simple Harmonic Motion

LEVEL 1

2. $\dfrac{3u}{8l}$

3. $\omega/3$

4. (*a*) $a_{block} = \dfrac{g}{2\sqrt{2}}$

 $a_{cylinder} = \dfrac{\sqrt{2}g}{3}$

(*b*) $\frac{3g}{5\sqrt{2}}$

5. $\cos^{-1}\frac{1}{\sqrt{1+\frac{2gh}{V_0^2}}\sqrt{1-\frac{h^2}{r_0^2}}}$

6. $\omega = \frac{3b\sqrt{2gh}}{2(b^2+c^2)}$

7. $\omega = 4\sqrt{\frac{g}{5a}}$

8. $\frac{8}{3}$

9. $\frac{mV_0^2}{2\sqrt{2}g}$

10. $\left(\frac{2}{L}\right)\sqrt{6gR(\theta\sin\theta+\cos\theta-1)}$

11. $\frac{1}{2\pi}\sqrt{\frac{4k}{m}}$

12. $\mu > \frac{2}{7}\tan\theta$

13. $2\pi\sqrt{\frac{2L}{3g}}$

16. $B_x = 2034$ N
$B_y = 4530$ N
$A_x = 3.39$ m/s^2
$A_y = -2.26$ m/s^2

LEVEL 2

1. $\frac{4\pi m}{M+2m}$

2. $1.074\sqrt{g/R}$

3. (a) $\mu_s = 0.299$
(b) $\theta = 54.6°$

4. $2\pi\sqrt{\frac{28R}{5g}}$

5. $2\pi\sqrt{\frac{m+I/R^2}{K}}$

6. $\pi\sqrt{\frac{ml}{F}}$

7. $\dot{\theta} = \sqrt{(3g\cos\alpha)/2a}$

8. $0.712\sqrt{ag}$

9. $m(a^2+k^2)\,\omega$,
$\frac{1}{2}\mathrm{m}\,(a^2+k^2)\,\omega^2 > mgh$

10. $O_x = \frac{15}{56}$ mg
$O_y = \frac{403}{448}$ mg

14. $\sqrt{\frac{4R(1-\cos\theta_0)g}{13}}$

15. $\omega^1 = \frac{4}{9a}\left(u-\sqrt{5ag}\right)$
$u = \left(3\sqrt{2}+\sqrt{5}\right)\sqrt{ag}$

17. $\frac{4}{5}\left(\frac{m+M}{M}\right)\frac{v_0 r\omega_0}{g}$

20. $\frac{1}{2\pi}\sqrt{\frac{\mu g}{l}}$

21. $2\pi\sqrt{\frac{m(k_1+k_2)}{4k_1k_2}}$

22. $f = \frac{1}{\pi}\sqrt{\frac{2k}{M}}$

25. 17 N, 26 N,
2 m/s², 3 m/s²

26. $\frac{12V}{7L}, \frac{7}{12}\sqrt{2gL}$

27. $\frac{2\sqrt{3}}{13}$ mg

LEVEL 3

1. $2\pi\sqrt{2l/5g}$

2. $4\pi\sqrt{m\left(\frac{1}{k_1}+\frac{1}{k_2}+\frac{1}{k_3}+\frac{1}{k_4}\right)}$

3. 0.012 turns

4. $\tan^{-1}\left(\frac{6}{\pi g}\right)$

6. 3 m/s² ↑
0.5 m/s² ↓

7. $r_1 = \frac{mv_0^2}{4Mg} + \sqrt{\left(\frac{mV_0}{4Mg}\right)^2 + \frac{mv_0^2 R}{4Mg}}$

8. $\frac{5V_0}{4r}$

9. $\sqrt{\frac{K(M+m)}{m(M+m\sin^2\alpha)}}$

11. $\left(\frac{1}{1\pm\sqrt{2n}}\right)\cdot L$

12. 1 m/s²

13. $v_0 = \frac{3}{2}\frac{\omega_0}{a}R^2 - \omega_0 R$

Solutions

Equilibrium

Level 1

Example 3

Solution

Clearly

$$\Sigma F_y = 200 \text{ kN} \uparrow$$
$$\Sigma F_x = 0$$
$$\Sigma M_B = 0$$
$$\Sigma F_x = 0$$
$$P \cos 60° - Q \cos 60° - R = 0$$

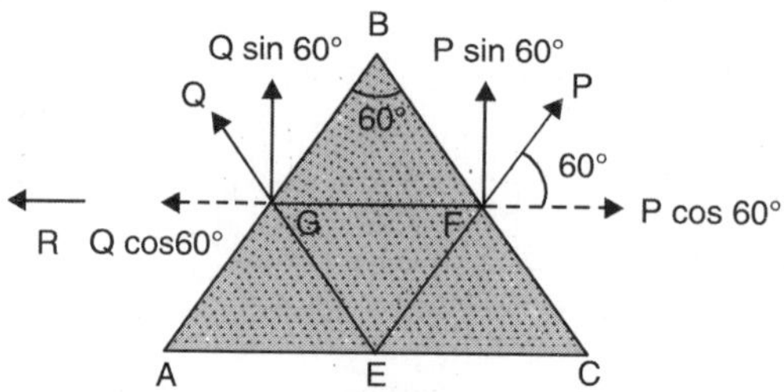

Fig. E3

$$P - Q = 2R \quad \ldots(i)$$
$$\Sigma F_y = 200 \text{ kN}$$
$$Q \sin 60° + P \sin 60° = 200$$
$$P + Q = 230.94 \quad \ldots(ii)$$
$$\Sigma M_B = 0$$
$$-P(a \sin 60°) + Q(a \sin 60°) + R(a \sin 60°) = 0$$
$$P - Q = R \quad \ldots(iii)$$

By equation (*i*) and (*iii*)

$$\mathbf{R = 0}$$

So,

$$P + Q = 230.94$$

$$\because P = Q$$
$$2P = 230.94$$
$$\mathbf{P = 115.47\ kN}$$
$$\mathbf{Q = 115.47\ kN} \qquad \textbf{...Ans.}$$

Example 4

Solution

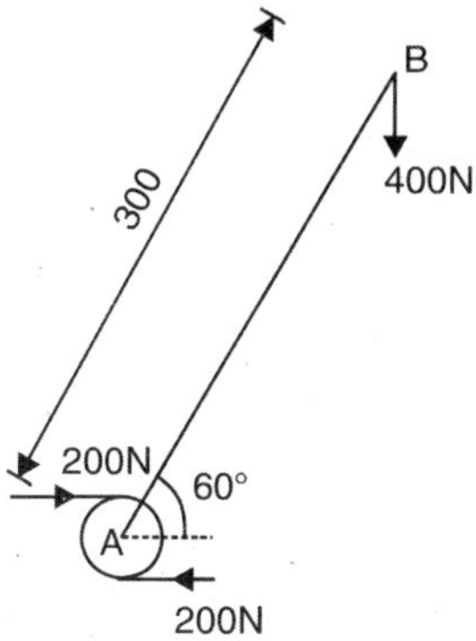

Fig. E4

Applying Varignon theorem about point *A*

$$Ry = \sum M_A$$
$$(400)\ (y) = 200 \times 60 + 200 \times 60 + 400 \times (300 \cos 60°)$$
$$\mathbf{y = 210\ mm.}$$

Example 5

Solution

Total clockwise couple produced by the drills is 24 Nm.

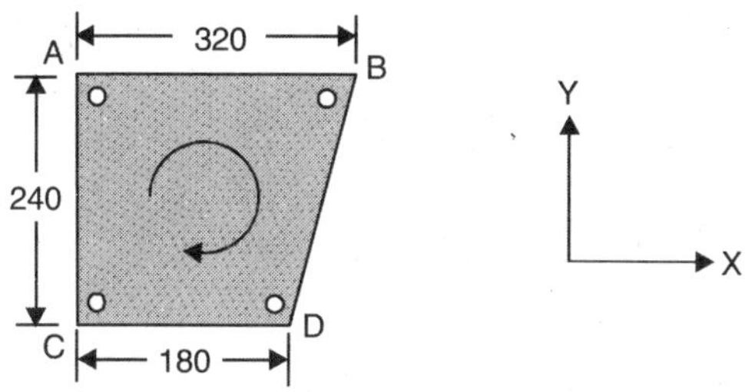

Fig. E5(a)

At point *A* and *C* forces parallel to *y*-axis do not form any couple. And

$$|A_x| = |C_x| = F$$

$$\text{Then} \quad F(0.24) = 24$$
$$F = 100 \text{ N}$$

For clockwise moment

$$\mathbf{A_x = 100\ N \rightarrow}$$
$$\mathbf{C_x = 100\ N \leftarrow}$$

Example 6

Solution

Apply Lami's theorem at *B*

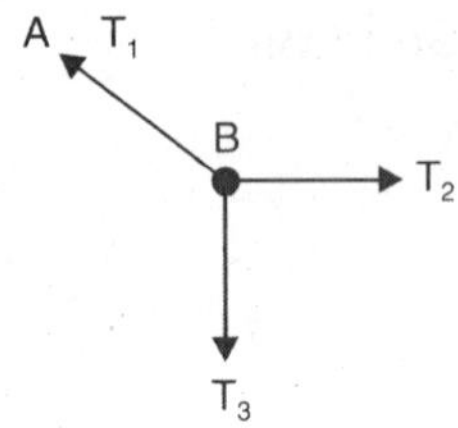

Fig. E6(a)

$$\frac{T_1}{\sin 60^\circ} = \frac{T_2}{\sin 150^\circ} = \frac{22}{\sin 150^\circ}$$
$$T_2 = 22 \text{ N}$$
$$T_1 = 38.1 \text{ N}$$

If W_C is weight at *C*

Then again by Lami's theorem at *C*

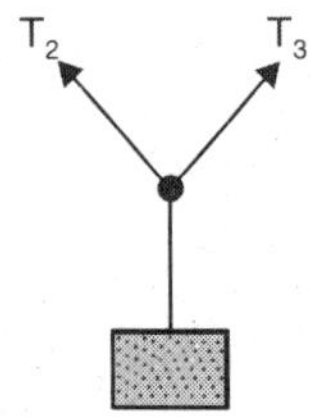

Fig. E6(b)

$$\frac{W_C}{\sin 95^\circ} = \frac{T_3}{\sin 120^\circ} = \frac{T_2}{\sin 145^\circ}$$
$$W_C = 38.2 \text{ N}$$
$$\mathbf{T_3 = 33.2\ N} \qquad \textbf{...Ans.}$$

Example 7

Solution

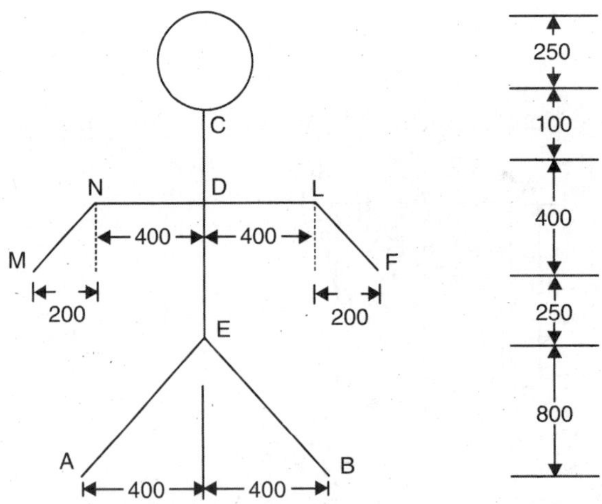

Fig. E7

For circle

$$L_1 = \pi\,(250) = 785.4 \text{ mm}$$
$$y_1 = 800 + 250 + 400 + 100 + 125 = 1675 \text{ mm}$$

For *CE*

$$L_2 = 250 + 400 + 100 = 750 \text{ mm}$$
$$y_2 = 800 + \frac{750}{2} = 1175 \text{ mm}$$

For *NL*

$$L_3 = 400 + 400 = 800 \text{ mm}$$
$$y_3 = 800 + 250 + 400 = 1450 \text{ mm}$$

For *MN* and *LE*

$$L_4 = 2\sqrt{(200)^2 + (400)^2} = 894.4 \text{ mm}$$
$$y_4 = 800 + 250 + 200 = 1250 \text{ mm}$$

For *AE* and *EB*

$$L_5 = 2\sqrt{(400)^2 + (800)^2} = 1788.9 \text{ mm}$$
$$y_5 = 400 \text{ mm}$$
$$\bar{y} = \frac{\Sigma L_i y_i}{\Sigma L_i}$$
$$\bar{\mathbf{y}} = \mathbf{1034.2\ mm}$$ **...Ans.**

Example 8

Solution

Here point B is taken as origin. Clearly x-coordinate of the center of mass of the system lie $r \cos\theta$ distance from B on the x-axis because the system is suspended from point A.

$$x = \frac{(r)\left(\frac{r\cos\theta}{2}\right) + (\pi r)\left(\frac{2r}{\pi}\right)}{r + \pi r} = \frac{r(\cos\theta + 4)}{2(1+\pi)}$$

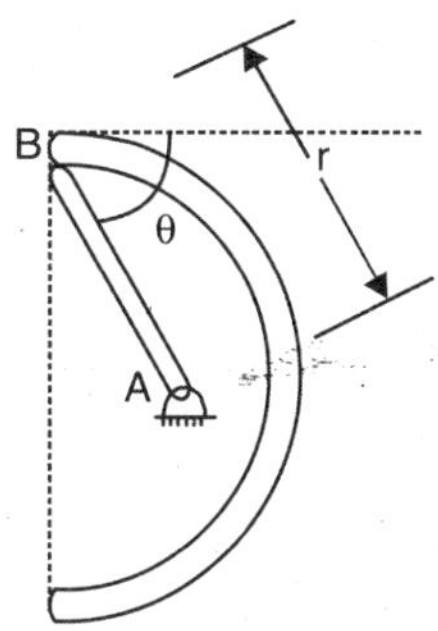

Fig. E8(a)

now,

$$r\cos\theta = \frac{r(\cos\theta + 4)}{2(1+\pi)}$$
$$\theta = \cos^{-1}\left(\frac{4}{1+2\pi}\right)$$
$$= \mathbf{56.69^\circ}$$ **...Ans.**

Example 9

Solution

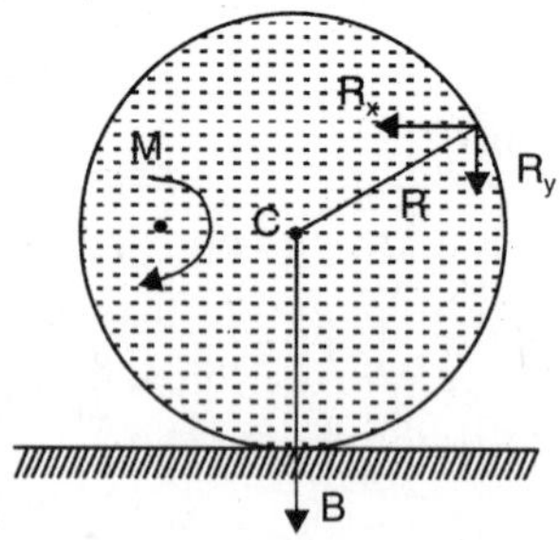

Fig. E9(a)

Take the moment about bottommost point of the wheel.

Suppose R is the normal force applied by the roller A.

$$Rx = R\cos 30°$$

$$Ry = R\sin 30°$$

$$\Sigma M_B = 0$$

$$-60 + R\left(\frac{\sqrt{3}}{2}\right)(450)\ (10^{-3}) - R\left(\frac{1}{2}\right)(150\sqrt{3})(10) = 0$$

$$\Rightarrow \qquad R\left(\frac{\sqrt{3}}{2}\right)(450 - 150) = 60\ (10^3)$$

$$R\left(\frac{\sqrt{3}}{2}\right)(300) = 60000$$

$$R = \frac{200 \times 2}{1.73} = \frac{40000}{173}$$

$$\approx \mathbf{231\ N}$$

...Ans.

Example 10

Solution

Drawing the free body diagram of pulley C

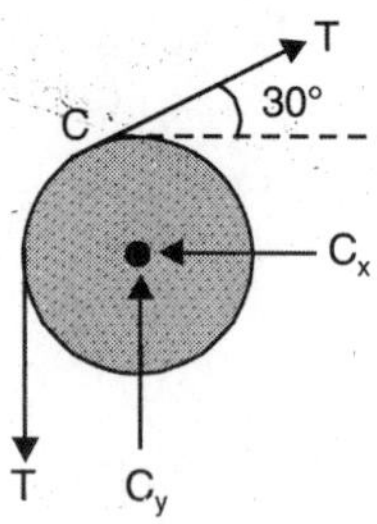

Fig. E10(a)

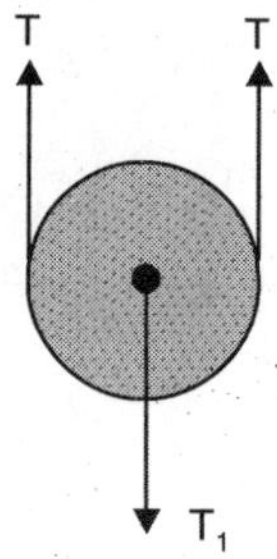

Fig. E10(b)

FBD of pulley B

For A

$$2T = T_1 \qquad \text{...}(i)$$

$$2T_1 = 1000$$

$$T_1 = 500 \text{ N} \quad \ldots(ii)$$

By (*i*) and (*ii*)

$$T = 250 \text{ N}$$

For pulley *C* apply condition of equilibrium

$$\Sigma F_x = 0$$

$$-C_x + T\cos 30° = 0$$

$$C_x = 250\cos 30° = 217 \text{ N}$$

$$\Sigma F_y = 0$$

$$C_y + T\sin 30° - T = 0$$

$$C_y + 250\sin 30° - 250 = 0$$

$$C_y = 125 \text{ N}$$

$$\mathbf{R_C = 250.4\ N} \quad \ldots\textbf{Ans.}$$

Example 15

Solution

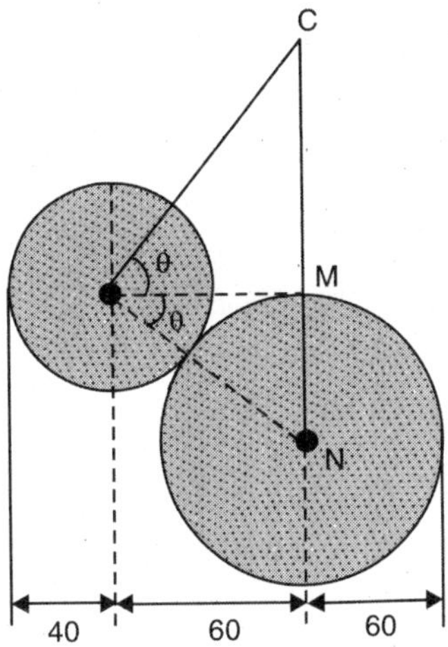

Fig. E15(a)

Clearly

$$\theta = \cos^{-1}\left(\frac{60}{40+60}\right) = 53.1°$$

$$\alpha = \tan^{-1}\left(\frac{20080}{60}\right) = 63.4°$$

$$MN = 100\sin 53.1° = 80 \text{ mm}$$

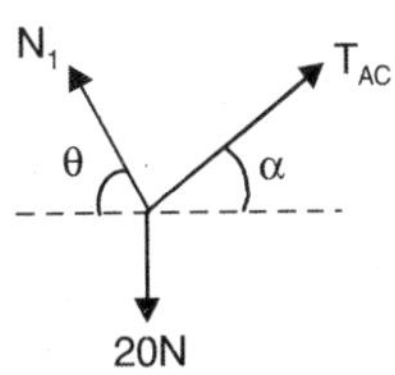

Fig. E15(b)

For cylinder A

Apply Lami's theorem

$$\frac{N_1}{\sin(90° + 63.4°)} = \frac{T_{AC}}{\sin(90° + 53.1°)} = \frac{20}{\sin(63.4°)}$$

we get

$$\mathbf{N_1 = 10\ N} \quad \textbf{...Ans.}$$
$$\mathbf{T_{AC} = 13.42\ N} \quad \textbf{...Ans.}$$

For cylinder B

Apply condition of equilibrium

$$\Sigma F_x = 0$$
$$10 \cos 53.1 - N_0 = 0$$
$$\mathbf{N_0 = 6N} \quad \textbf{...Ans.}$$
$$\Sigma F_y = 0$$
$$T_{BC} - 40 - 10 \sin 53.1° = 0$$
$$\mathbf{T_{BC} = 48\ N} \quad \textbf{...Ans.}$$

Example 16

Solution

For $\theta = 30°$ (taking D as origin)

For ABC

$$X_1 = \frac{r \sin \pi/6}{\pi/6} = \frac{3r}{\pi}$$

For BDE

$$X_2 = \frac{l + r}{2} - r = \frac{lr}{2}$$

$$\bar{x} = \frac{L_1 x_1 + L_2 x_2 + \dots}{L_1 + L_2 + \dots}$$

$$\bar{x} = 0$$

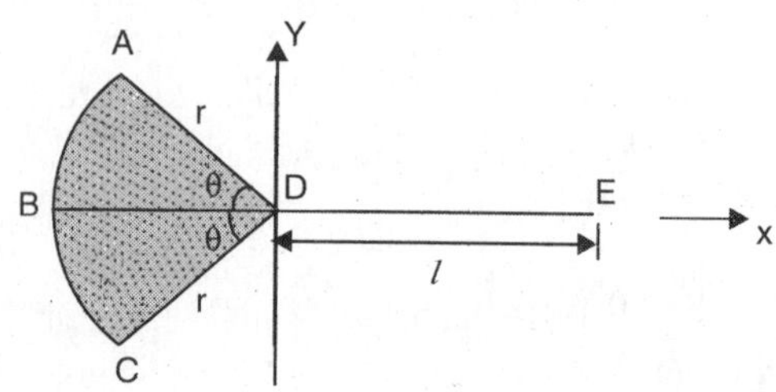

Fig. E16(a)

$$-r\,\pi/3\,(3r/\pi) + \left(\frac{lr}{2}\right)(l + r) = 0$$

$$l = \sqrt{3}\,\mathbf{r} \qquad \textbf{...Ans.}$$

Example 17

Solution

It is given that

$$\bar{y} = r$$

$$V_1 = \text{Volume of cone} = \frac{\pi r^2 h}{3}$$

$$V_2 = \text{Volume of hemisphere} = \frac{2\pi r^3}{3}$$

$$y_1 = r + h/4$$

$$y_2 = r - 3r/8$$

By centroid formula

$$\bar{y} = \frac{V_1 y_1 + V_2 y_2}{V_1 + V_2} = \frac{\frac{\pi r^2 h}{3}[r + h/4] + \frac{2\pi r^3}{3}[r3r/8]}{\frac{\pi r^2 h}{3} + \frac{2\pi r^3}{3}}$$

$$\mathbf{r} = \frac{\mathbf{h(r + h/4) + 2r(r3r/8)}}{\mathbf{h + 2r}} \Rightarrow \mathbf{h} = \sqrt{3}\,\mathbf{r} \qquad \textbf{...Ans.}$$

Example 18

Solution

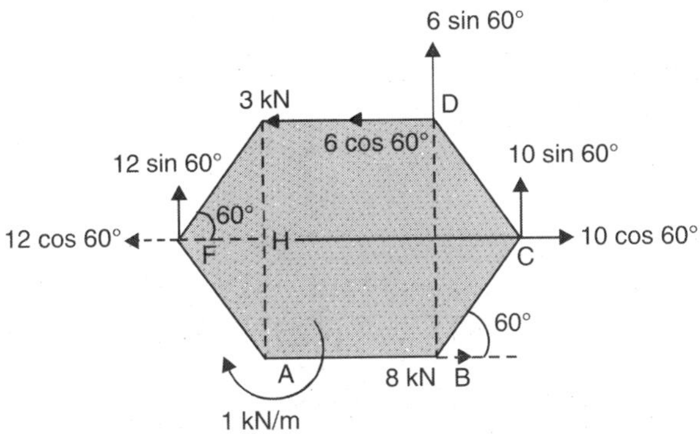

Fig. E18

Let x be the side of polygon.

$$\Sigma M_A = 0$$

$$10 \cos 60° \times (x \sin 60°) - 10 \sin 60° \times (x + x \cos 60°) - 6 \cos 60° \times (2x \sin 60°) - 6 \sin 60° \times x - 3 \times (2x \sin 60°) + 1 = 0$$

$$\mathbf{x = 41.2\ m}$$ **...Ans.**

Example 19

Solution

A_1 = Area of circular sector

$$= (30)^2 \left(35 \times \frac{\pi}{180}\right) = 549.8 \text{ cm}^2$$

A_2 = Area of quarter circle

$$= \frac{\pi}{4} \times (30)^2 = 706.86 \text{ cm}^2$$

A_3 = Area of circular sector

$$= (30)^2 \left(15 \times \frac{\pi}{180}\right) = 235.62 \text{ cm}^2$$

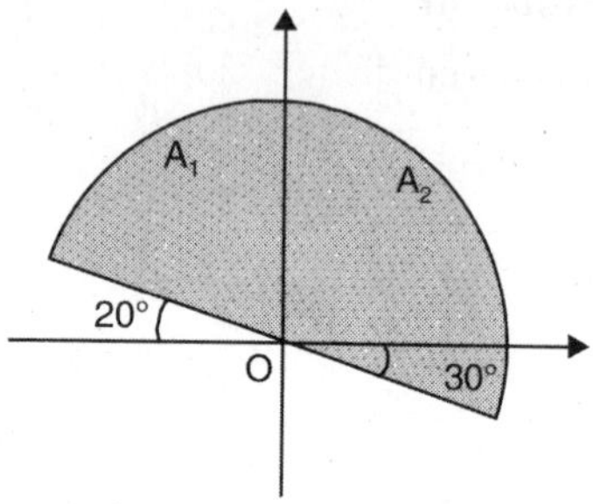

Fig. E19 (a)

A_1	A_2	A_3
$x_1' = \frac{2r\sin\alpha}{3\alpha}$ $= \frac{2 \times 30\sin 35°}{3 \times \left(\frac{35 \times \pi}{180}\right)}$ $= 18.78$ cm $x_1 = x_1' \sin 35°$ $= 18.78 \sin 35°$ $= 10.77$ cm Similarly $y_1 = 15.38$ cm	$x_2 = \frac{4r}{3\pi} = \frac{4 \times 30}{3\pi} = 12.73$ cm $y_2 = 12.73$ cm	$x_3' = \frac{2r\sin\alpha}{3\alpha}$ $= \frac{2 \times 30\sin 15°}{3 \times \left(\frac{15 \times \pi}{180}\right)}$ $= 19.77$ cm $x_3 = x_1' \cos 15°$ $= 19.77 \cos 15°$ $= 19.1$ cm $y_3 = 5.117$ cm

$$\bar{x} = \frac{(549.8)(10.77) + (706.86)(12.73) + (235.62)(19.1)}{(549.8 + 706.86 + 235.62)} = 5.08 \text{ cm}$$

$$\bar{y} = 10.89 \text{ cm}$$

$$\mathbf{G(\bar{x}, \bar{y}) = (5.08, 10.89)}$$ **...Ans.**

Example 20

Solution

FBD of the ring

Here T, N denotes the tension in the thread and reaction offered by wire on the ring

In ΔAOB

$$\angle BAO = \angle ABO = \theta$$

apply Lami's theorem at point B

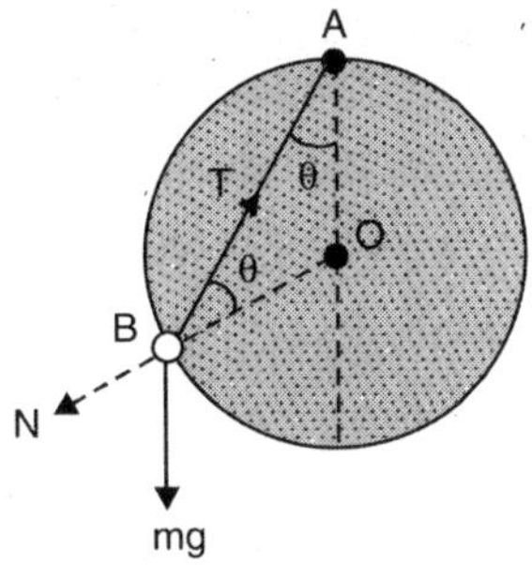

Fig. 20(a)

$$\frac{T}{\sin 2\theta} = \frac{mg}{\sin(180°\theta)}$$

$$T = \frac{mg}{\sin\theta}\sin 2\theta$$

$$= \mathbf{2\ mg\ cos\ \theta} \qquad \text{...Ans.}$$

Similarly we can get

$$\mathbf{N = mg} \qquad \text{...Ans.}$$

Level 2

Example 1

Solution

Each box can extend maximum amount over the next.

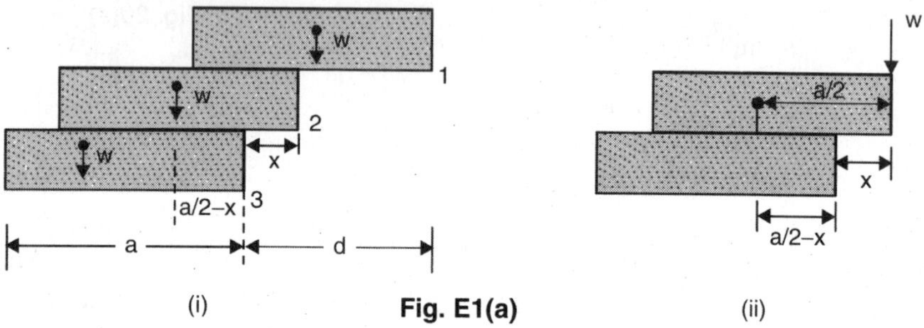

Fig. E1(a)

$$\Sigma M_B = 0$$

$$w \times x - w\,(a/2 - x) = 0$$

$$x = a/4$$

$$\text{Hence,} \quad l = \frac{a}{2} + x = \frac{a}{2} + \frac{a}{4} = \mathbf{\frac{3a}{4}}$$

Example 2

Solution

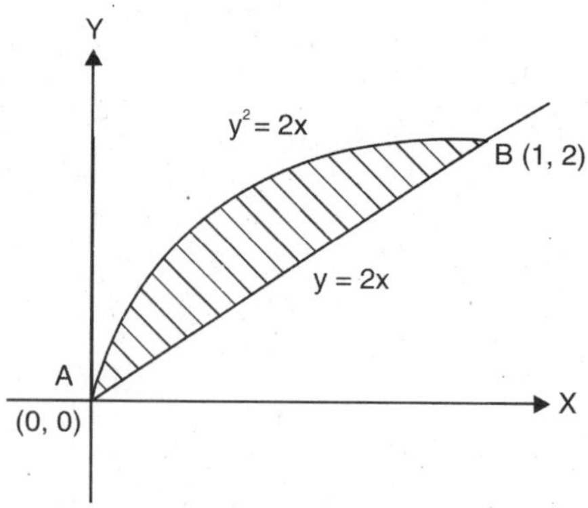

Fig. E2(a)

Clearly this is two-dimensional geometry so we have to use double integration for finding the centroid of the graph.

$$\bar{x} = \frac{\iint x dx dy}{\iint dx dy}$$

and

$$\bar{y} = \frac{\iint y dx dy}{\iint dx dy}$$

Consider an arbitrary strip in the intersected region parallel to y-axis. Then

$$\bar{x} = \frac{\int_0^1 \int_{2x}^{2\sqrt{x}} x\, dx\, dy}{\int_0^1 \int_{2x}^{2\sqrt{x}} dx\, dy} = \frac{2 \times 3}{15 \times 1} = \frac{2}{5}$$

$$\bar{y} = \frac{\int_0^1 \int_{2x}^{2\sqrt{x}} y\, dx\, dy}{\int_0^1 \int_{2x}^{2\sqrt{x}} dx\, dy} = \frac{1 \times 3}{3 \times 1} = 1$$

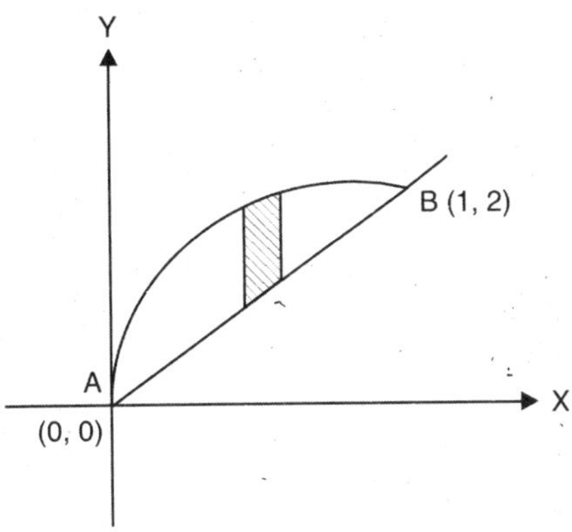

Fig. E2(b)

$\Rightarrow$ $$(\bar{x}, \bar{y}) = \left(\frac{2}{5}, 1\right)$$ **...Ans.**

Example 4

Solution

Suppose m is the mass of each small sphere
Then clearly

$$2N_0 \cos\theta = 3mg \quad \ldots(i)$$

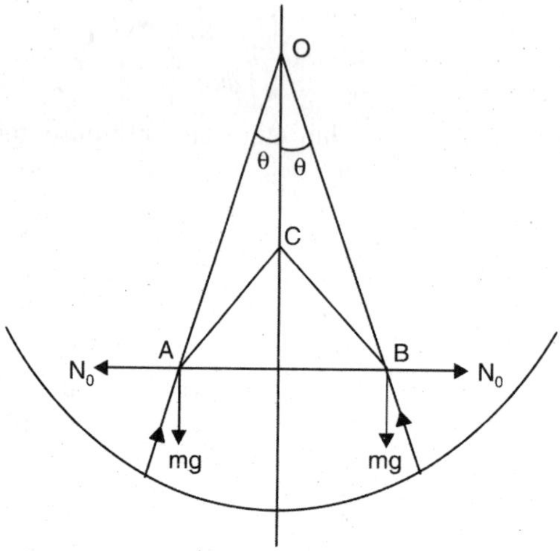

Fig. E4(a)

for cylinder having center C

$$2T\cos 30° = mg$$

$$\sqrt{3}\,\mathbf{T} = mg \quad \ldots(ii)$$

for cylinder having center A

$$T\cos 60° + N_1 = N_0 \sin\theta \quad \ldots(iii)$$

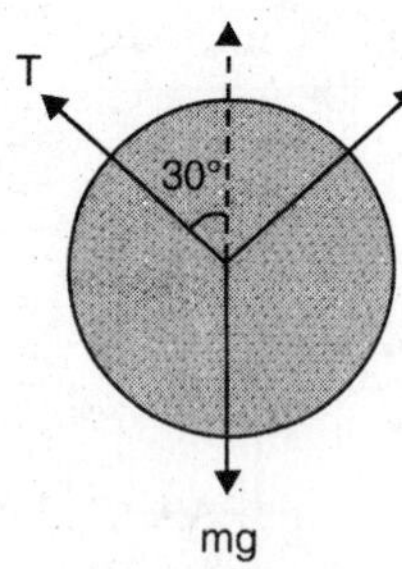

Fig. E4(b)

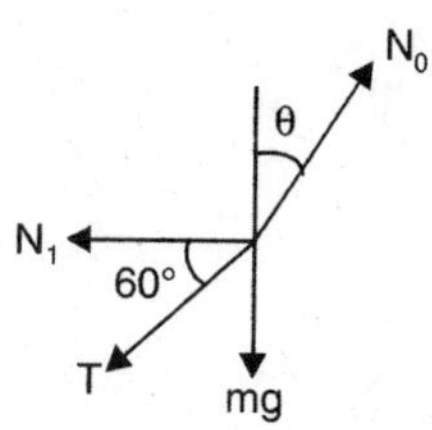

Fig. E4(c)

According to the problem
Here N_1 represents pressure between the lower spheres.

$$\text{So,} \quad N_1 > 0$$

$$T \cos 60° \le N_0 \sin \theta \qquad \ldots(iv)$$

By using (*i*) and (*ii*) in (*iv*)

We get

$$\frac{mg}{2\sqrt{3}} \le \frac{3mg}{2} \tan\theta$$

$$\tan \theta \ge \frac{1}{3\sqrt{3}}$$

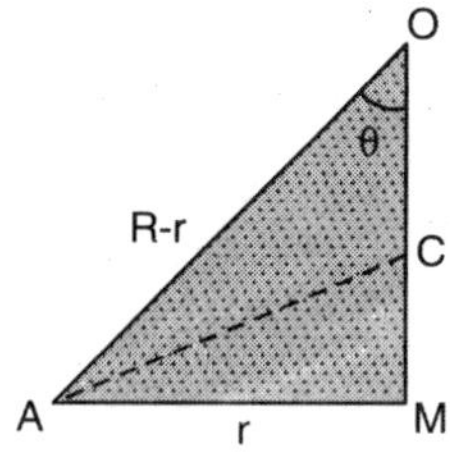

Fig. E4(d)

clearly from ΔOAM

$$\tan \theta = \frac{r}{\left[(Rr)^2 - r^2\right]^{1/2}}$$

$$\frac{r}{\left[(Rr)^2 - r^2\right]^{1/2}} \ge \frac{1}{3\sqrt{3}}$$

squaring both side

we get $$\frac{r^2}{(Rr)^2 - r^2} \ge \frac{1}{27}$$

$$\mathbf{R \le r(1 + 2\sqrt{7})} \qquad \textbf{...Ans.}$$

Example 5

Solution

Let

$$\lambda = \text{mass per unit length}$$

$$m = \text{mass of ladder}$$

According to the problem λ increases uniformly

$$\lambda = \lambda_0 + ax \qquad \ldots(i)$$

$$a = \text{positive constant}$$

$$\lambda_0 = \text{Linear mass density of the upper end B}$$

$$\text{now when } x = l$$

$$\lambda = 2\lambda_0$$

$$\text{then } a = \frac{\lambda_0}{l}$$

using a's value in (*i*)

we get

$$\lambda = \lambda_0 \left(1+\frac{x}{l}\right)$$

now,

$$m = \int_0^l \lambda\, dx = \int_0^l \lambda_0 \left(1+\frac{x}{l}\right) dx = \frac{3}{2}(\lambda_0 l)$$

$$x_C = \frac{\int_0^l \lambda\, x\, dx}{\int_0^l \lambda\, dx} = \frac{\int_0^l \lambda_0 \left(x+\frac{x^2}{l}\right) dx}{\frac{3}{2}(\lambda_0 l)}$$

$$\mathbf{x_C} = \left(\frac{\mathbf{5}}{\mathbf{9}}\right) l$$ **...Ans.**

Example 6

Solution

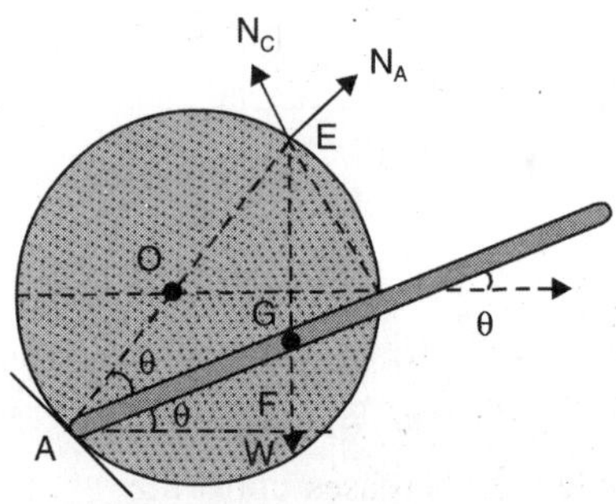

Fig. E6(a)

For the equilibrium condition these three forces must be concurrent

In ΔAGF $AF = 1.5\, R \cos \theta$

In ΔAEF $AF = 2\, R \cos 2\theta$

So, $1.5\, R\cos \theta = 2R\cos 2\theta$

$\therefore\ \theta = \mathbf{23.2°}$ **...Ans.**

Example 7

Solution

Clearly curve is symmetrical about $y = x$ axis so

$$\bar{y} = \bar{x}$$

so, for $\bar{x}$

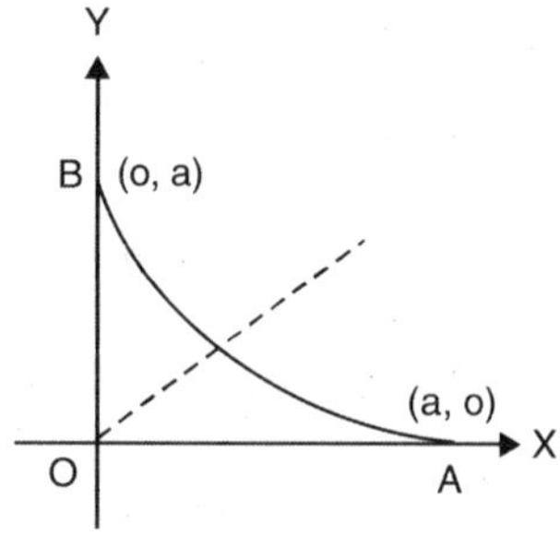

Fig. E7

$$\bar{x} = \frac{\int_0^a xy\,dx}{\int_0^a x\,dx}$$

$$\int_0^a xy\,dx = \int_0^a x\left(a^{2/3} - x^{2/3}\right)^{3/2} dx$$

by putting $x = a \sin^3 \theta$

we get

$$\int_0^a xy\,dx = 3a^3 \int_0^{\pi/2} \sin^5\theta \cos^4\theta\, d\theta = \frac{8}{105}a^3$$

similarly for $\int_0^a y\,dx$

$$= \int_0^a \left(a^{2/3} - x^{2/3}\right)^{3/2} dx$$

by putting $x = a \sin^3\theta$
we get

$$\int_0^a y\,dx = \frac{3\pi a^2}{32}$$

now

$$\bar{x} = \frac{256a}{315\pi}$$

$$\Rightarrow \quad (\bar{x}, \bar{y}) = \left(\frac{256a}{315\pi}, \frac{256a}{315\pi}\right) \quad \text{...Ans.}$$

Example 8

Solution

Clearly according to the figure

$$DE = a\sqrt{2}$$

In ΔCDF

$$DF = R \sin \theta_2$$

$$R \sin \theta_2 = \frac{a\sqrt{2}}{2} = \frac{a}{\sqrt{2}}$$

$$\sin \theta_2 = \frac{20}{\sqrt{2} \times 100} = 0.1414$$

$$\theta_2 = 8.13°$$

$$\sin \theta_3 = \frac{a+c}{R}$$

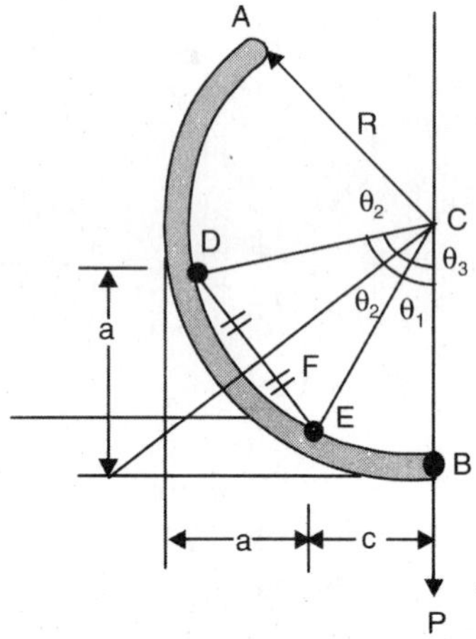

Fig. E8(a)

$$\sin \theta_1 = \frac{c}{R}; \cos \theta_1 = \frac{(R^2c^2)^{1/2}}{R}$$

$$\sin \theta_3 = \sin (16.26 + \theta_1)$$

$$\frac{a+c}{R} = \sin 16.26° \cos \theta + \cos 16.26° \sin \theta$$

$$\frac{a+c}{R} = 0.28 \left(\frac{\sqrt{R^2c^2}}{R}\right) + (0.96) \left(\frac{c}{R}\right)$$

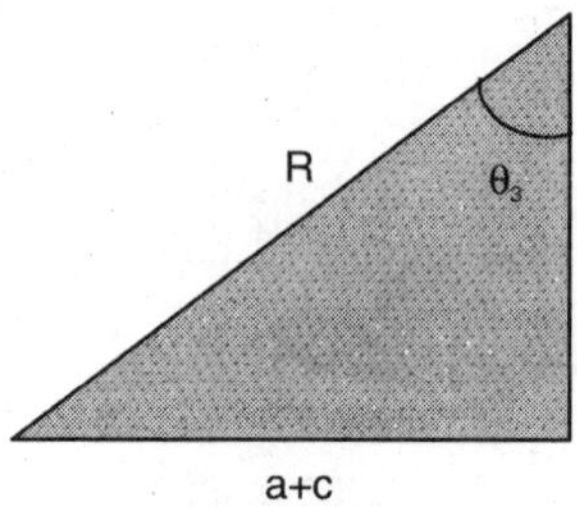

Fig. E8(b)

we get

$$(0.08)\, c^2 + (1.6)\, c - 384 = 0$$

after solving for c

we get

$$\mathbf{c = 60\ mm} \qquad \text{...Ans.}$$

Example 10

Solution

Area of shaded region = Area of semicircle ABO – (Area of quarter circle ONM + Area of bigger triangle AMN + Area of smaller triangle MLC)

$$= \frac{\pi(8)^2}{2} - \left(\frac{\pi(4)^2}{4} + \frac{1}{2}(4)(12) + \frac{1}{2}(6)(2)\right)$$

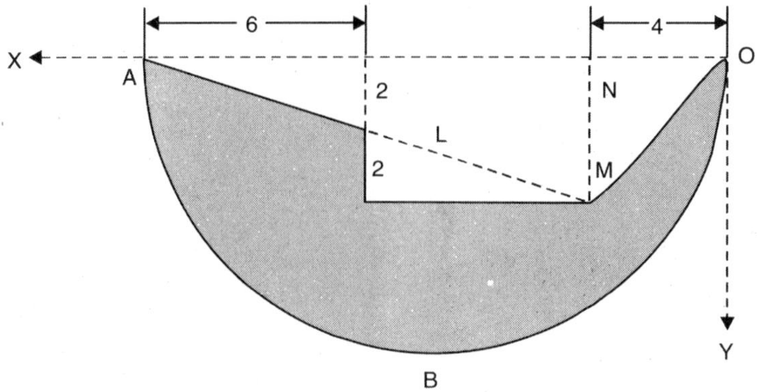

Fig. E10(a)

$$= 100.53 - (12.54 + 24 + 6) = 57.96$$

Now $x_{(ABO)} = 8$

$$x_{(OMN)} = 4 - \frac{4(4)}{3\pi} = 2.302$$

$$x_{(AMN)} = 4 + \left(\frac{12}{3}\right) = 8$$

$$x_{(MLC)} = 4 + \frac{2}{3}(6) = 8$$

$$y_{(ABO)} = \frac{4(8)}{3\pi} = 3.4$$

$$y_{(OMN)} = \frac{4(4)}{3\pi} = 1.7$$

$$y_{(AMN)} = \frac{4}{3} = 1.3$$

$$y_{(MLC)} = 2 + \frac{2}{3}(2) = 3.3$$

$$\bar{x} = \frac{\sum A_i x_i}{\sum A_i}$$

$$= \frac{(100.53)(8) - [(12.5)(2.302) + (24)(8) + (6)(8)]}{[100.53 - (12.57 + 24 + 6)]}$$

$$\bar{x} = 9.24$$

similarly

$$\bar{y} = 4.62$$

When lamina is suspended from A, AG line make α (suppose) with the line AB, here G denote center of gravity

$$\text{Sought angle } \alpha = \tan^{-1}\left(\frac{\overline{y}}{\overline{x}}\right) = \tan^{-1}\left(\frac{4.62}{9.24}\right)$$

$$= \mathbf{26.6^\circ} \qquad \textbf{...Ans.}$$

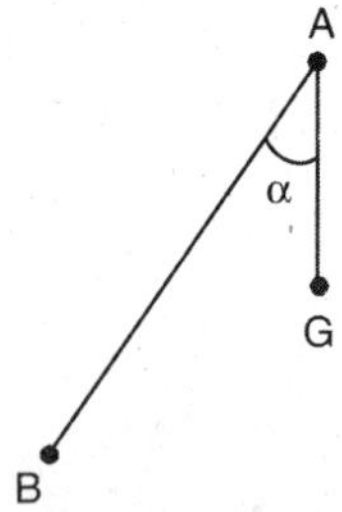

Fig. E10(b)

Example 12

Solution

$$X_1 = 12.5 \text{ cm}$$
$$X_2 = 20 \text{ cm}$$
$$X = 65 - x_2 - x_1 = 65 - 20 - 12.5 = 32.5 \text{ cm}$$
$$y_1 = 21.65 \text{ cm}$$
$$y_2 = 34.64 \text{ cm}$$
$$y = y_2 - y_1$$
$$= (34.64 - 21.65) = 13 \text{ cm}$$

for finding the C.G of composite figure

AB	BC	CD	AD
$x_{AB} = 6.25$ cm	$x_{BC} = 12.5 + \frac{32.5}{2} = 28.75$ cm	$x_{CD} = 12.5 + 32.5 + \frac{30}{2} = 55$ cm	$x_{AD} = \frac{65}{2} = 32.5$ cm

$$\overline{x} = \frac{\Sigma L_i X_i}{\Sigma L_i}$$
$$\overline{x} = 33.18 \text{ cm}$$
$$y_{AB} = 10.825 \text{ cm}$$
$$y_{BC} = 28.15$$
$$y_{CD} = 17.32$$
$$y_{AD} = 0$$
$$\overline{y} = \frac{\Sigma L_i X_i}{\Sigma L_i}$$

$$= \frac{25 \times 10.825 + 35 \times 28.15 + 40 \times 17.32 + 65 \times 0}{25 + 35 + 40 + 65}$$

$$= \mathbf{11.81\ cm}$$

Example 13

Solution

Let AB is a horizontal line and $A'B'$ is the diameter of the semi cylindrical shell of mass m_1.

For an equilibrium

$$\Sigma M_0 = 0$$

$$- m_1\, g \cos\theta \times r + m_2\, g \sin\theta \times \frac{2r}{\pi} = 0$$

$$\tan\theta = \frac{\pi m_1}{2m_2}$$

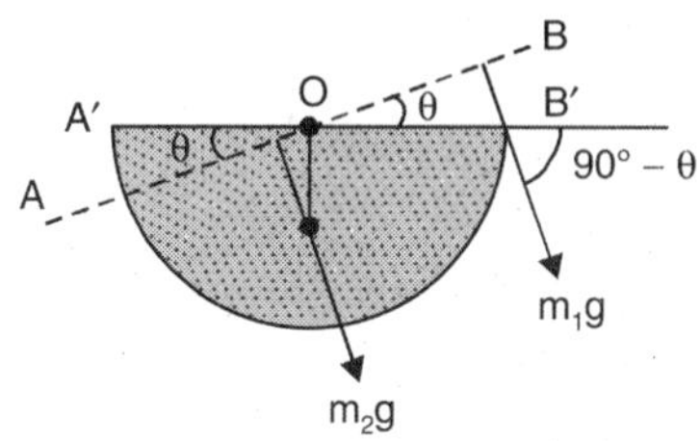

Fig. E13(a)

$$\theta = \tan^{-1}\frac{\pi \mathbf{m}_1}{2\mathbf{m}_2}$$ **...Ans.**

Example 14

Solution

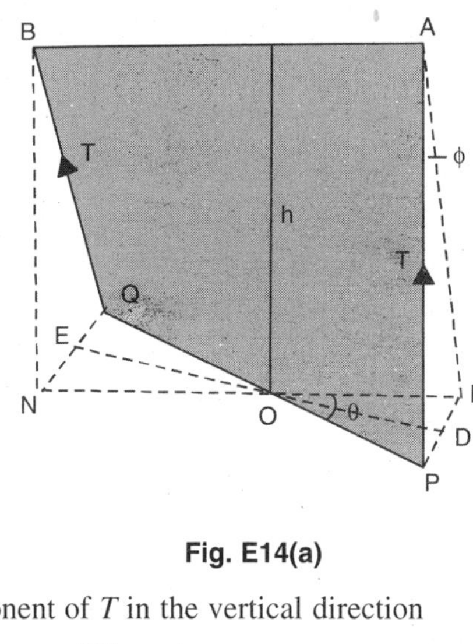

Fig. E14(a)

Taking the component of T in the vertical direction

$$2T \cos\phi = w \qquad ...(i)$$

Suppose moment M is applied on the structure

$$\text{So, } M = T \sin \phi \times DE \quad \text{...(ii)}$$

$$PM = h \tan \phi$$

$$DE = \left(\frac{1}{2}\right) (ab/h) \sin \theta \cot \phi \quad \text{...(iii)}$$

Using (*i*), (*ii*) and (*iii*)
We get,

$$\mathbf{M} = \left(\frac{\mathbf{w}}{\mathbf{4}}\right) \mathbf{(ab/h) \sin \theta} \quad \text{...Ans.}$$

Example 15

Solution

FBD of the water-containing region OBD for *mg*

$$mg = \rho g V$$

$$= (1)(9.8)\left[\frac{\pi(4)^2}{4}\right](30)$$

$$= 3700 \text{ kN}$$

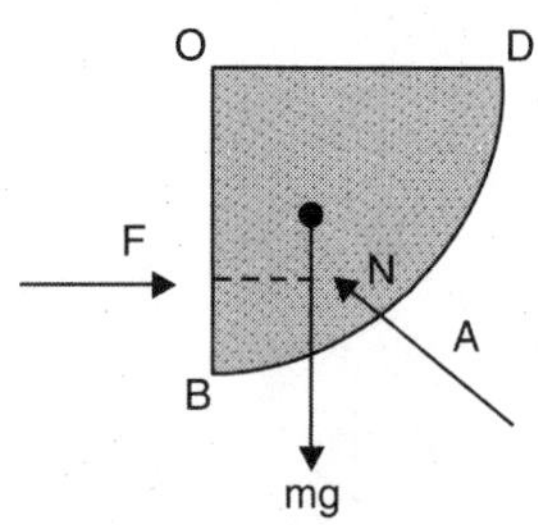

Fig. E15(a)

for *F*

$$A = b \times R = (30)(4) = 120 \text{ m}^2$$

$$F = \rho g h A$$

$$= (1)(9.8)\left(\frac{4}{2}\right)(120)$$

$$= 2350 \text{ kN}$$

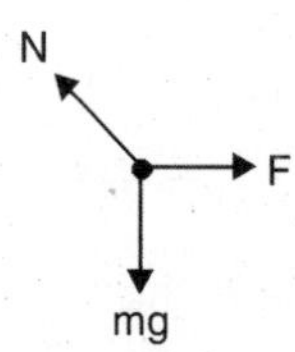

Fig. E15(b)

Clearly

Normal force at point *A* is resultant of forces *mg* and *F*

so,

$$N = \sqrt{(2350)^2 + (3800)^2}$$

$$= \mathbf{4380 \text{ kN}} \quad \text{...Ans.}$$

Example 16

Solution

Let *G* be the middle point of rod *AB* and *H* is the point on the cylinder. Rod touches the cylinder at point *H*.

$$OM \perp XY$$

$$OM = AC + OC \sin\theta$$
$$= l\cos\theta + OC\sin\theta \qquad \ldots(i)$$
$$GC = l\sin\theta = CH\sec\theta = (r - OC)\sec\theta$$

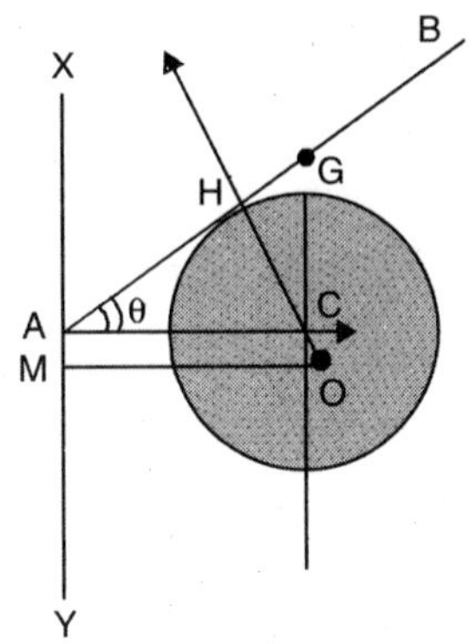

Fig. E16(a)

By eliminating *OC*

$$l\sin\theta\cos\theta = r - OC$$
$$= r - (C - l\cos\theta)\,\text{cosec}\,\theta \qquad \text{using } (i)$$
$$\boldsymbol{l\cos^2\theta = c - r\sin\theta}$$

Level 3

Example 1

Solution

Co-ordinate of center of gravity of the area of a curve $r = f(\theta)$ bounded by $\theta = \alpha$ and $\theta = \beta$.

$$\bar{x} = \frac{\int_{\alpha}^{\beta} \left(\frac{2}{3} r\cos\theta\right) \cdot \frac{1}{2} r^2 d\theta}{\int_{\alpha}^{\beta} \frac{1}{2} r^2 d\theta}$$

Similarly for y

$$\bar{y} = \frac{\int_{\alpha}^{\beta} \left(\frac{2}{3} r\cos\theta\right) \cdot \frac{1}{2} r^2 d\theta}{\int_{\alpha}^{\beta} \frac{1}{2} r^2 d\theta}$$

Not for lemniscate

$$r^2 = a^2 \cos 2\theta$$

Curve is mirror image of X-axis so

$$\bar{y} = 0$$

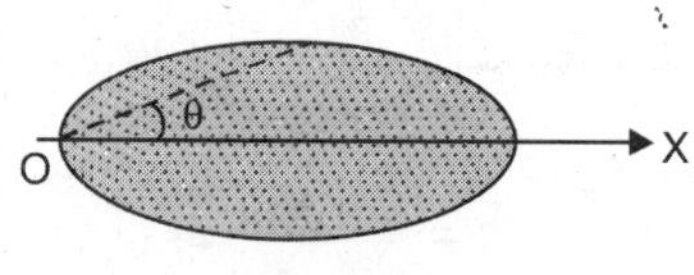

Fig. E1

for
$$\bar{x} = \frac{\int_{-\pi/4}^{\pi/4} \left(\frac{2}{3} r\cos\theta\right) \cdot \frac{1}{2} r^2 d\theta}{\int_{-\pi/4}^{\pi/4} \frac{1}{2} r^2 d\theta}$$

$$= \frac{\int_{-\pi/4}^{\pi/4} \frac{1}{3} a^3 \cos^{3/2} 2\theta \cos\theta \, d\theta}{\int_{-\pi/4}^{\pi/4} \frac{a^2}{2} \cos 2\theta d\theta}$$

$$\int_{-\pi/4}^{\pi/4} a^2 \cos 2\theta d\theta = \frac{a^2}{2}$$

For numerator

$$\text{Put } \sqrt{2} \sin\theta = \sin\alpha$$

$$\sqrt{2} \cos\theta \, d\theta = \cos\alpha \, dx$$

Then we get

$$= \frac{a^3}{3\sqrt{2}} \int_{-\pi/2}^{\pi/2} \cos^4 \alpha \, d\alpha$$

$$= \frac{2}{3\sqrt{2}} (a^3) \cdot \left(\frac{\pi}{2}\right)\left(\frac{1}{2}\right)\left(\frac{3}{4}\right)$$

$$= \frac{\pi a^3}{8\sqrt{2}}$$

$$\bar{x} = \frac{\pi a^3 \times 2}{8\sqrt{2} \times a^2} = \frac{\pi a}{4\sqrt{2}}$$

$$\Rightarrow \left(\frac{\pi a}{4\sqrt{2}}, 0\right)$$ ...**Ans.**

Example 2

Solution

Clearly due to symetricity one can easily say that $\bar{z} = 0$. For this type of problem first of all we calculate the centroid of the small element strip of the volume and then integrate it with proper limit.

Take the small element strip having thickness dx and radius r (assume)

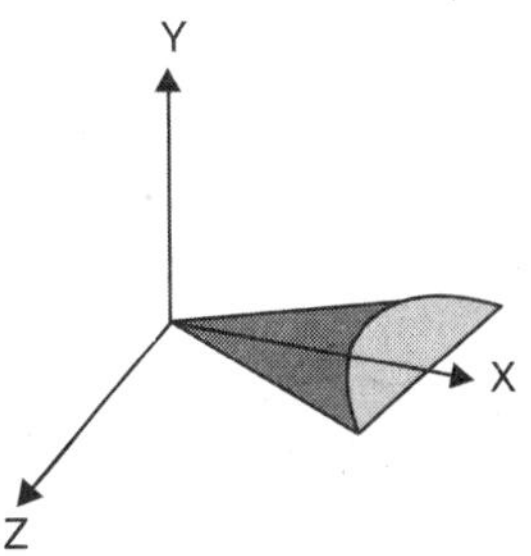

Fig. E2(a)

$$dv = \frac{1}{2}(\pi r^2)\ dx \qquad \ldots(i)$$

By similar triangle

$$R = \frac{a}{h}(x) \qquad \ldots(ii)$$

$$\bar{x} = \frac{\int x dv}{\int dv}$$

and

$$\bar{y} = \frac{\int y dv}{\int dv}$$

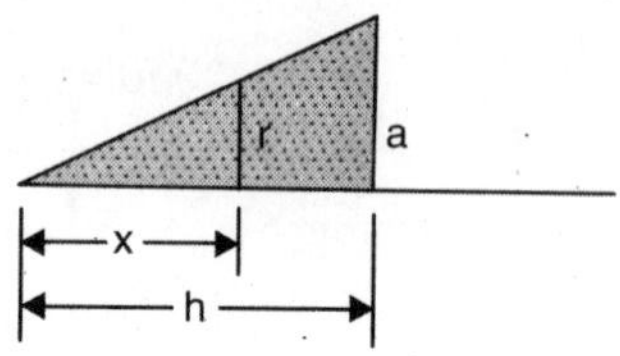

Fig. E2(b)

$(\bar{x}\ \bar{y})$ shows centroid of the strip

By using (*ii*)

Now
$$\int \bar{x}\, dv = \int_0^h x\left(\frac{1}{2}\pi r^2\right) dx$$

$$= \frac{\pi a^2 h^2}{8}$$

Similarly

$$\int \bar{y}\, dv = \int_0^h \frac{4r}{3\pi}\left(\frac{1}{2}\pi r^2\right) dx$$

$$= \frac{a^3 h}{6}$$

$$\int dv = \int_0^h \frac{1}{2}\pi r^2 dx = \frac{\pi a^2 h}{6}$$

by using (*i*)

so,

$$\bar{x} = \frac{\int x dV}{\int dV} = \frac{\left(\dfrac{\pi a^2 h^2}{8}\right)}{\left(\dfrac{\pi a^2 h}{6}\right)}$$

$$= \frac{3}{4}(h)$$

$$\bar{y} = \frac{\int y dV}{\int dV} = \frac{\frac{a^3 h}{6}}{\frac{\pi a^2 h}{6}} = \frac{a}{\pi}$$

$$\Rightarrow \qquad (\bar{\mathbf{x}}, \bar{\mathbf{y}}) = \left(\frac{\mathbf{3h}}{\mathbf{4}}, \frac{\mathbf{a}}{\pi}\right) \qquad \textbf{...Ans.}$$

Example 3

Solution

According to the problem statement mass of hinge is considerable let it be m_0.

At Point C

Apply Lami's theorem at C

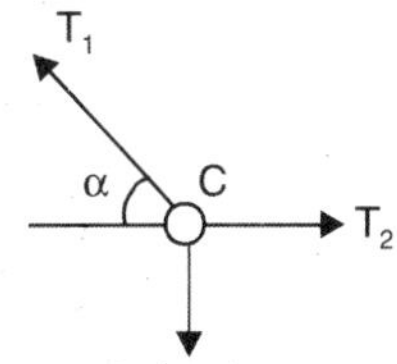

Fig. E3(a)

$$\frac{T_2}{\cos\alpha} = \frac{(m+m_0)g}{\sin\alpha}$$

$$T_2 \sin\alpha = (m + m_0)\, g \cos\alpha \qquad ...(i)$$

At point D

Suppose line $a - b$ is perpendicular to T_3.

$$T_2 \sin\alpha = m_0\, g \cos\alpha + R\cos\beta \qquad ...(ii)$$

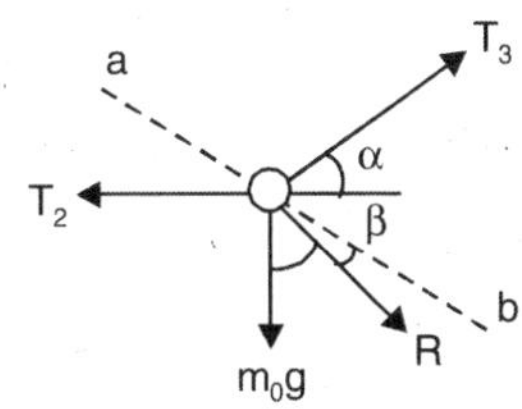

Fig. E3(b)

By equations (i) and (ii) we get

$$R = \frac{mg\cos\alpha}{\cos\beta}$$

$$R = \frac{mg\cos\alpha}{\cos\beta}$$

For R_{min} = cos β should be max

$$\cos\beta = 1$$

$$R_{min} = mg\cos\alpha = mg\cos(80°)$$

$$= \frac{\mathbf{mg}}{\mathbf{2}} \qquad \textbf{...Ans.}$$

Example 4

Solution

According to the centroid method

Three forces F_1, F_2, F_3 are acting along OA_1, OA_2, and OA_3

G is the centroid of triangle formed by A_1, A_2, A_3

We can write as

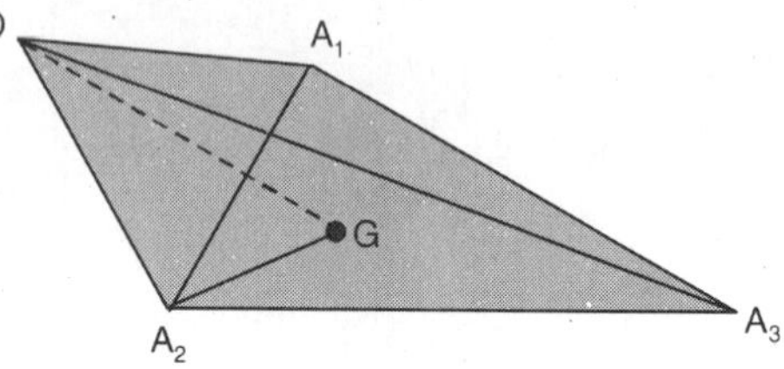

Fig. E4(a)

$$\frac{F_1}{k_1 OA_1} = \frac{F_2}{k_2 OA_2} = \frac{F_3}{k_3 OA_3}$$

$$= \frac{R}{(k_1 + k_2 + k_3)OG}$$

Here R denote resultant along OG, now come on towards problem

T_1 = Tension in the string OA

T_2 = Tension in the string OB

T_3 = Tension in the string OC.

$$\frac{T_1}{l_1} = \frac{T_2}{l_2} = \frac{T_3}{l_3} = \frac{W}{3(OG)}$$

OG = distance from fixed point O to centroid G of the triangle.

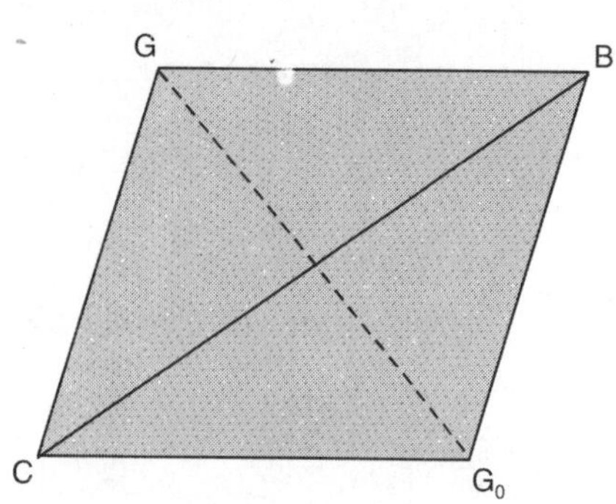

Fig. E4(b)

Now according to Lagrange's formula

$$\Sigma mOA^2 = \Sigma mGA^2 + OG^2 \Sigma m$$

so we can write as

$$l_1^2 + l_2^2 + l_3^2 = AG^2 + BG^2 + CG^2 + 3OG^2 \qquad \ldots(ii)$$

Concentrate on triangle BGC

Construct a parallelogram $BGCG_0$ as shown in figure

$$(GG_0)^2 + (BC)^2 = 2\,[(BG)^2 + (GC)^2]$$

$$(AG)^2 + (BC)^2 = 2\,[(BG)^2 + (GC)^2]$$

Similarly

$$(BG)^2 + (AC)^2 = 2\,[(CG)^2 + (AG)^2]$$

$$(CG)^2 + (AB)^2 = 2\,[(AG)^2 + (BG)^2] \qquad \ldots(iii)$$

$$(AG)^2 + (BG)^2 + (CG)^2 = \frac{a^2 + b^2 + c^2}{3} \qquad \ldots(iv)$$

therefore

$$9(OG)^2 = 3(l_1^2 + l_2^2 + l_3^2) - a^2 - b^2 - c^2 \qquad \ldots(v)$$

now put (v) in (i)
we get the result as

$$\mathbf{T_1} = \frac{\mathbf{Wl_1}}{\left[\mathbf{3\,(121 + 122 + 123)\; a^2b^2c^2}\right]^{1/2}}$$

$$\mathbf{T_2} = \frac{\mathbf{Wl_2}}{\left[\mathbf{3(121 + 122 + 123)a^2b^2c^2}\right]^{1/2}}$$

$$\mathbf{T_3} = \frac{\mathbf{Wl_3}}{\left[\mathbf{3(121 + 122 + 123)a^2b^2c^2}\right]^{1/2}} \qquad \textbf{...Ans.}$$

Example 8

Solution

According to **Pappus-Guldinus** theorem

$$A = 2\pi\bar{y}L \qquad \ldots(i)$$

Where L is the length of the line that is revolved for generating a surface.
In problem

$$\bar{x} = 2r - \frac{2r}{\pi}$$

$$= 2r\left(1-\frac{1}{\pi}\right) \qquad \ldots(ii)$$

$$L = \frac{\pi r}{2} \qquad \ldots(iii)$$

Use (*ii*) and (*iii*) in (*i*)
We get,

$$\mathbf{A = 2\pi r^2 (\pi - 1)} \qquad \ldots\textbf{Ans.}$$

Example 9

Solution

According to the problem rod must be in equilibrium in any position. The equilibrium can be only neutral.

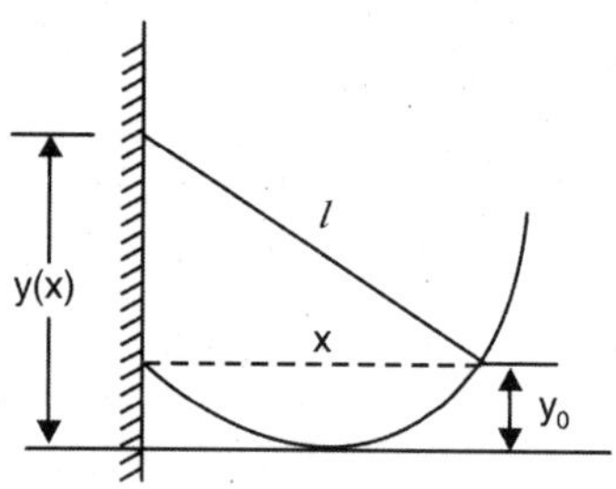

Fig. E9(a)

We can write as

$$l^2 = (y(x) - y_0)^2 + x^2$$

$$y_0 = y(x) \pm \sqrt{l^2 - x^2}$$

Assume ordinate of the CM is zero

$$\frac{y_0 + y(x)}{2} = 0$$

$$y(x) = \pm\sqrt{\frac{l^2 - x^2}{2}}$$

Now we can write as a general function y(x) as

$$\mathbf{y(x) = K - \sqrt{\frac{l^2 - x^2}{2}}}$$

Where *K* is an arbitrary constant.

Example 10

Solution

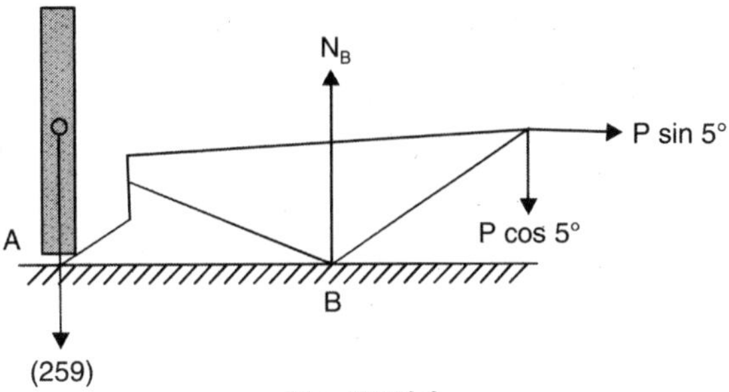

Fig. E10(a)

Taking moment about point B

$$\sum M_B = 0$$

$$(25)\,(9.8)\,(0.09) - P\,(\sin 5°)\,(0.036) - P\,(\cos 5°)\,(0.06) = 0$$

$$\mathbf{P = 351\ N} \qquad \textbf{...Ans.}$$

Friction

Level 1

Example 1

Solution

$$\mu = \tan 25^\circ = 0.466$$

Free body diagram of column supported by vertical position

$$\Sigma F_y = 0$$

$$N_A = 2000\text{N}$$

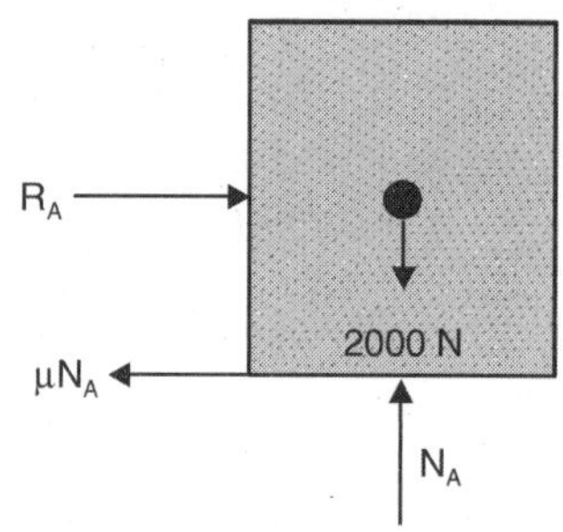

Fig. E1(a)

Free body diagram of wedges

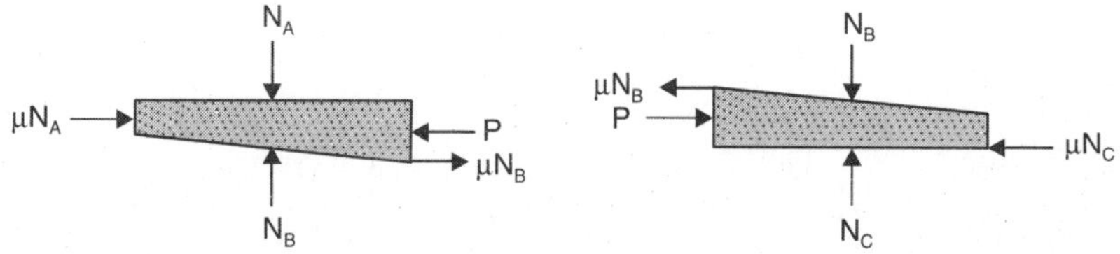

Fig. E1(b)

For upper wedge equation diagram (E, D).

$$\Sigma F_y = 0$$

$$N_B \cos 5^\circ - N_A - \mu N_B \sin 5^\circ = 0$$

$$N_B \cos 5^\circ - N_A - 0.466 N_B \sin 5^\circ = 0$$

$$N_B = 2092.97 \text{ N}$$

$$\Sigma F_x = 0$$

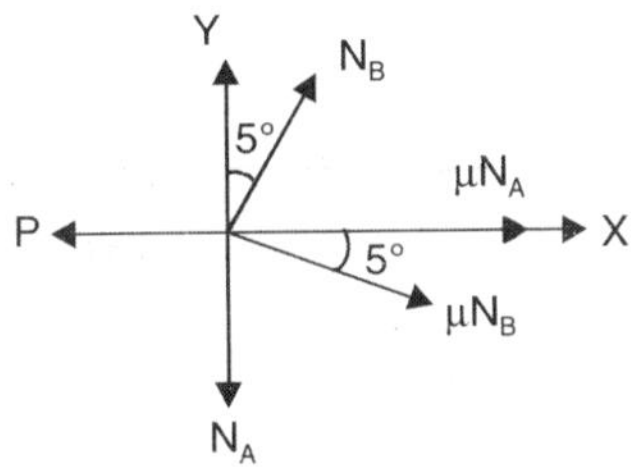

Fig. E1(c)

$$P = \mu N_A + \mu\ N_B \cos 5° + N_B \sin 5°$$

Hence, $\mathbf{P = 2086.03\ N}$ **...Ans.**

Example 2

Solution

Due to symmetricity of the blocks *B* and *C*, consider equal reaction N_A will act on the wedge

$$\alpha = \tan^{-1}\left(\frac{15}{80}\right) = 10.62°$$

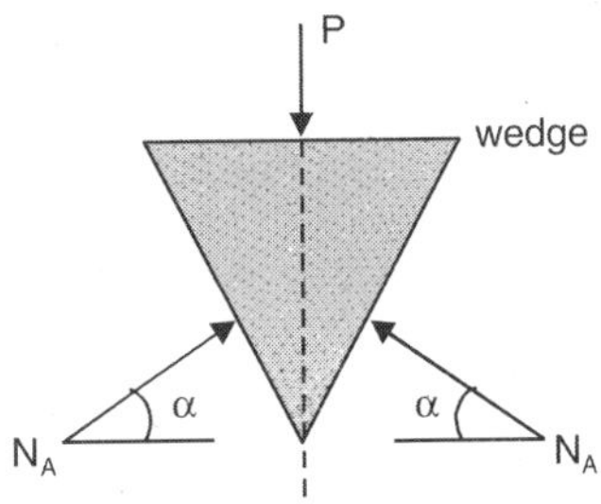

Fig. E2(a)

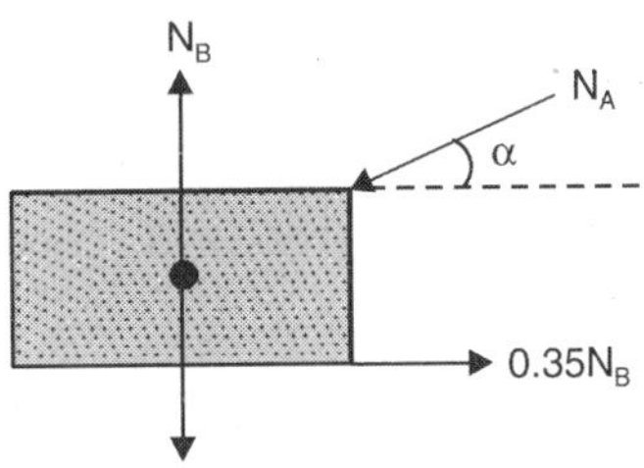

Fig. E2(b)

For block *B* or *C*

$$\Sigma F_x = 0$$

$$0.35\ N_B - N_A \cos 10.62° = 0$$

$$N_B = 2.808\ N_A \quad \ldots(i)$$

$$\Sigma F_y = 0$$

$$N_B - N_A \sin 10.62° - 51\ (9.81) = 0$$

Using equation (*i*)

$$N_A\ (2.808 - \sin 10.62°) = 51\ (9.81)$$

$$N_A = 190.688\ \text{N} \quad \ldots(ii)$$

For wedge

$$\Sigma F_y = 0$$

$$2N_A \sin 10.60° - P = 0$$

Using equation (*ii*)

$$2 \times 190.688 \sin 10.62° - P = 0$$

$$\mathbf{P = 70.285\ N}$$ **...Ans.**

Example 3

Solution

For clockwise rotation of pulley breaking torque must be anticlockwise for this movement.

$$T_1 > T_2$$

$$\beta = \text{lap angle} = 180° + 45°$$

$$= 225°$$

$$= 3.927 \text{ radians}$$

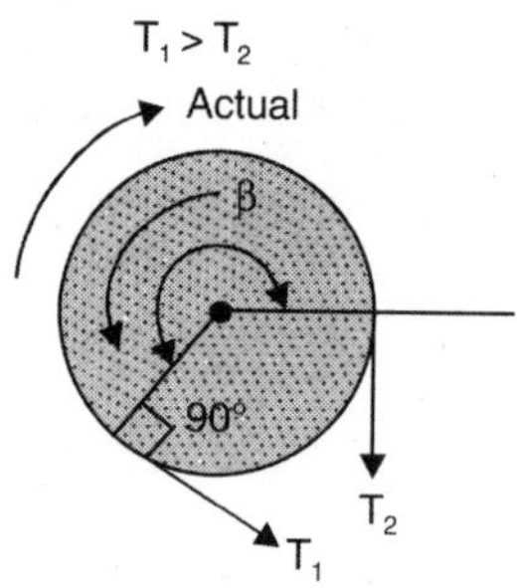

Fig. E3(a)

$$\frac{T_1}{T_2} = e^{\mu\beta}$$

$$\frac{T_1}{T_2} = e^{0.3 \times 3.927} \Rightarrow T_1 = 3.428T_2$$

Breaking torque

$\Rightarrow$

$$(T_1 - T_2)\,R = M$$

$$T_2\,(3.428 - 1)\,(0.4) = 100$$

$$T_2 = 103 \text{ N}$$

$$\text{And } T_1 = 353 \text{ N}$$

FBD of *AC*

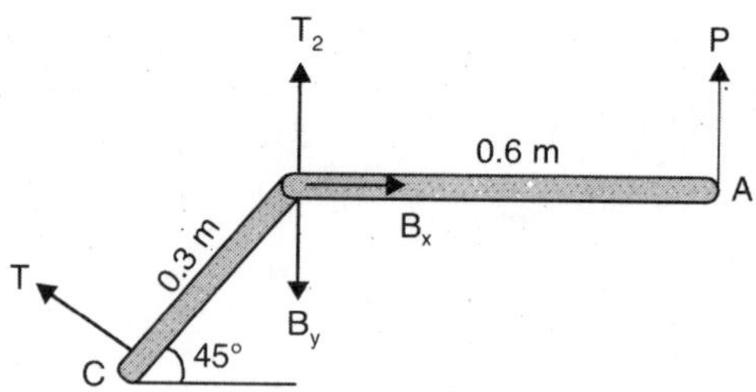

Fig. E3(b)

$$\Sigma M_B = 0$$

$$P\,(0.6) - (353)\,(0.3) = 0$$

$$\mathbf{P = 176.5\ N}$$ **...Ans.**

Example 4

Solution

$$N_D = N_B = N$$

$$\mu N_D = \mu N_B = \mu N$$

$$\Sigma M_B = 0$$

(neglecting the small deviation of window from vertical)

FBD of window

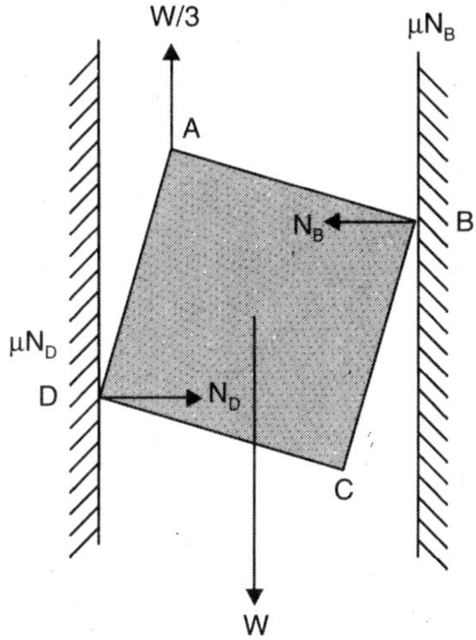

Fig. E4(a)

$$\frac{Wa}{2} - W\,a/2 + \mu Na - Nb = 0$$

$$\mu = \mathbf{b/a} \qquad \text{...Ans.}$$

Example 5

Solution

The resultant N of this distribution acts at some point A. For maintaining equilibrium it must pass through the center of the disc. Assume r is the radius of pulley and AB is the horizontal distance between point A and B.

Taking moment about A and equate it with zero.

$$\Sigma M_A = 0$$

$$P\,(r) - mg\,(AB) = 0$$

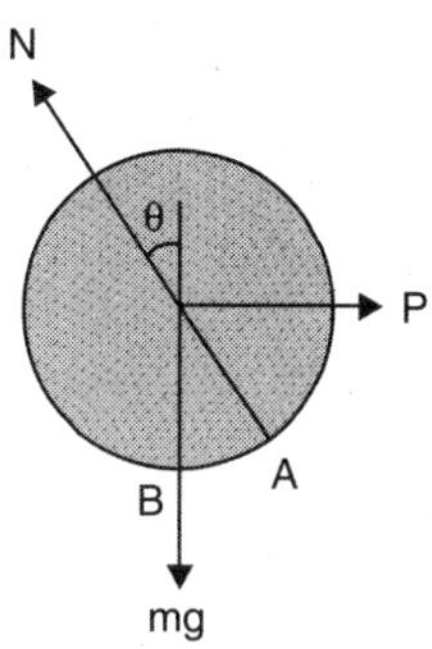

Fig. E5(a)

$$\frac{P}{mg} = \frac{AB}{r} = \frac{OB}{r} = \tan\phi = \mu$$

$$\mu = \frac{\mathbf{P}}{\mathbf{mg}} \qquad \textbf{...Ans.}$$

Example 6

Solution

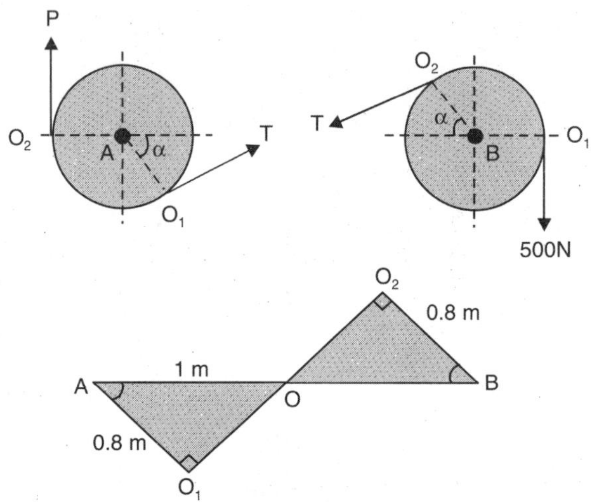

Fig. E6(a)

Clearly force P is required to hold the weight of 500 N

$$\text{So, } P < 500 \text{ N}$$

$$\cos\alpha = \frac{0.8}{1}$$

$$\alpha = 37°$$

For pulley B

$$\beta = 180° - \alpha = 180° - 37° = 143° = 2.5 \text{ rad}$$

$$\frac{500}{T} = e^{(0.3)\,(2.5)}$$

$$\therefore T = 236.18 \text{ N}$$

for pulley A

$$\frac{T}{P} = e^{(0.4)\,(2.5)}$$

$$\frac{236.18}{P} = e$$

$$\mathbf{P = 86.88\ N} \qquad \textbf{...Ans.}$$

Example 7

Solution

FBD of block *A*

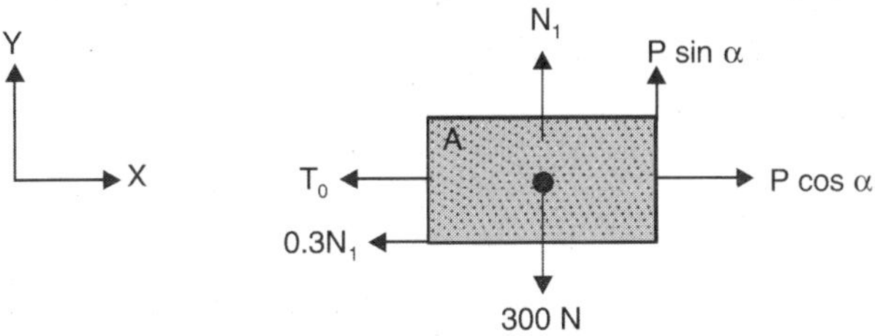

Fig. E7(a)

FBD of block *B*

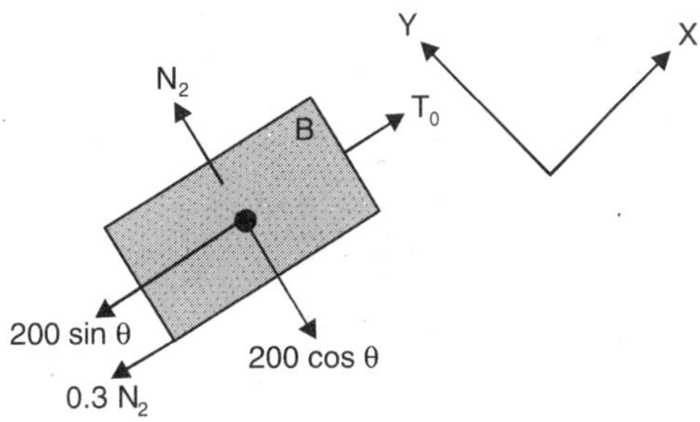

Fig. E7(b)

For block *A*

$$\Sigma F_x = 0$$

$$P\cos\alpha - T_0 - 0.3\,N_1 = 0 \qquad \ldots(i)$$

$$\Sigma F_y = 0$$

$$P\sin\alpha + N_1 - 300 = 0 \qquad \ldots(ii)$$

For block *B*

$$\Sigma F_x = 0$$

$$T_0 - 200\sin\theta - 0.3\,N_2 = 0 \qquad \ldots(iii)$$

$$\Sigma F_y = 0$$

$$N_2 - 200\cos\theta = 0 \qquad \ldots(iv)$$

Using above all these equations we get

$$P = \frac{258}{(\cos\alpha + 0.3\sin\alpha)} \quad \ldots(v)$$

For $P\big|_{min}$, $(\cos\alpha + 0.3\sin\alpha)\big|_{max}$

$$\frac{d(\cos\alpha + 0.3\sin\alpha)}{d\alpha} = 0$$

$$\text{we get } \tan\alpha = 0.3$$

$$\alpha = 16.7°$$

$$P_{min} = \frac{258}{\cos(16.7°) + 0.3\sin(16.7°)}$$

$$P_{min} = \mathbf{247.12\ N} \quad \ldots\textbf{Ans.}$$

Example 8

Solution

FBD of cylinder

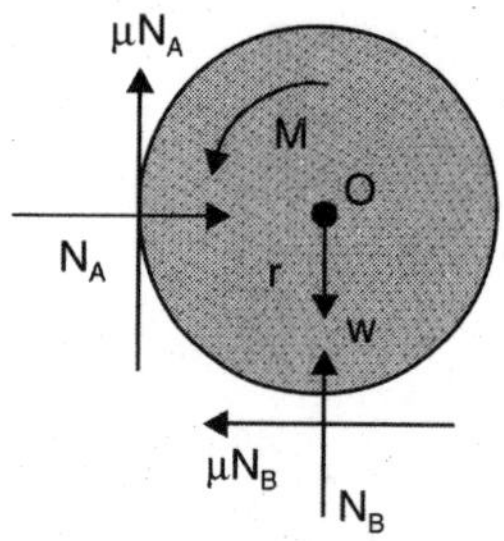

Fig. E8(a)

Apply condition of equilibrium

$$\Sigma F_x = 0$$

$$N_A - \mu N_B = 0 \quad \ldots(i)$$

$$\Sigma F_y = 0$$

$$N_B - W + \mu N_A = 0 \quad \ldots(ii)$$

$$\Sigma M_0 = 0$$

$$-(\mu N_A)(r) - (\mu N_B)(r) + M = 0 \quad \ldots(iii)$$

using all above equations we get

$$\mathbf{M} = \frac{\mu \mathbf{W r}(1+\mu)}{(1+\mu^2)} \quad \ldots\textbf{Ans.}$$

Example 12

Solution

FBD of plank

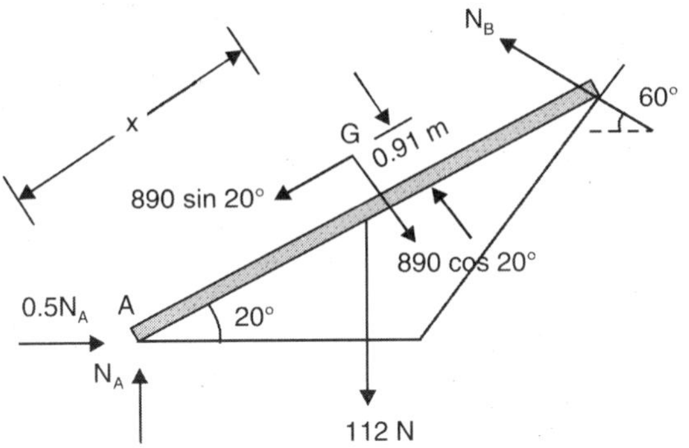

Fig. E12(a)

Apply equation of equilibrium

$$\Sigma F_x = 0$$

$$0.5\, N_A - N_B \cos 60° = 0 \qquad \ldots(i)$$

$$\Sigma F_y = 0$$

$$N_A - 890 - 112 + N_B \sin 60° = 0 \qquad \ldots(ii)$$

$$\Sigma M_A = 0$$

$$(890 \sin 20°)\,(0.91) - (890 \cos 20°)\,(x) - (112)\,(2.29 \cos 20°)$$
$$+ (N_B \sin 60°)\,(4.58 \cos 20°) + (N_B \cos 60°)\,(4.58 \sin 20°) = 0 \qquad \ldots(iii)$$

using all above equations
we get

$$\mathbf{x = 2.94\ m} \qquad \textbf{...Ans.}$$

Example 13

Solution

FBD for rod AB

$$\Sigma M_A = 0$$

$$-\,6g\,(OG) - (0.4\, N_2 \cos 10°)\,(AM) - (0.4\, N_2 \sin 10°)\,(BM)$$
$$-\,(N_2 \cos 80°)\,(AM) + (N_2 \sin 80°)\,(BM) = 0$$

$$\because \left(OG = \frac{2r}{\pi}\right)$$

$$AM = BM = 0.3\ \text{m}$$

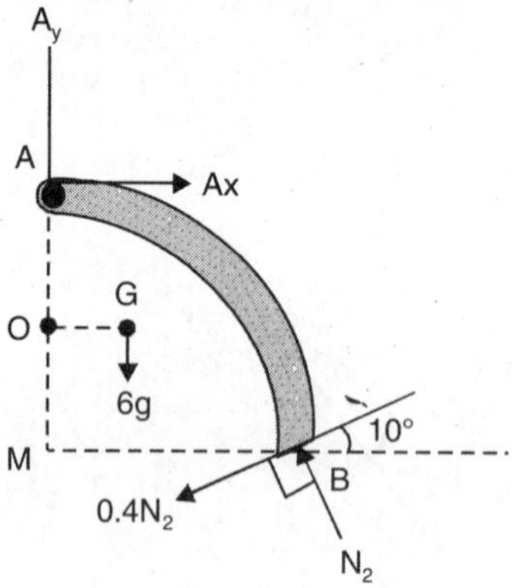

Fig. E13(a)

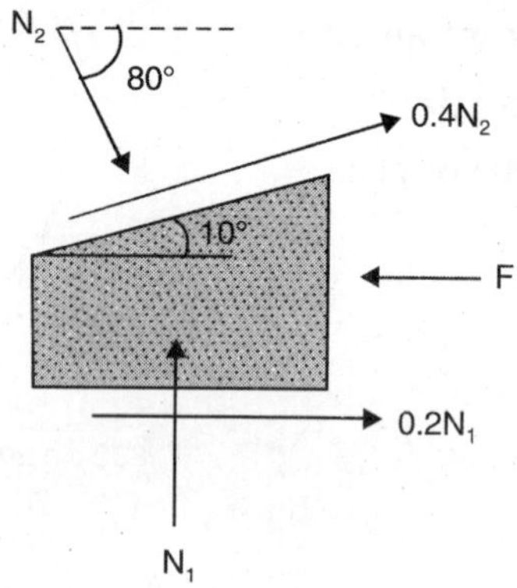

Fig. E13(b)

$$\text{We get, } N_2 = 108.07 \text{ N}$$

For wedge

$$\Sigma F_x = 0$$

$$N_1 + 0.4\, N_2 \sin 10° - N_2 \sin 80° = 0 \qquad \ldots(i)$$

$$\Sigma F_y = 0$$

$$-F + 0.2\, N_1 + 0.4\, N_2 \cos 10° + N_2 \cos 80° = 0 \qquad \ldots(ii)$$

Putting the value of N_2

In (*i*) and (*ii*)

We get

$$\mathbf{F = 81.11\ N} \qquad \ldots\text{Ans.}$$

Example 14

Solution

(*a*) FBD of block

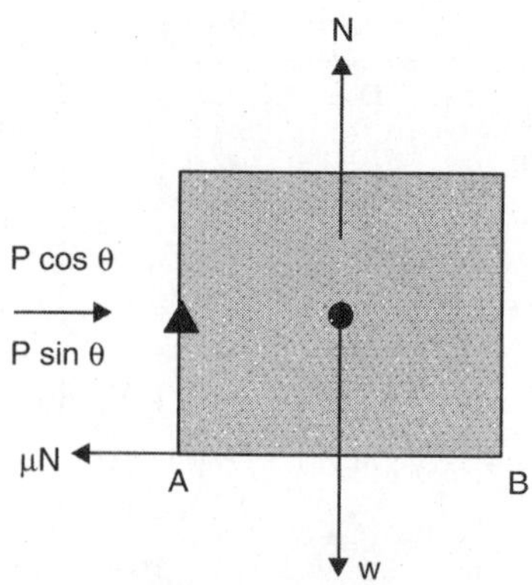

Fig. E14(a)

$$\Sigma F_x = 0$$

$$p \cos\theta - \mu N = 0 \quad \ldots(i)$$

$$\Sigma F_y = 0$$

$$P \sin\theta + N - W = 0 \quad \ldots(ii)$$

From equation (*i*) and (*ii*)
We get

$$\frac{P}{W} = \frac{\mu}{\cos\theta + \mu\sin\theta}$$

So,

$$\frac{P}{W} < \frac{\mu}{\cos\theta + \mu\sin\theta}$$

(*b*) For tipping condition
Clockwise moment about point *B* should be less than anticlockwise moment about point *B*

$(P\cos\theta)(h) + (P\sin\theta)(2a) \le W(a)$

$$\frac{P}{W} \le \frac{a}{h\cos\theta + 2a\sin\theta}$$

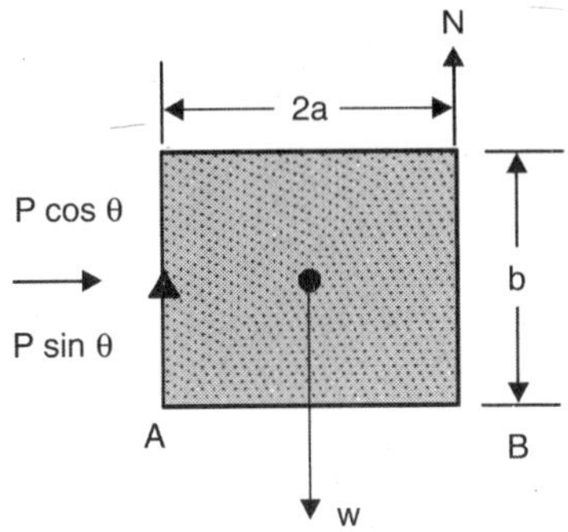

Fig. E14(b)

Example 16

Solution

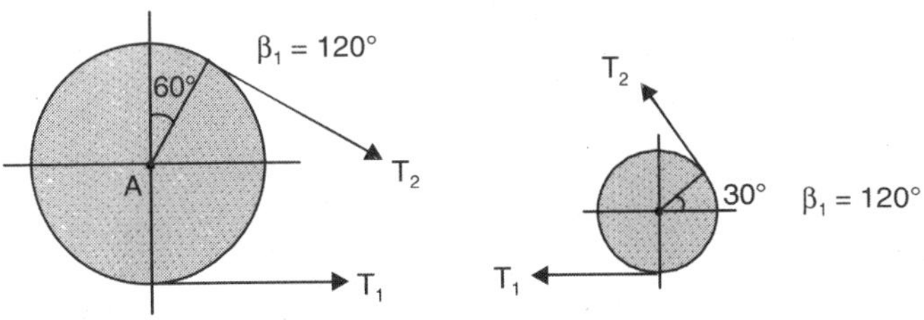

Fig. E16(a)

$\beta_1 > \beta_2$ and μ is same for both pulleys

so, slippage will occur at pulley *B* for pulley *B*, $\dfrac{T_2}{T_1} = e^{\mu\beta_2}$

$$\frac{600}{T_1} = e^{0.25} \times 2\pi/3$$

$$T_1 = 355.43 \text{ N}$$

For pulley A

Torque exerted by belt on pulley

$$M = (T_2 - T_1)\, r$$
$$= (600 - 355.43)\ .08$$
$$= \mathbf{19.57\ Nm} \qquad \textbf{...Ans.}$$

Example 17

Solution

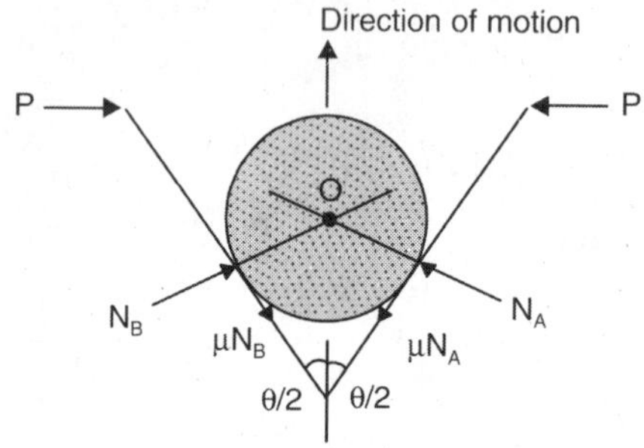

Fig 17(a)

Applying condition of equilibrium

$$\Sigma F_x = 0$$
$$N_B \cos \theta/2 + \mu\ N_B \sin \theta/2 \ - \mu N_C \cos \theta/2 - N_C \cos \theta/2 \ = 0$$
$$\Sigma F_y = 0$$
$$N_B \sin \theta/2 + \mu\ N_B \cos \theta/2 \ - \mu N_C \cos \theta/2 + N_C \sin \theta/2 \ = 0$$
$$\Sigma M_0 = 0$$
$$\mu N_C \times r - \mu N_B \times r = 0$$
$$N_B = N_C \qquad ...(iii)$$

We get θ = **33.4°** **...Ans.**

Example 18

Solution

(*a*) All coasters are locked

$$\Sigma F_x = 0$$
$$P - \mu N_A - \mu\ N_B = 0 \qquad ...(i)$$
$$\Sigma F_y = 0$$
$$N_A + N_B \ - W = 0$$

Here ($W = 60g$). ...(*ii*)

By (*i*) and (*ii*) we can get

$$\mathbf{P = 176.6\ N}$$

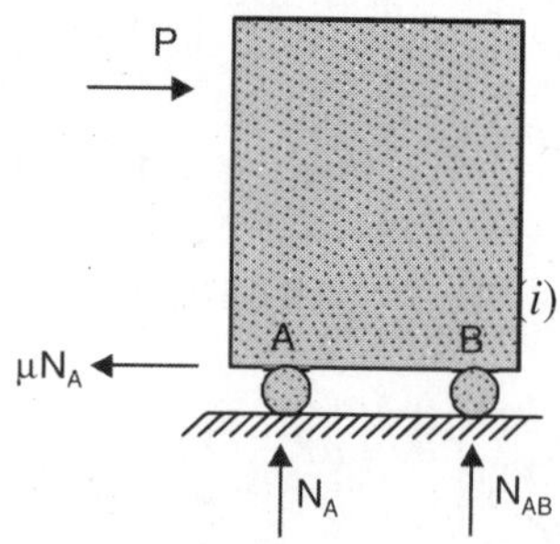

Fig. E18(a)

(*b*) Casters locked at *B* and free at *A*.

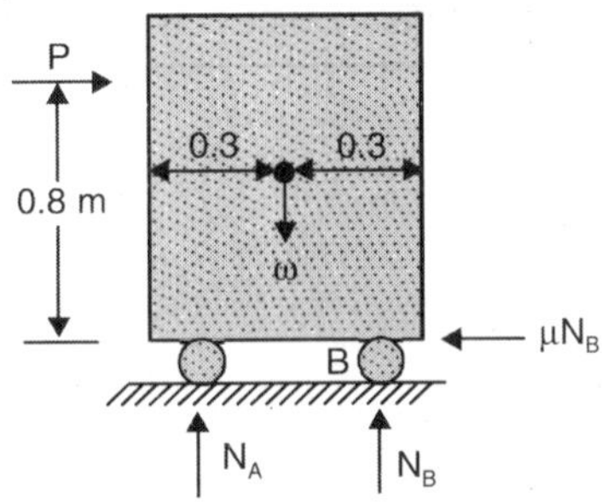

Fig. E18(b)

$$\Sigma F_x = 0$$

$$P - \mu N_B = 0, \quad P = 0.3\ N_B$$

$$\Sigma M_A = 0$$

$$P \times 0.8 + (60 \times 9.81) \times 0.3 - N_B \times 0.6 = 0$$

$$\mathbf{P = 147.3\ N}$$ **...Ans.**

(*c*) Casters locked at *A* and free at *B*

By using

$$\Sigma F_x = 0$$

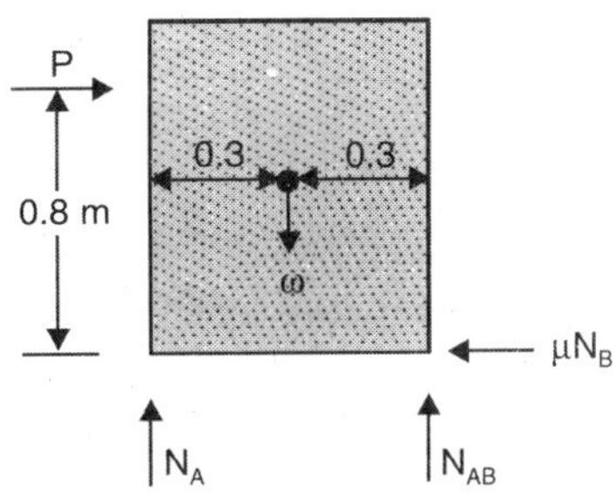

Fig. E18(c)

$$\Sigma M_B = 0$$

$$\mathbf{P = 63.1\ N}$$ **...Ans.**

Level 2

Example 2

Solution

Consider triangle ΔAEB

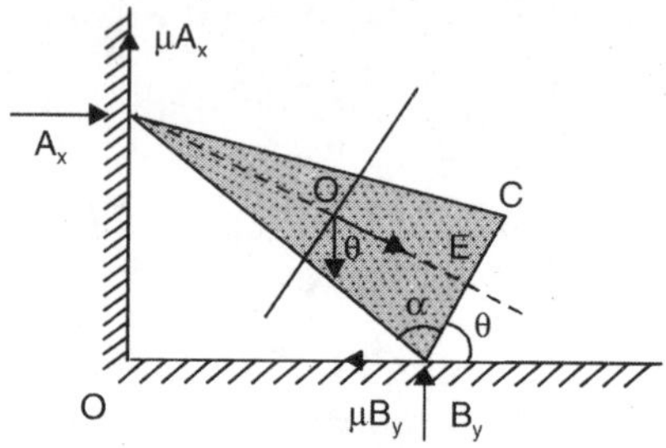

Fig. E2(a)

$$\angle ABE = \alpha$$

$$\tan\alpha = \frac{AE}{BE} = \frac{h}{l/4}$$

$$(AE)^2 + (BE)^2 = (AB)^2$$

$$(l/4)^2 + (h)^2 = l^2$$

$$\therefore h = 0.968\, l$$

$$\text{so } \tan\alpha = \frac{0.968l}{l/4} = \frac{0.968}{0.25}$$

$$\alpha = 75.5°$$

from figure

$$\Sigma F_x = 0$$

$$A_x - \mu B_y = 0$$

$$A_x = \mu B_y \qquad \ldots(i)$$

$$\Sigma F_y = 0$$

$$\mu A_x + B_y - W = 0$$

$$\mu^2 B_y + B_y - W = 0 \qquad \text{Using } (i)$$

$$B_y = 0.9412 \text{ W}$$

Triangle ABO

$$OA = (0.968 \cos\theta + 0.25 \sin\theta)\, l$$

$$OB = (-0.25 \cos\theta + 0.968 \sin\theta)\, l$$

$$\Sigma M_A = 0$$

$$w \sin\theta \times AD + \mu B_y \times OA - B_y \times OB = 0$$

$$AD = \frac{2}{3}(h) = \frac{2}{3} \times 0.968\, l = 0.6453\, l$$

Sought result θ = **65.5°** **...Ans.**

Example 3

Solution

Consider an element part of the thread.

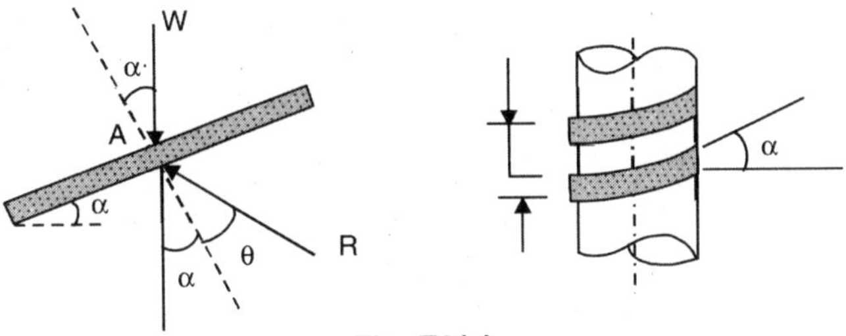

Fig. E3(a)

Concentrate at point A

$$W = R\cos(\alpha + \phi) \quad \ldots(i)$$

$$\sum M_0 = M - R\sin(\alpha + \phi)\cdot r = 0$$

$$M = R\sin(\alpha + \phi)\cdot r \quad \ldots(ii)$$

Using (*i*) and (*ii*)
We get

$$M = Wr\tan(\alpha + \phi)$$

$$\text{Where } \alpha = \tan^{-1}\left(\frac{L}{2\pi r}\right)$$

$$\mathbf{M = Wr\tan\left[\tan^{1}\left(\frac{L}{2\pi r}\right) + \phi\right]} \quad \textbf{...Ans.}$$

Example 4

Solution

FBD of the block B
Applying equation of equilibrium

$$\Sigma F_x = 0$$

$$T_B + \mu N_B + \mu N_A - 4905\sin 40° = 0 \quad \ldots(i)$$

$$\Sigma F_y = 0$$

$$N_B - N_A - 4905\cos 40° = 0 \quad \ldots(ii)$$

FBD of fixed drum

$$\beta = \text{lap angle} = \pi = 3.14,\ \mu = 0.2$$

$$T_B > T_A$$

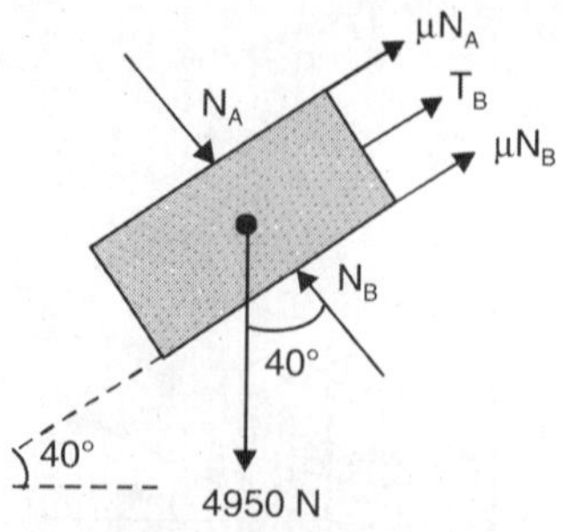

Fig. E4(a)

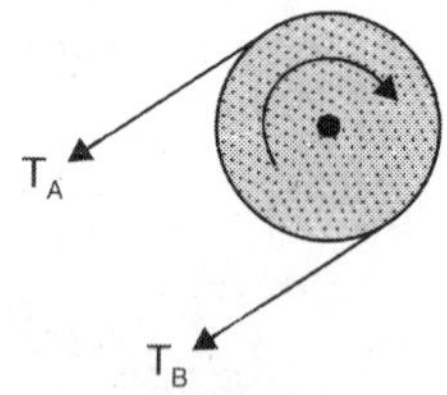

Fig. E4(b)

$$\frac{T_B}{T_A} = e^{(0.2 \times 3.14)}$$

$$T_B = 1.874\ T_A \qquad \ldots(iii)$$

FBD of block A

Applying equation of equilibrium

$$\Sigma F_x = 0$$

$$\mu N_A + (\text{mg}) \sin 40° - T_A = 0 \qquad \ldots(iv)$$

$$\Sigma F_y = 0$$

$$N_A - (\text{mg}) \cos 40° = 0 \qquad \ldots(v)$$

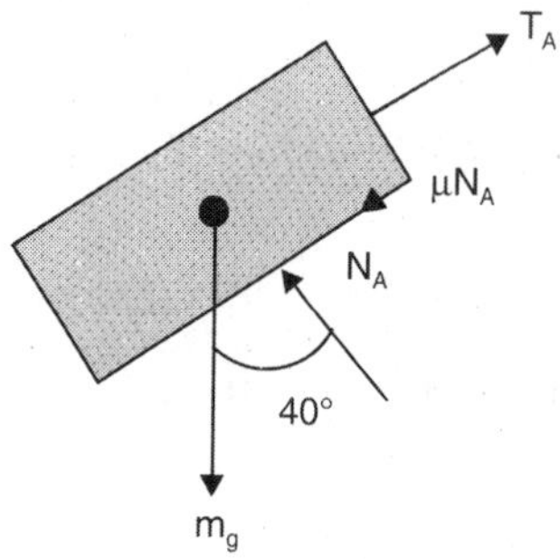

Fig. E4(c)

By these equations we can get

m = 135.25 kg **...Ans.**

Example 5

Solution

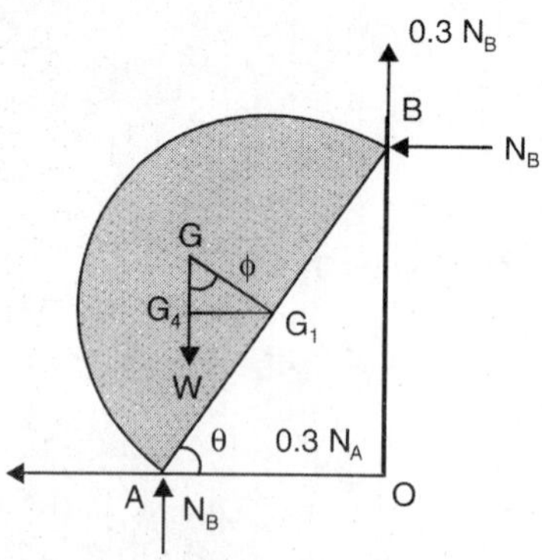

Fig. E5(a)

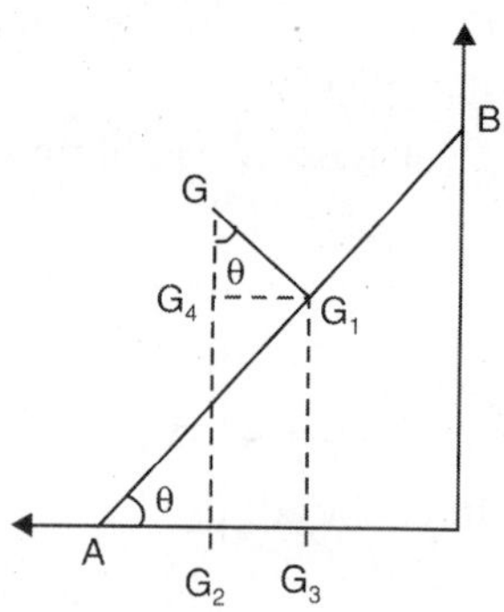

Fig. E5(b)

$$GG_1 = \frac{4r}{3\pi}$$

Taking moment about point A

$$\Sigma M_A = 0$$

$$-N_B \times OB - 0.3\ N_B \times OA - W \times AG_2 = 0$$

by geometry

$$OB = 2r \sin \theta,\ OA = 2r\cos\theta$$

$$AG_2 = \left(r\cos\theta - \frac{4r}{3\pi}\sin\theta\right)$$

By equation (i)

$$N_B = \frac{w\left(\cos\theta \frac{4}{3\pi}\sin\theta\right)}{(2\sin\theta + 0.6\cos\theta)} \qquad \ldots(i)$$

Applying equation of equilibrium

$$\Sigma F_y = 0$$

$$N_A - W + 0.3\ N_B = 0 \qquad \ldots(ii)$$

$$\Sigma F_x = 0$$

$$0.3\ N_A - N_B = 0 \qquad \ldots(iii)$$

By using these equations we can get result.

Hence $\theta = \mathbf{40.6°}$ **...Ans.**

Example 6

Solution

Clearly impending motion of end A must be towards left

FBD of the rod AB

$$R_A = \text{Resultant force at point } A$$

$$= \sqrt{N_A^2 + (\mu_s N_A)^2} = N_A\sqrt{1+\mu_s^2}$$

$$\tan\phi = \mu_s$$

$$\therefore \phi = \tan^{-1}\mu_s = \tan^{-1}(.35) = 19.29°$$

now concentrate on point D

$$\alpha = 90° - \frac{\theta}{2} \qquad \ldots(i)$$

Lami's theorem at P Fig.6(b)

$$\frac{T}{\sin(180°\phi)} = \frac{W}{\sin(\alpha + \phi)}$$

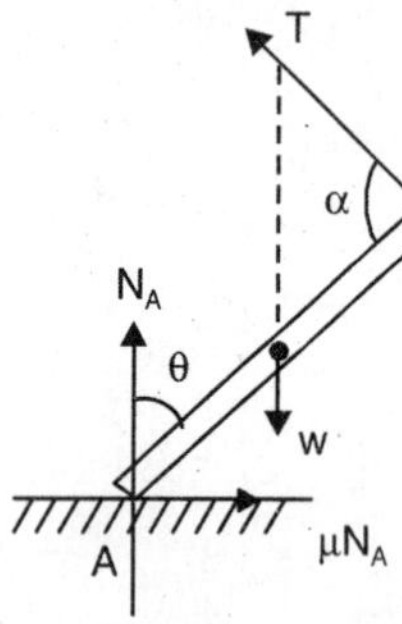

Fig. E6(a)

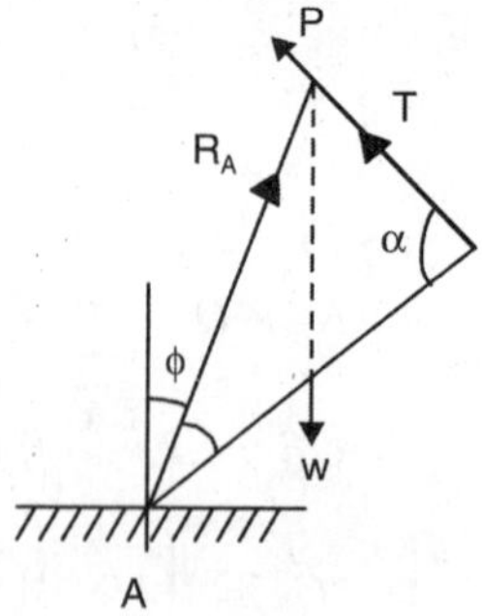

Fig. E6(b)

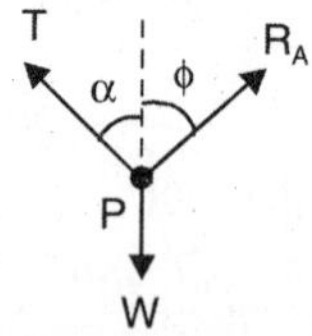

Fig. E6(c)

$$T = \frac{W \sin \phi}{\sin(\alpha + \phi)}$$

For $$T|_{\min},\ \alpha + \phi = \frac{\pi}{2}$$

$$90° - \frac{\theta}{2} + \phi = 90° \quad (\text{using } (i))$$

$$\frac{\theta}{2} = \phi,\ \theta = 2\phi$$

$$\theta = 2\,(19.29°) = 35.58°$$

now,

$$T_{\min} = \frac{W \sin(19.29°)}{\sin 90°}$$

$$\mathbf{T_{min} = 0.33\ W}$$ **...Ans.**

Example 8

Solution

FBD of ladder

By conditions of equilibrium

$$\Sigma F_x = 0$$

$$\mu_S N_A + \mu_S N_C \cos\theta - N_C \sin\theta = 0 \qquad ...(i)$$

$$\Sigma F_y = 0$$

$$N_A - W + N_C \cos\theta + \mu_S N_C \sin\theta = 0 \qquad ...(ii)$$

From (*i*) and (*ii*)

$$W = N_C \sin\theta \left(\frac{1 + u2S}{u_S}\right) \qquad ...(iii)$$

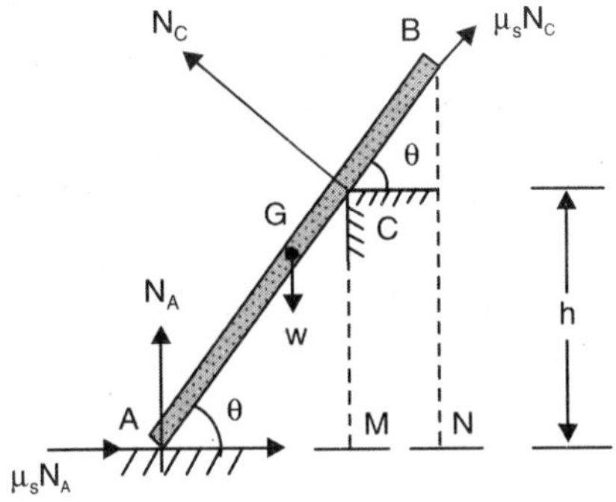

Fig. E8(a)

$$\Sigma M_A = 0$$

$$N_C\,(AC) - W\left(\frac{L}{2}\cos\theta\right) = 0$$

By geometry
We get

$$W = \frac{2N_C h}{L\sin\theta\cos\theta} \qquad \ldots(iv)$$

Now equating (*iii*) and (*iv*) we get

$$f(\theta) = \cos\theta - \cos^3\theta = \frac{2\mu_S h}{(1+\mu 2S\cdot L)}$$

For ladder not to slip at any angle

$$f'(\theta) = 0$$

$$\text{We get } \cos\theta = \frac{1}{\sqrt{3}}$$

$$f(\theta) = \frac{2}{\sqrt{27}}$$

$$\frac{2}{\sqrt{27}} < \frac{2\mu_S h}{(1+\mu_S)^2\cdot L}$$

$$\frac{\mathbf{\mu_S h}}{\mathbf{(1+\mu 2S)L}} > \frac{\mathbf{1}}{\mathbf{\sqrt{27}}} \qquad \ldots\textbf{Ans.}$$

Example 10

Solution

As size of pulley is negligible

$$AA' = x$$
$$BB' = 10 - x$$

Here $T_1 > T_2$

$$\frac{T_1}{T_2} = e^{\mu\beta}$$
$$\beta = \pi, \mu = 0.5$$
$$\frac{x(0.5)}{(10-x)(0.5)} = e^{(0.5)\,(\pi)}$$

We get **x = 8.28 m** **...Ans.**

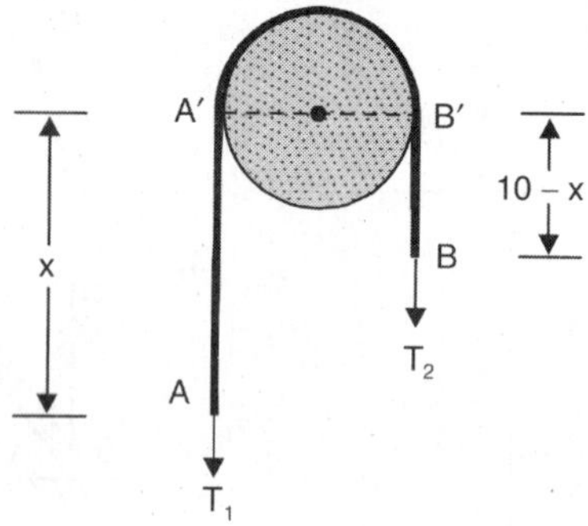

Fig. E10(a)

Example 11

Solution

FBD of pulley

Clearly

$$T_2 > T_1$$
$$\frac{T_2}{T_1} = e\mu^k \cdot \beta$$
$$\frac{T_2}{T_1} = e^{0.25} \times {}^{(3\pi/2)}$$
$$T_2 = 3.25\ T_1 \qquad ...(i)$$

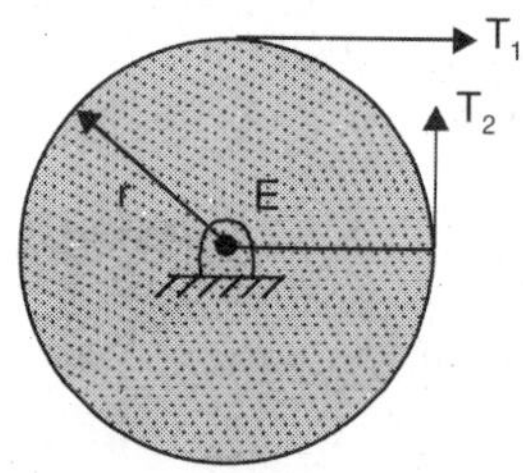

Fig. E11(a)

FBD of lever *ABD*

$$\Sigma M_C = 0$$
$$T_1 \times 80 + T_2 \times 80 - 100 \times 320 = 0$$
$$T_1 + T_2 = 400 \qquad ...(ii)$$

From (*i*) and (*ii*)

We get

$$T_1 = 94.12 \text{ N}$$
$$T_2 = 305.89 \text{ N}$$

Now

Breaking torque

$= (T_2 - T_1)\,r$

$= (305.89 - 94.12)\ 0.15$

= 31.765 Nm **...Ans.**

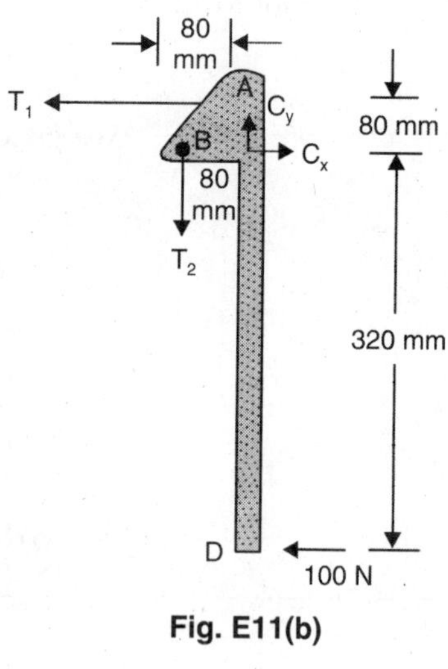

Fig. E11(b)

Example 12

Solution

When force P is applied to wedge B, it exerts pressure on C to move rightward corresponding block D will move

FBD of wedge B FBD of C and D

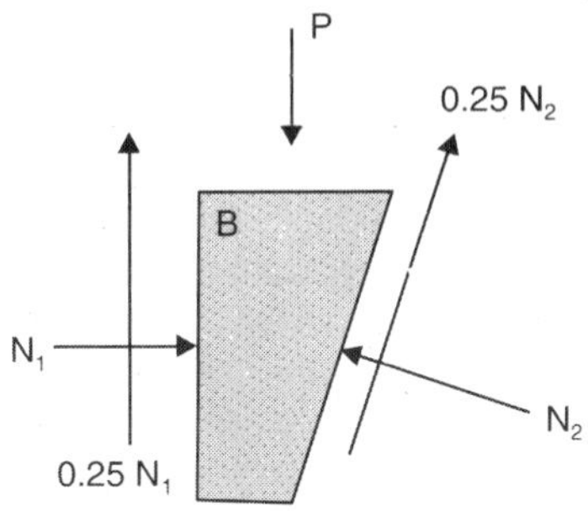

Fig. E12(a)

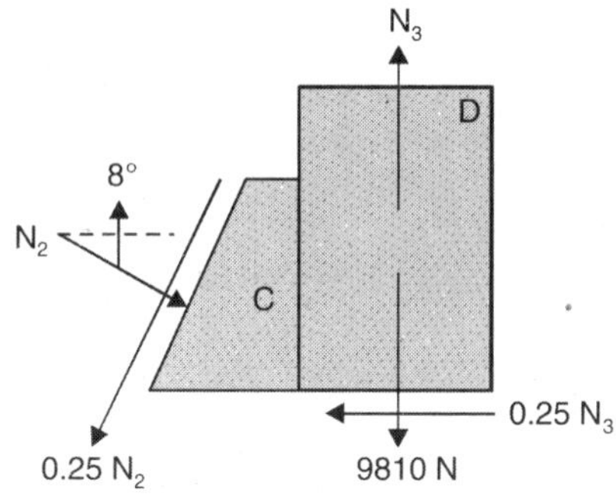

Fig. E12(b)

Equation diagram (E D)

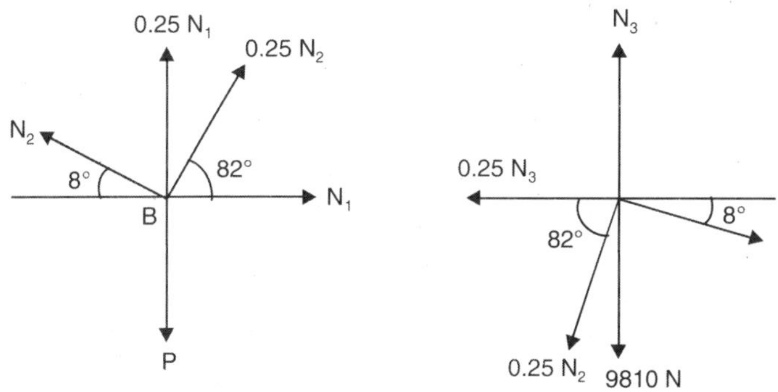

Fig. E12(c)

For B

$$\Sigma F_x = 0$$

$$N_1 + 0.25\ N_2 \cos 82° - N_2 \cos 8° = 0 \quad \ldots(i)$$

$$\Sigma F_y = 0$$

$$0.25\ N_2 \sin 82° + 0.25\ N_1 + N_2 \sin 8° - P = 0$$

For C and D

$$\Sigma F_x = 0$$

$$N_2 \cos 8° - 0.25\ N_3 - 0.2\ N_2 \cos 82° = 0 \quad \ldots(ii)$$

$$\Sigma F_y = 0$$

$$N_3 - 0.25\,N_2 \sin 82° - N_2 \sin 8° - 9810 = 0 \qquad \ldots(iii)$$

Using above all equations

$$\mathbf{P = 1796.7\ N} \qquad \textbf{...Ans.}$$

Example 13

Solution

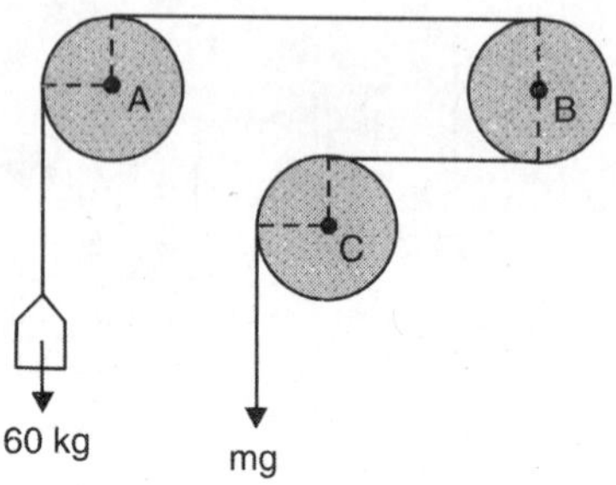

Fig. E13(a)

For A

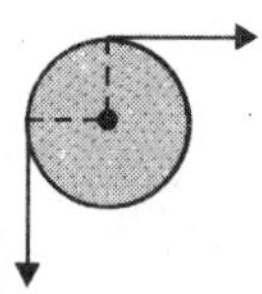

Fig. E13(b)

For B

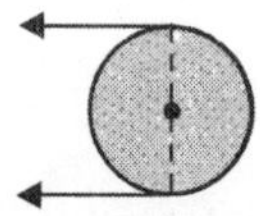

Fig. E13(c)

For C

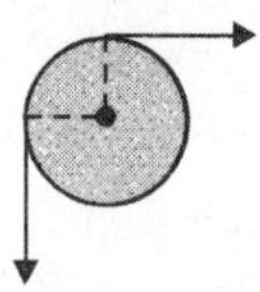

Fig. E13(d)

(*a*) The smallest weight *m* for equilibrium

$$60(g) > mg$$

$$\beta = \pi/2 + \pi + \pi/2 = 2\pi \text{ radian}$$

$$\frac{60g}{mg} = e^{(0.25)\,(2\pi)}$$

$$\therefore m = 12.47 \text{ kg}$$

(*b*) According to the question pipe *B* is rotated slowly in anticlockwise direction so clearly,
for pipe *A*

$$\frac{T_1}{60g} = e^{(0.25)\,(\pi/2)}$$

$$\mathbf{T_1 = 870.828\ N} \qquad \textbf{...Ans.}$$

For pipe *B*
Now use μ_k

$$\frac{T_1}{T_2} = e^{\mu_k \beta} = e^{(0.2)\,(\pi)}$$

$$\frac{870.828}{T_2} = e^{(0.2)\,(3.14)}$$

$$\mathbf{T_2 = 464.569\ N} \qquad \textbf{...Ans.}$$

For pipe *C*

$$\frac{mg}{T_2} = e^{\mu_s \beta}$$

$$\frac{mg}{464.569} = e^{(0.25)\,(\pi/2)}$$

$$\mathbf{m = 70.2\ kg} \qquad \textbf{...Ans.}$$

Level 3

Example 1

Solution

Force P is applied perpendicular to the rod AB, so it will rotate rod AB in anticlockwise sense, clearly friction will oppose in the clockwise sense

Suppose $AC = l_0$

$$\Sigma M_C = 0$$

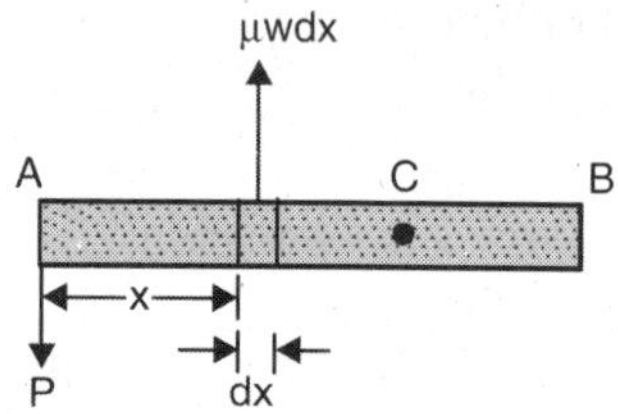

Fig. E1(a)

$$\int_0^{l_0} \mu Wx\,dx - \int_{l_0}^{l} \mu Wx\,dx = 0$$

$$\frac{1}{2}\left(\mu W l_0^2\right) = \frac{1}{2}\mu W\left(l^2 - l_0^2\right)$$

$$\sqrt{2}\,l_0 = l$$

$$\boldsymbol{l_0 = \frac{l}{\sqrt{2}}} \qquad \text{...Ans.}$$

So, least value of P to just move the rod is

$$P = \int_0^{l_0} \mu W\,dx - \int_{l_0}^{l} \mu W\,dx$$

$$= \mu W\left(\sqrt{2} - 1\right)l$$

$$\mathbf{P = \mu W l\left(\sqrt{2} - 1\right)} \qquad \text{...Ans.}$$

Example 3

Solution

By each peg reaction is equal to

$$N = N_A = N_B = N_C = \frac{W}{3}$$

There is an instantaneous center of rotation L inside the triangle so slipping takes place simultaneously at A, B and C.

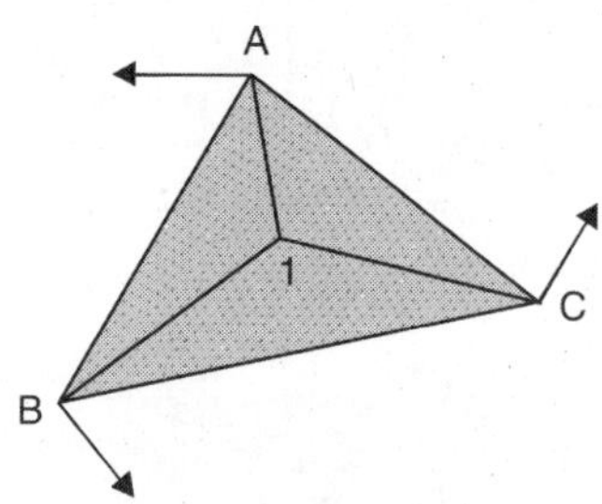

Fig. E3

$$F = F_A = F_B = F_B = F_C = m\frac{W}{3}$$

The couple required is the sum of moments about I of the frictions

$$M = I\,(F_A + I_B + I_C)$$

For the obtuse angle triangle I may be pass outside the triangle.

Turning about one corner let A resultant of F/s along AB and I along AC is $2\,F\cos A/2$ and this can only be in equilibrium with a force less than F.

$$\cos A/2 < 1/2$$
$$A > 120$$

∴ It follows that lamina will turn about a corner if the angle at that corner is greater than 120°

$$\text{if } A > 120°$$
$$\text{Then } M = F\,(AB + AC)$$

Sought result $\mathbf{M} = \dfrac{\mu \mathbf{W}}{\mathbf{3}}\,(\mathbf{AB} + \mathbf{AC})$ **...Ans.**

Example 5

Solution

Suppose that the rod is just about to slide down the plane AC and up the plane CB.

Here R_A and R_B denotes the resultant force by the inclined plane at A and B on the rod

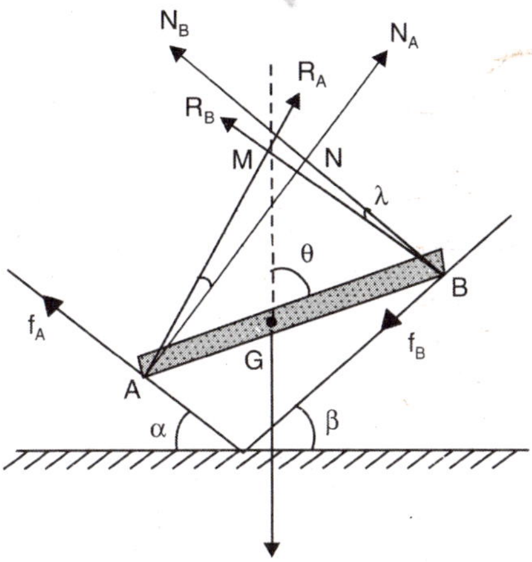

Fig. E5(a)

θ = Angle made by rod AB with vertical

Clearly
in ΔAMG

$$\frac{\sin MAG}{\sin AMG} = \frac{MG}{AG} \qquad \ldots(i)$$

in ΔBMG

$$\frac{\sin MBG}{\sin BMG} = \frac{MG}{BG} \qquad \ldots(ii)$$

$$\because AG = BG$$

then by equations (*i*) and (*ii*)

$$\frac{\sin(\theta\alpha + \lambda)}{\sin(\alpha\lambda)} = \frac{\sin(\theta + \beta + \lambda)}{\sin(\beta + \lambda)}$$

which gives us

$$\theta = \tan^{-1}\left[\frac{2\sin(\alpha - \lambda)\sin(\beta + \lambda)}{\sin(2\lambda + \beta\alpha)}\right] \qquad \textbf{...Ans.}$$

Example 6

Solution

(*ii*) FBD of rod *AB* in ΔAOB

$$\tan\theta = \frac{OB}{OA}$$

in Δ ALC

$$\tan\phi = \frac{Al}{CL} = \frac{(L/2)\cos\theta}{L\sin\theta}$$

$$2\tan\phi = \cot\theta = \tan\left(\frac{\pi}{2} - \theta\right)$$

$$2\tan(26.56°) = \tan\left(\frac{\pi}{2} - \theta\right)$$

$$\theta = 45° \qquad \textbf{...Ans.}$$

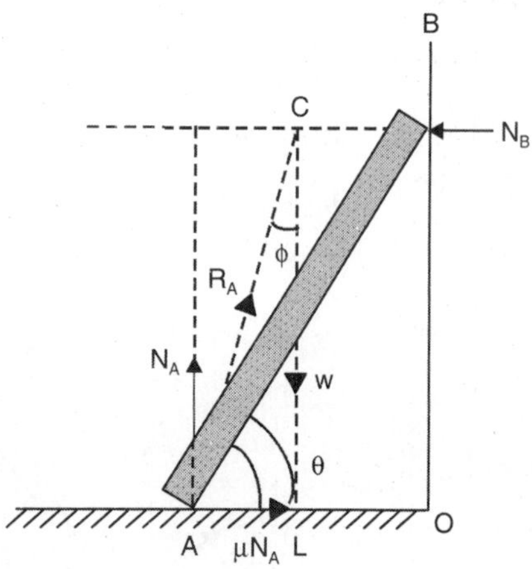

Fig. E6

Analysis of Structures

Level 1

Example 1

Solution

FBD of the *X* frame is

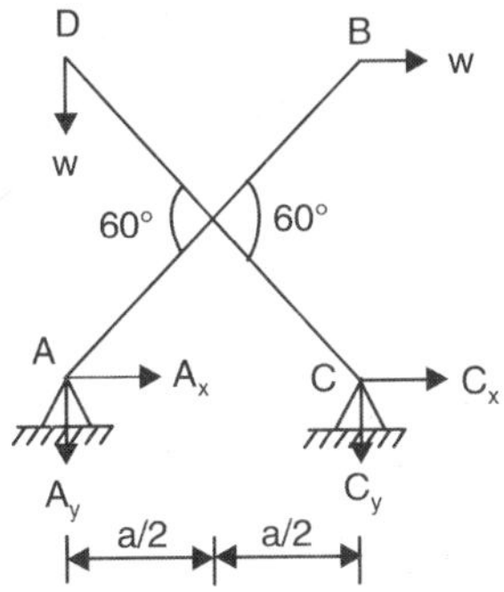

Fig. E1(a)

Taking moment about *A*

$$\Sigma M_A = 0$$

$$-Cy \times a - w \times BC = 0 \qquad \ldots(i)$$

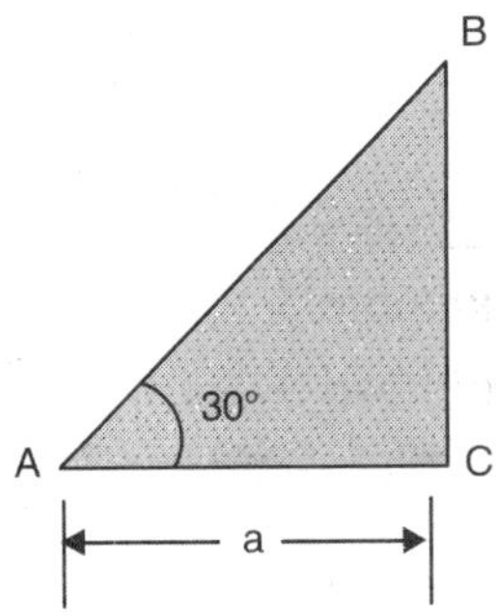

Fig. E1(b)

$$\tan 30^\circ = \frac{BC}{AC}$$

$$BC = a/\sqrt{3}$$

$$W \times \frac{a}{\sqrt{3}} = -Cy \times a$$

$$\therefore \mathbf{Cy} = \left(\mathbf{w}/\sqrt{\mathbf{3}}\right)\uparrow \qquad \textbf{...Ans.}$$

$$\Sigma F_x = 0$$

$$+A_x + Cx - w = 0 \qquad \ldots(ii)$$

$$\Sigma F_y = 0$$

$$-Ay + Cy - w = 0 \qquad \ldots(iii)$$

$$Ay = -(w - Cy) = Cy - w = \frac{w}{\sqrt{3}} - w$$

$$\mathbf{Ay} = \left(\frac{\sqrt{3}-1}{\sqrt{3}}\right)\mathbf{w}\uparrow \qquad \text{...Ans.}$$

Concentrate on rod *AB*

$$\Sigma M_E = 0$$

$$W \times \frac{a}{2\sqrt{3}} - A_x \times \frac{a}{2\sqrt{3}} + A_y + \frac{a}{2} = 0$$

$$\frac{w}{\sqrt{3}} - \frac{A_x}{\sqrt{3}} + A_y = 0$$

$$\frac{w}{\sqrt{3}} - \left(\frac{\sqrt{3}-1}{\sqrt{3}}\right)w = \frac{A_x}{\sqrt{3}}$$

$$w\left(1 - \sqrt{3} + 1\right) = A_x$$

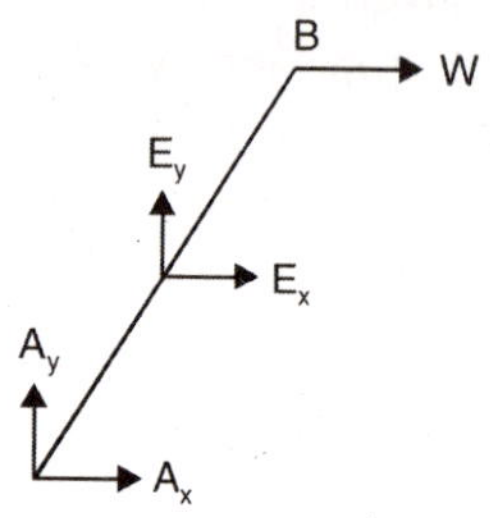

Fig. E1(c)

$$A_x = (2 - \sqrt{3})\ W \qquad \text{...Ans.}$$

Putting the A_x value in equation (*ii*)
We get

$$\mathbf{C_x} = \left(\sqrt{3} - 1\right)\mathbf{W} \qquad \text{...Ans.}$$

Example 3

Solution

Consider rod *AB*

$$\Sigma M_B = 0$$

$$-A_y \times 4 + 50 \times 2 = 0$$

$$A_y = 25 \text{ kN}$$

Consider rod *CD*

$$\Sigma M_B = 0$$

$$-30\cos 30° \times 2 + 60 + D_x \times 7 = 0$$

$$D_x = 9/7 \text{ kN}$$

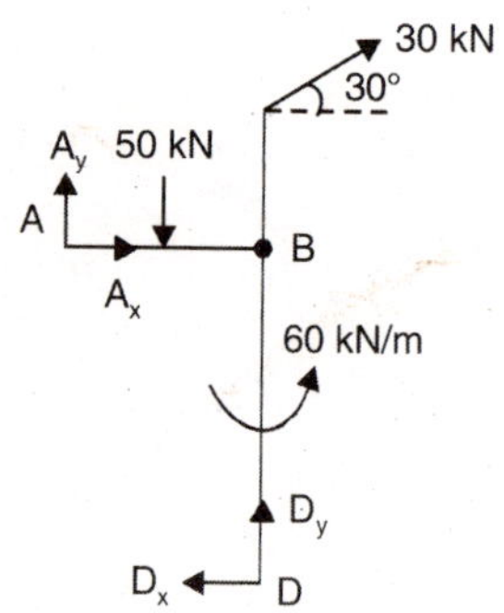

Fig. E3(a)

Similarly

$$\mathbf{A_x = 24.83\ kN} \qquad \text{...Ans.}$$

$$\Sigma F_y = 0$$

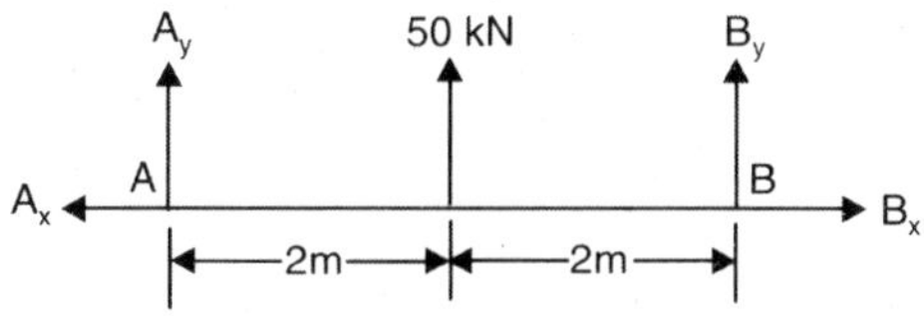

Fig. E3(b)

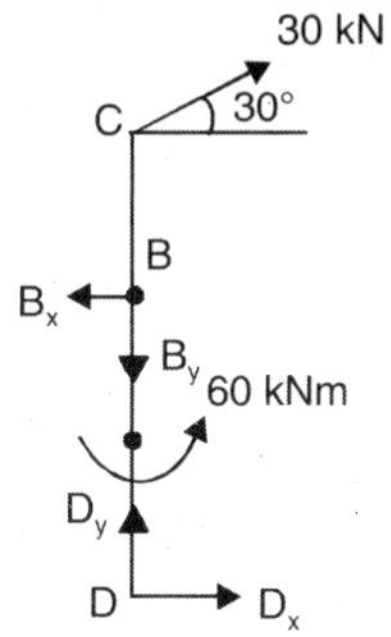

Fig. E3(c)

$$A_y + D_y + 30 \sin 30° - 50 = 0$$

$$\mathbf{D_y = 10\ kN} \uparrow \qquad \textbf{...Ans.}$$

$$R_A = \sqrt{A_x^2 + A_y^2} = 35.24 \text{ kN}$$

$$R_D = \sqrt{D_x^2 + D_y^2} = 10.07 \text{ kN}$$

Example 4

Solution

FBD of full beam *AC*

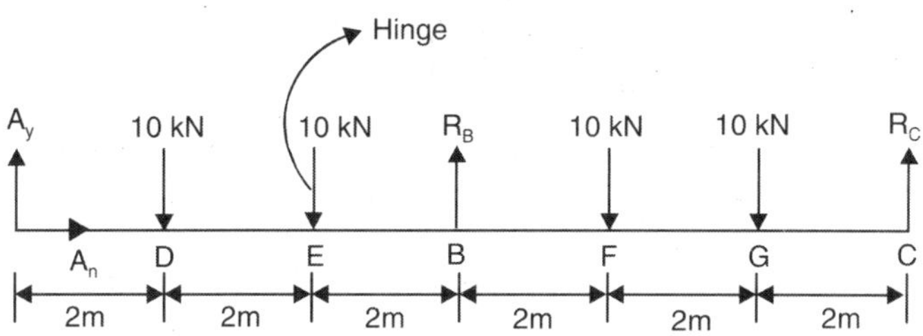

Fig. E4(a)

For *AE*

$$\Sigma M_E = 0 \quad + 10 \times 2 - A_y \times 4 = 0$$

$$\mathbf{A_y = 5\ kN} \qquad \textbf{...Ans.}$$

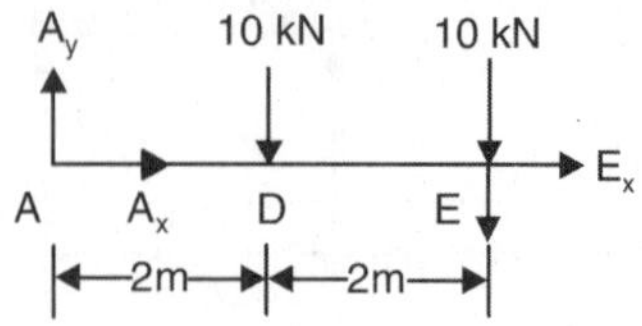

Fig. E4(b)

$$\Sigma F_y = 0$$

For AC

$$A_y - 40 + R_C + R_B = 0$$

$$\mathbf{R_B = -5 - 5 + 40 = 30\ kN}$$ **...Ans.**

For complete beam AC

$$\Sigma M_B = 0$$

$$-A_y \times 6 + 10 \times 4 + 10 \times 2 - 10 \times 2 - 10 \times 4 + R_C \times 6 = 0$$

$$\mathbf{R_C = 5\ kN}$$ **...Ans.**

Example 5

Solution

FBD for joint A

$$\Sigma F_x = 0$$

$F_{AB} \cos 58.1 - 8.4 = 0$

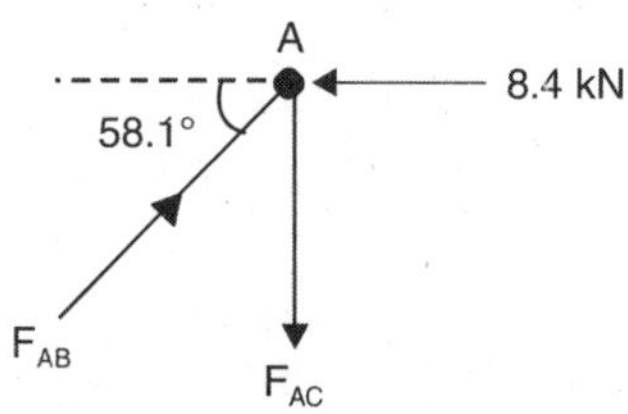

Fig. E5(a)

$$F_{AB} = 15.9 \text{ kN}$$

(force is acting positive towards the point A), so compressive

$$\Sigma F_y = 0$$

$$F_{AB} \sin 58.1° - F_{AC} = 0$$

$$F_{AC} = 13.5 \text{ kN}$$

(force is acting away from the point A) so 'tensile'.
Like this we get the result as
For joint C

$$\mathbf{F_{CD} = 15.9\ kN\ (tensile)}$$

$$\mathbf{F_{BC} = 16.8\ kN\ (comp)}$$

For joint D

$$\mathbf{F_{BD} = 13.5\ kN\ (comp)}$$ **...Ans.**

Example 6

Solution

Draw the free body diagram of *BD* and *AB* independently

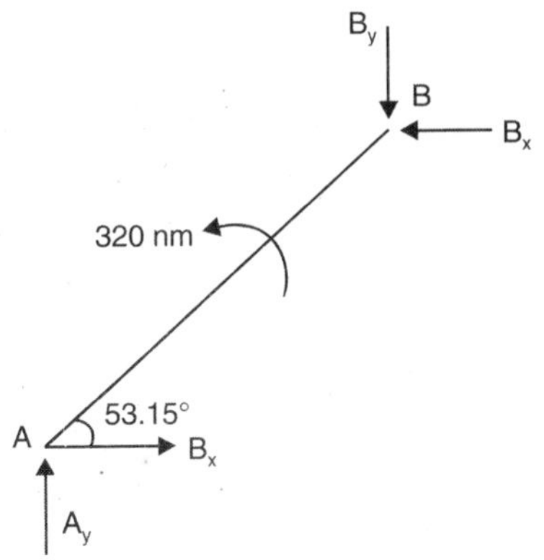

Fig. E6(a)

For member *AB*

$$\Sigma F_x = 0$$

$$A_x - B_x = 0 \qquad \ldots(i)$$

$$\Sigma F_y = 0$$

$$A_y - B_y = 0 \qquad \ldots(ii)$$

$$\Sigma M_A = 0$$

$$320 + B_x \times (5 \sin 53.13°) - By \times (5 \cos 53.13°) = 0 \qquad \ldots(iii)$$

For member *BD*

$$\Sigma F_x = 0$$

$$R_D + B_x - T \cos 30° = 0 \qquad \ldots(iv)$$

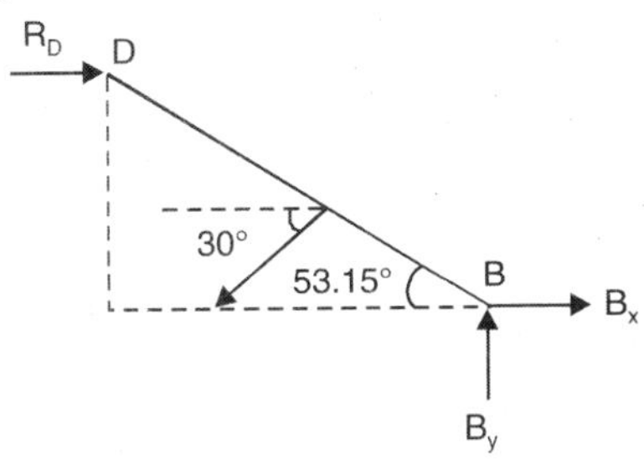

Fig. E6(b)

$$\Sigma F_y = 0$$

$$By - T \sin 30° = 0 \qquad \ldots(v)$$

$$\Sigma M_B = 0 \text{ gives}$$

$$R_D\,(10\sin 53.13°) - (T\cos 30°)\,(5\sin 53.13°) - (T\sin 30°)\,(\cos 53.13°)\,(5) = 0 \quad \ldots(vi)$$

Using these six equations, we get desired result

$$T = 461.54 \text{ N}$$
$$R_D = 286.62 \text{ N} \rightarrow$$
$$B_x = 113.08 \text{ N}$$
$$B_y = 230.77 \text{ N}$$
$$A_y = 230.77 \text{ N}$$
$$A_x = 113.08 \text{ N}$$
$$R_B = \sqrt{B_x^2 + B_y^2}$$
$$= \sqrt{(113.08)^2 + (230.77)^2}$$
$$\approx \mathbf{257\ N} \quad \ldots\textbf{Ans.}$$
$$\phi = \tan^{-1}\left(\frac{230.77}{113.08}\right) = 63.9° \quad \ldots\textbf{Ans.}$$

Example 7

Solution

Taking moment about point D

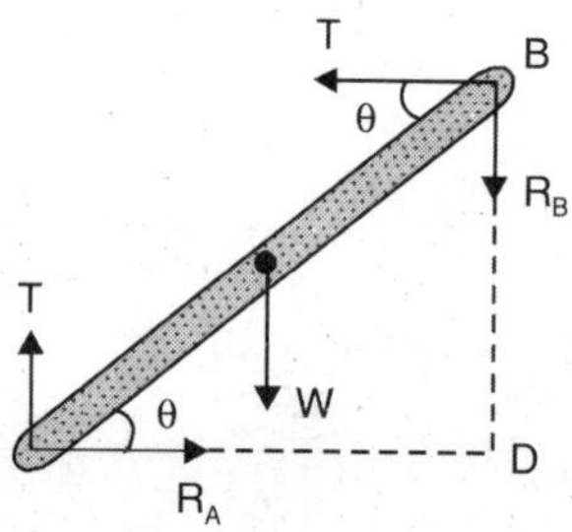

Fig. E7

$$\Sigma M_D = 0$$
$$w\left(\frac{L}{2}\right)\cos\theta + T(L\sin\theta) - T(L\cos\theta) = 0$$
$$T = \frac{W}{2(L\tan\theta)}$$

Put $T = 3\,W$

$$\theta = \mathbf{39.8°} \quad \ldots\textbf{Ans.}$$

Example 8

Solution

Consider joint D

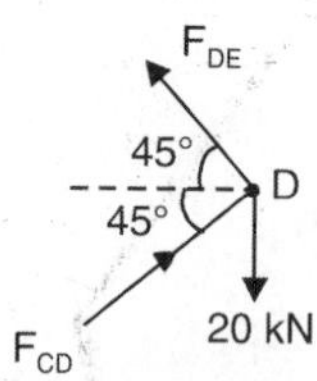

Fig. E8(a)

$$\Sigma F_x = 0$$
$$F_{DE} = F_{CD}$$
$$\Sigma F_y = 0$$

$F_{DE} \sin 45° + F_{CD} \sin 45° - 20 = 0$

$F_{CD} = 14.14$ kN (comp.)

$\mathbf{F_{DE} = 14.14}$ **kN (tensile)** **...Ans.**

Consider joint *E*

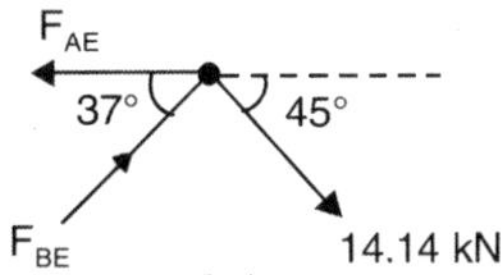

Fig. E8(b)

$\Sigma F_y = 0$

$-14.14 \sin 45° + F_{BE} \sin 37° = 0$

$F_{BE} = 16.67$ kN (comp.)

$\Sigma F_x = 0$

$- F_{AE} + 16.67 \cos 37° + 14.14 \cos 45° = 0$

$\mathbf{F_{AE} = 23.33}$ **kN (tensile)** **...Ans.**

Now apply same technique for joint *C*, *A* and *B*, we get the desire result

$\mathbf{F_{AC} = 16.67}$ **kN (tensile)**

$\mathbf{F_{BC} = 23.33}$ **kN (comp.)**

$\mathbf{A_x = 36.67}$ **kN ←**

$\mathbf{A_y = 10}$ **kN ↑**

$\mathbf{B_x = 36.67}$ **kN →**

$\mathbf{B_y = 10}$ **kN ↑** **...Ans.**

Example 9

Solution

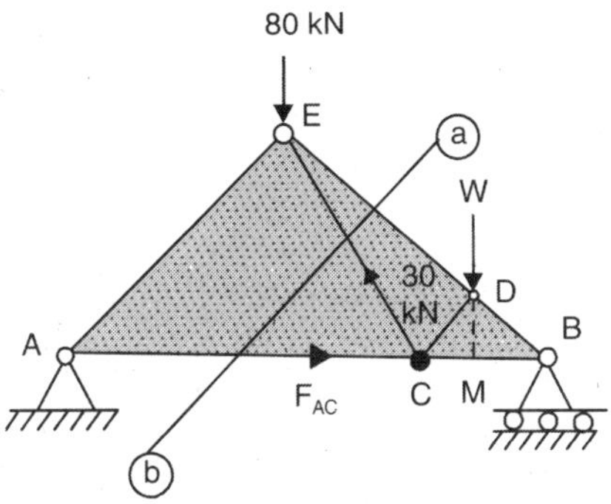

Fig. E9(a)

Applying condition of equilibrium to the right of section *a-b*

$\Sigma M_B = 0$

$W(BM) - 30 \sin 60° (CB) = 0$

$w (3) - 30 (\sin 60°) (4) = 0$

$\mathbf{w = 34.64}$ **kN** **...Ans.**

For complete truss

$$\Sigma M_B = 0$$

$$R_A(8) - 80\,(6) - 34.64\,(3) = 0$$

$$\mathbf{R_A = 73\ kN} \uparrow$$

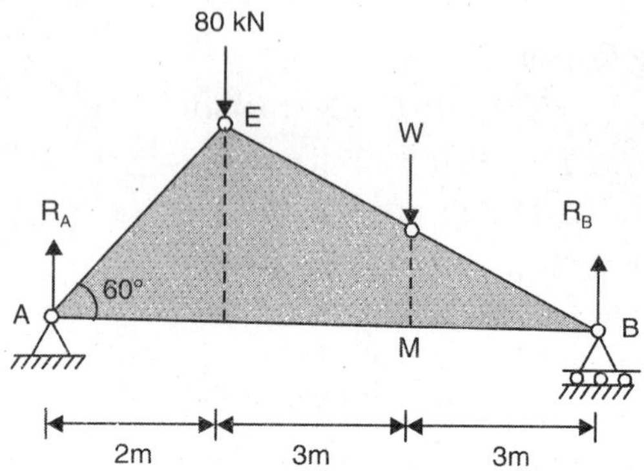

Fig. E9(b)

At joint A of truss

$$\Sigma F_y = 0$$

$$\mathbf{F_{AE} = 84.29\ kN\ (Comp)}$$

$$F_{AC} = \mathbf{42.15\ kN\ (Tensile)}$$ **...Ans.**

Level 2

Example 1

Solution

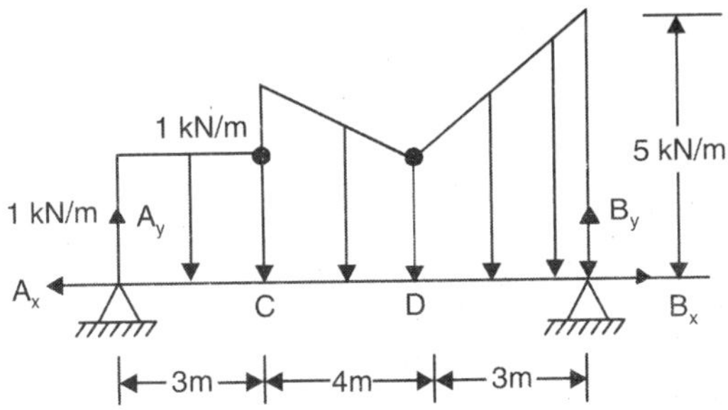

Fig. E1(a)

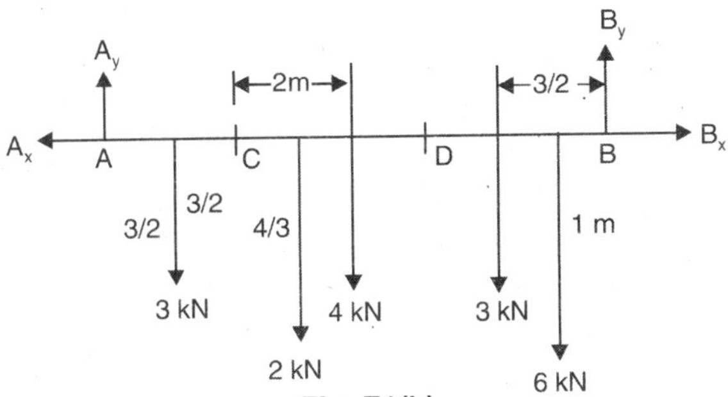

Fig. E1(b)

$$\Sigma M_A = 0$$

$$3\times\left(\frac{3}{2}\right)+2\times\left(\frac{13}{3}\right)+4\times(5)+3\times\left(\frac{17}{2}\right)+6\times(9)-B_y\times 10=0$$

$$\mathbf{B_y = 11.29\ kN}$$

$$\Sigma F_y = 0$$

$$A_y - 3 - 2 - 4 - 3 - 6 + B_y = 0$$

$$\mathbf{A_y = 6.74\ kN}$$

There is no horizontal force component acting along the AB so

$$\mathbf{A_x = B_x = 0} \qquad \textbf{...Ans.}$$

Example 2

Solution

Area under such parabola is $\frac{1}{3}bh$.

$$\text{Where } b = \text{base of the parabola}$$
$$h = \text{height of the parabola}$$

Here
$$A = \frac{1}{3}bh = \frac{1}{3} \times 4 \times 1500 = 2000$$

Centroid of parabola lies $\frac{3}{4}(b)$ distance from zero intensity point

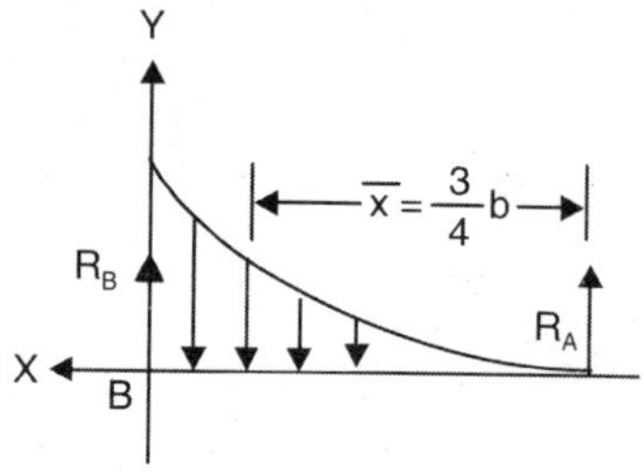

Fig. E2(a)

$$\Sigma M_A = 0$$
$$2000 \times \left(\frac{3}{4} \times 4\right) - 4 \times R_B = 0$$
$$\mathbf{R_B = 1500\ N} \uparrow$$
$$\mathbf{R_A = 500\ N} \uparrow$$

...Ans.

Example 3

Solution

FBD of the cable ABCDEF

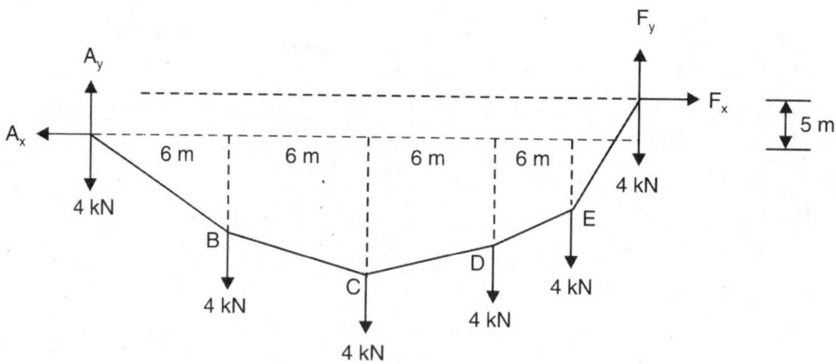

Fig. E3(a)

Applying condition of equilibrium

$$\Sigma F_x = 0$$
$$F_x - A_x = 0 \qquad ..(i)$$

$$\Sigma F_y = 0$$

$$A_y + F_y - 24 = 0 \qquad \ldots(ii)$$

$$\Sigma M_A = 0$$

$$5\,(F_x) + 4\,(6) + 4\,(12) + 4\,(18) + 4\,(24) + 4\,(30) - 30\,(F_y) = 0$$

$$F_x = 6\,F_y - 72 \qquad \ldots(iii)$$

Sag distance is known at point *c* so we can take moment about this point to the right of *C*.

$$\Sigma M_C = 0$$

$$4(6) + 4\,(12) + 4(18) + 12(F_x) - 18\,(F_y) = 0$$

$$F_x = 1.5\,F_y - 12 \qquad \ldots(iv)$$

Now concentrate to the right of *D*

$$\Sigma M_D = 0$$

$$F_x(h) + 4\,(6) + 4\,(12) - F_y(12) = 0 \qquad \ldots(v)$$

using these five Equation we can get

$$F_x = 8 \text{ KN} \rightarrow F_y = 13.33 \text{ KN} \uparrow$$

$$A_x = 8 \text{ KN} \leftarrow A_y = 10.667 \text{ KN} \uparrow$$

T_{max} is observed by value of θ_{max} at *A* or *F*.

$$\theta_A = \tan^{-1}\left(\frac{6.667}{8}\right) = 39.8°$$

$$\theta_F = \tan^{-1}\left(\frac{9.333}{8}\right) = 49.4°$$

$$\theta_A < \theta_F \text{ so } T_A < T_F$$

$$T_F = T_{max} = \sqrt{(9.33)^2 + (8)^2} = \mathbf{12.29\ KN} \qquad \textbf{...Ans.}$$

Example 4

Solution

Due to symmetrical loading

$$R_A = R_H = 16 \text{ kN} \uparrow$$

Concentrate right part of *DE* of structure

We can calculate value of θ as

$$\theta = \tan^{-1}\left[\frac{6.72}{(10.54 + 12.5)}\right] = 16.26°$$

$$\therefore \alpha = 180° - 2\,(90 - \theta) = 32.52°$$

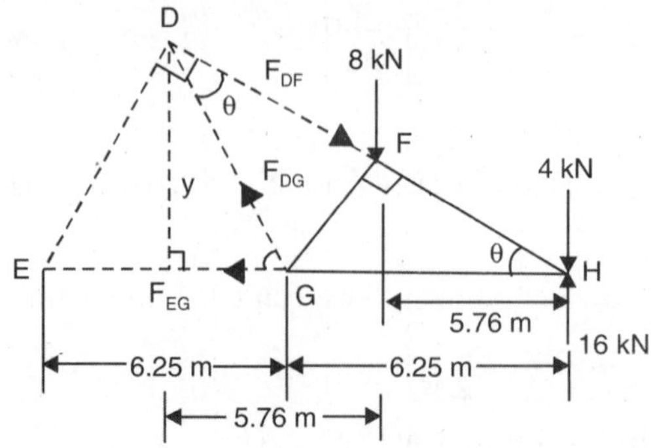

Fig. E4(a)

Now $\Sigma M_H =$

$$F_{DG} (\sin \alpha) (6.25) - 8 (5.76) = 0$$

$$\mathbf{F_{DG} = 13.71 \text{ kN (tensile)}} \quad \text{...Ans.}$$

Apply

$$\Sigma M_D = 0$$

$$8(5.76) + 4 (5.76 \times 2) - 16 (5.76 \times 2) + F_{EG} (y) = 0$$

$$y = 3.36 \text{ m by trigonometry}$$

$$F_{EG} = 27.43 \text{ kN (tensile)}$$

$$\text{Now apply } \Sigma M_G = 0$$

$$\Sigma F_y = 0 \text{ at joint } H$$

$$\Sigma F_x = 0 \text{ at joint } H$$

$$\Sigma F_y = 0 \text{ at joint } G$$

we can get our desired result

$$\mathbf{F_{DF} = 40.62 \text{ kN (comp)}}$$

$$\mathbf{F_{HF} = 42.86 \text{ kN (comp)}}$$

$$\mathbf{F_{GH} = 41.14 \text{ kN (tensile)}}$$

$$\mathbf{F_{FG} = 7.86 \text{ kN (comp.)}} \quad \text{...Ans.}$$

Example 5

Solution

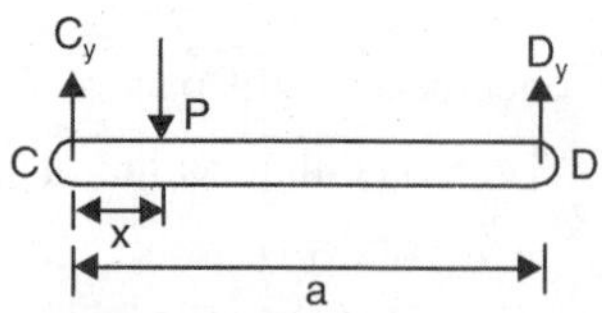

Fig. E5(a)

For FBD member *CD*

$$\Sigma M_C = P(x) - D_y (a) = 0$$

$$D_y = P\left(\frac{x}{a}\right)\uparrow$$

For complete system

$$\Sigma M_A = 0$$

$$\Sigma F_y = 0$$

$$R_B = P\left(\frac{x}{a}\right)\uparrow$$

$$R_A = \frac{P(ax)}{a}\uparrow$$

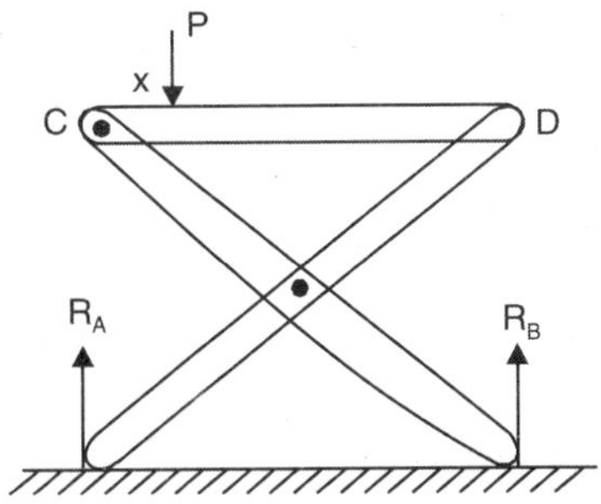

Fig. E5(b)

For FBD member *AD*

$$\Sigma M_E = 0$$

$$-D_x\left(\frac{a}{2}\right) - D_y\left(\frac{a}{2}\right) - R_A\left(\frac{a}{2}\right) = 0$$

$$-D_x = R_A + D_y$$

$$= P\left(\frac{a-x}{a}\right) + P\left(\frac{x}{a}\right) = P$$

$$D_x = -P$$

$$\Sigma F_y = 0$$

$$-E_y + R_A - D_y = 0$$

$$P\left(\frac{ax}{a}\right) + P\left(\frac{x}{a}\right) = E_y$$

$$P\left(1 - \frac{2x}{a}\right) = E_y$$

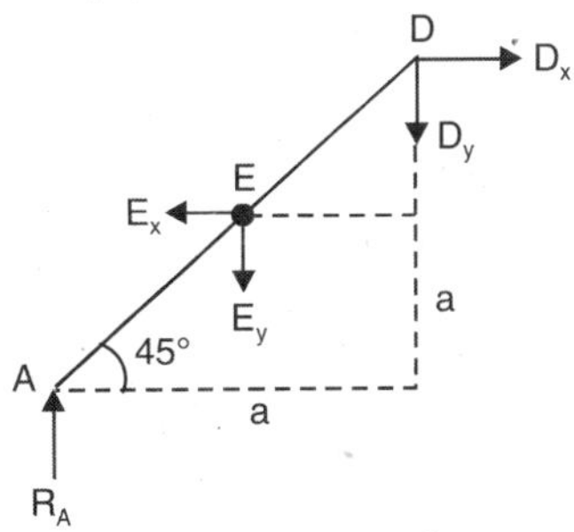

Fig. E5(c)

$$E_y = P\left(\frac{a-2x}{a}\right)$$

Force on point *E*

$$R_E^2 = E_x^2 + E_y^2 = P^2 + \frac{P^2}{a^2}(a-2x)^2$$

$$= P^2 + \frac{P^2}{a^2}\left(a^2 - 4ax + 4x^2\right)$$

$$\mathbf{R_E = P\sqrt{1 + \left(1 \frac{2x}{a}\right)^2}} \qquad \text{...Ans.}$$

R_E to be maximum

$$\frac{dR_E}{dx} = 0 \text{ gives}$$

$$x = 0 \text{ or } x = a$$

and $$\mathbf{R_{E(max)}} = \mathbf{P\sqrt{2}}$$ **...Ans.**

Example 6

Solution

For just stable without load moments about other rail must be balanced.

$$Q \cdot x - 200\,(3 + 2) = 0 \qquad \ldots(i)$$

For just stable with load moments about inner rail must be balanced.

$$Q\,(3 + x) - 200\,(2) - 400\,(2 + 8) = 0 \qquad \ldots(ii)$$

By equations (*i*) and (*ii*) we get the result as

$$Q = 1133.33 \text{ kN}$$

$$X = 0.882 \text{ m}$$ **...Ans.**

Example 7

Solution

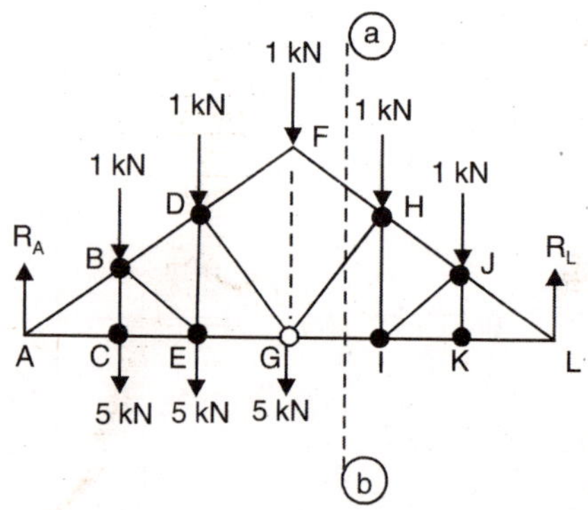

Fig. E7(a)

For overall structure

$$\Sigma M_A = 0 \text{ gives}$$

$$R_L = 7.5 \text{ kN}$$

We are only concentrating the structure right of *a-b* because it includes all members that are subjected in problem statement. Concentrating force in the member FH.

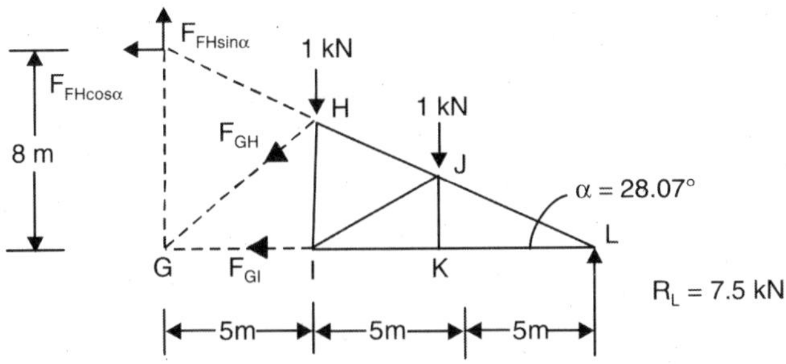

Fig. E7(b)

$$\text{Taking } \Sigma M_G = 0$$

$$(F_{FH} \cos \alpha)(8) - (1)(5) - (1)(10) + (7.5)(15) = 0$$

$$\mathbf{F_{FH} = -13.81 \text{ kN}} \qquad \text{...Ans.}$$

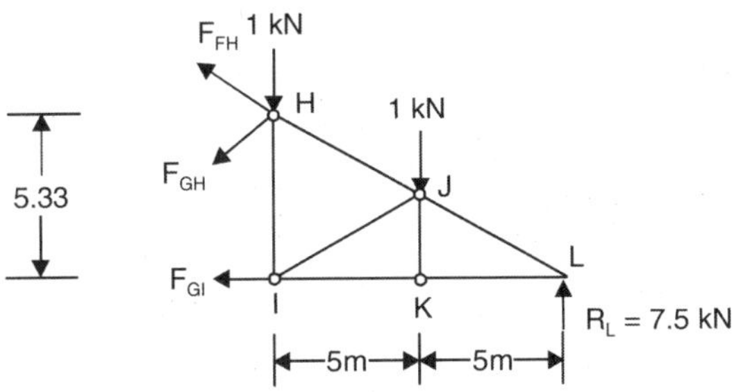

Fig. E7(c)

$$\Sigma M_H = 0$$

$$F_{GI} = 13.13 \text{ kN} \qquad \text{...Ans.}$$

Example 8

Solution

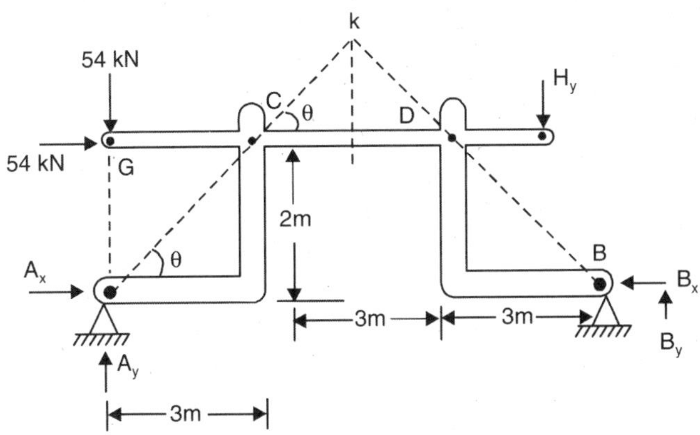

Fig. E8(a)

Applying equation of equilibrium for the frame

$$\Sigma M_K = 0$$
$$\Sigma M_A = 0$$
$$\Sigma F_y = 0$$
$$\mathbf{H_y = 66\ KN \downarrow}$$
$$\mathbf{A_y = 42\ KN \uparrow}$$
$$\mathbf{B_y = 78\ KN \uparrow}$$
$$\mathbf{B_x = 117\ KN \rightarrow}$$

...Ans.

Example 12

Solution

Here in the problem figure load is distributed on both side of pinned point. After opening the pinned point beam *AB* and *BC* will look like this

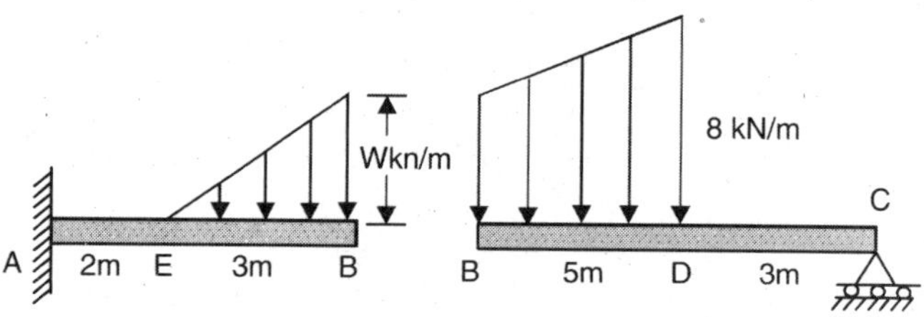

Fig. E12(a)

Now we have to find the value of *W* kN/m

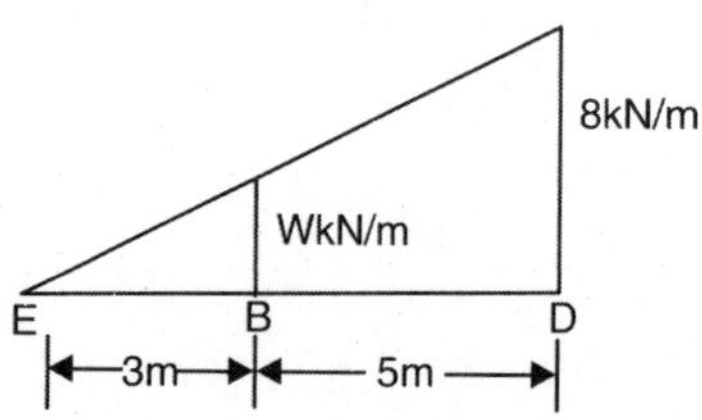

Fig. E12(b)

From similar triangle concept

$$\frac{W}{3} = \frac{8}{8}$$
$$\therefore W = 3 \text{ kN/m}$$

FBD of part *AB*

and for part *BC*

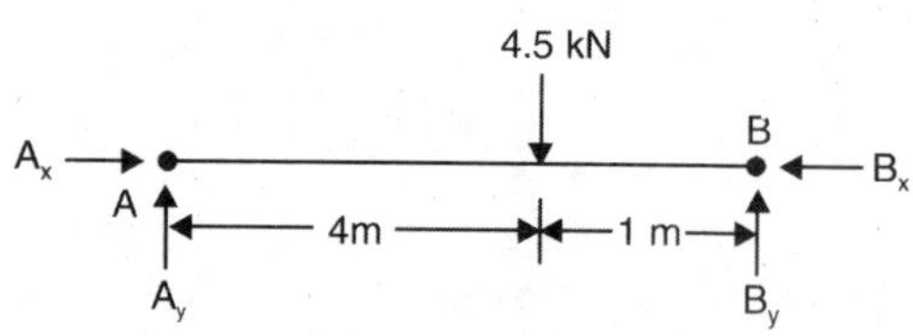

Fig. E12(c)

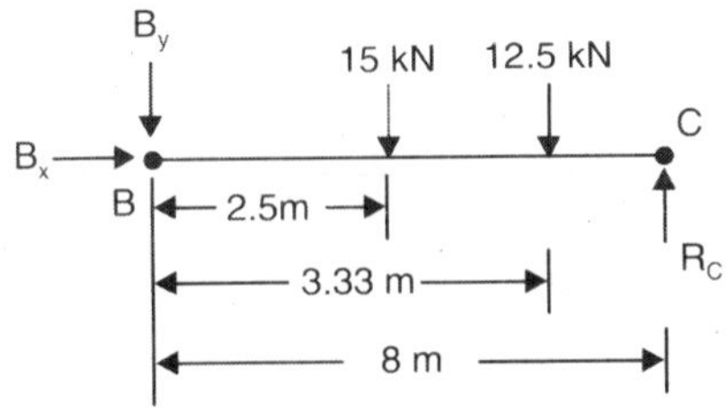

Fig. E12(d)

Concentrate on beam BC

Applying equations of equilibrium

$$\Sigma F_x = 0$$
$$B_x = 0$$
$$\Sigma F_y = 0$$
$$-B_y - 15 - 12.5 + R_C = 0 \qquad \ldots(i)$$
$$\Sigma M_B = 0$$
$$-(15 \times 2.5) - (12.5 \times 3.33) + (R_C \times 8) = 0 \qquad \ldots(ii)$$

for beam AB

$$\Sigma F_x = 0$$
$$A_x = B_x$$
$$\Sigma F_y = 0$$
$$A_y - 4.5 + B_y = 0 \qquad \ldots(iii)$$
$$\Sigma M_A = 0$$
$$M - (4.5 \times 4) + (B_y \times 5) = 0 \qquad \ldots(iv)$$

Using all above equation

We get

$$A_x = B_x = 0$$
$$R_C = 9.89 \text{ kN} \uparrow$$
$$B_y = -17.61 \text{ kN}$$
$$M = 106.05 \text{ kN.m}$$
$$\mathbf{A_y = 22.11 \text{ kN} \uparrow} \qquad \textbf{...Ans.}$$

Example 13

Solution

FBD of the entire beam

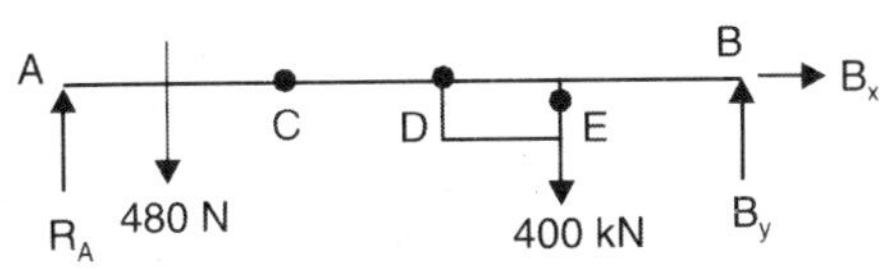

Fig. E13(a)

Applying condition of equilibrium

$$\Sigma M_A = 0$$

$$\text{By } (32) - 480\,(6) - 400\,(22) = 0$$

$$\therefore B_y = 365 \text{ N}$$

$$\Sigma F_x = 0$$

$$B_x = 0$$

$$\Sigma F_y = 0$$

$$R_A - 480 - 400 + B_y = 0$$

$$R_A - 480 - 400 + 365 = 0$$

$$\therefore R_A = 515 \text{ N}$$

Now for shear force diagram

<table>
<tr><th>A to C</th><th>C to D</th></tr>
<tr><td>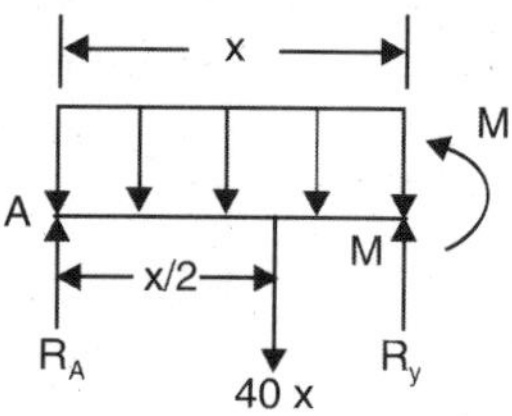

Fig. 13(b)</td><td>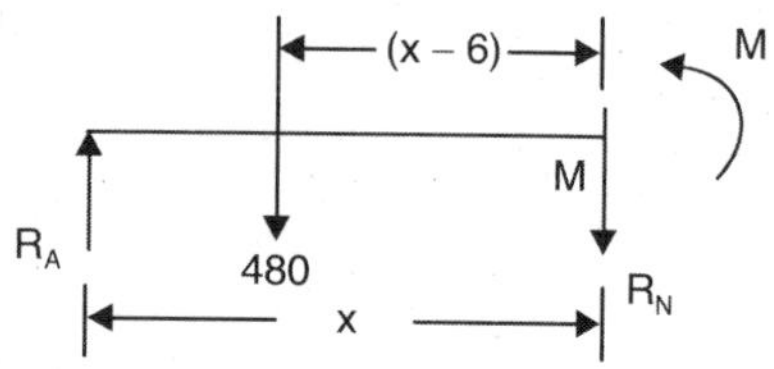

Fig. 13(c)</td></tr>
<tr><td>$\Sigma F_y = 0$
$515 - 40\,x - R_y = 0$
$R_y = 515 - 40x$ ...(i)</td><td>$\Sigma F_y = 0$
$515 - 480 - R_N = 0$
$R_N = 35$ N ...(iii)</td></tr>
<tr><td>$\Sigma M_M = 0$
$M + (40\text{x})\left(\dfrac{x}{2}\right) - 515\,x = 0$
$M = 515\,x - 20x^2$ ...(iii)</td><td>$\Sigma M_N = 0$
$M + 480\,(x - 6) - 515\,x = 0$
$M = (2880 + 35x)$ ...(iv)</td></tr>
</table>

For *D* to *B*

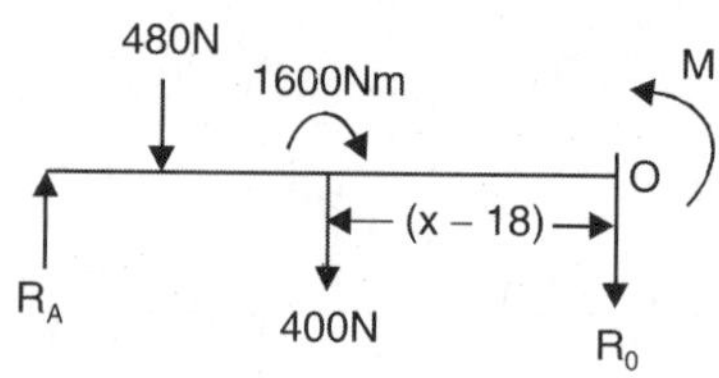

Fig. E13(d)

$$\Sigma F_y = 0$$

$$515 - 480 - 400 - R_0 = 0$$

$$R_0 = -365 \text{ N} \qquad \ldots(v)$$

$$\sum M_0 = 0$$

$$M + (400)(x - 18) + 480(x - 6) - 1600 - 515x = 0$$

$$M = (11680 - 365x) \qquad \ldots(vi)$$

Now shear force	**Bending moment**
$F_s(x) = \begin{cases} (515 - 40x)\text{N} & 0 < x < 12 \\ 35\text{N} & 12 < x < 18 \\ -365 \text{ N} & 18 < x < 32 \end{cases}$	$M_g(x) = \begin{cases} (515 - 20x^2) & 0 < x < 12 \\ (2880 + 35x) & 12 < x < 18 \\ (11680 - 365x) & 18 < x < 32 \end{cases}$
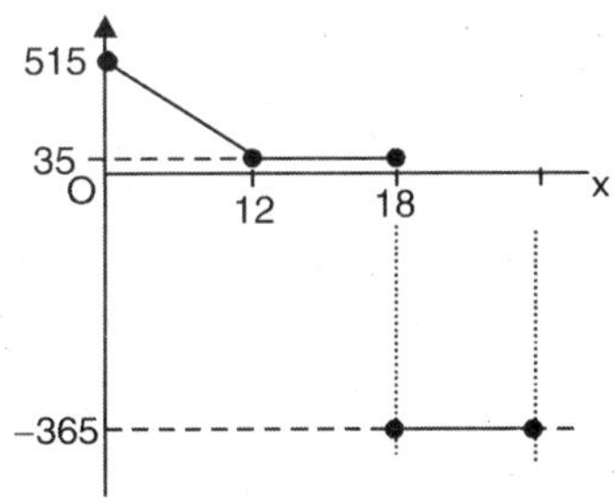 **Fig. E13(e)**	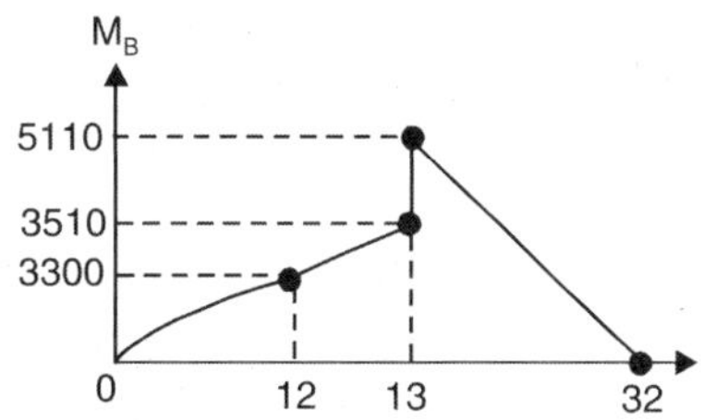 **Fig. E13(f)**

Level 3

Example 2

Solution

$$R_A = 40.06 \text{ kN}\downarrow \quad R_B = 198.19 \text{ kN}$$
$$R_D = 172.02 \text{ kN}\uparrow \quad R_E = 33.67 \text{ kN}$$
$$R_F = 14.67 \text{ kN}\uparrow \quad W = 122.13 \text{ kN}\downarrow$$

Example 4

Solution

FBD of whole beam *ABDE*

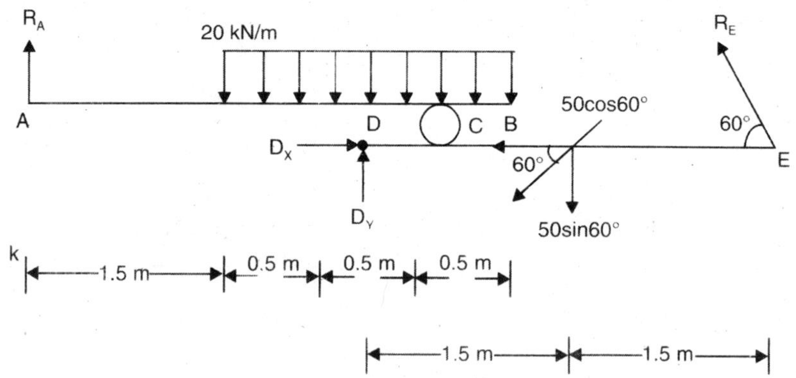

Fig. E4(a)

For *AB*

$$\Sigma M_A = 0$$

$$R_C \times (2.5) - 20 \times \left(1.5 + \left(\frac{1.5}{2}\right)\right) = 0$$

$$R_C = 27 \text{ KN} \uparrow$$

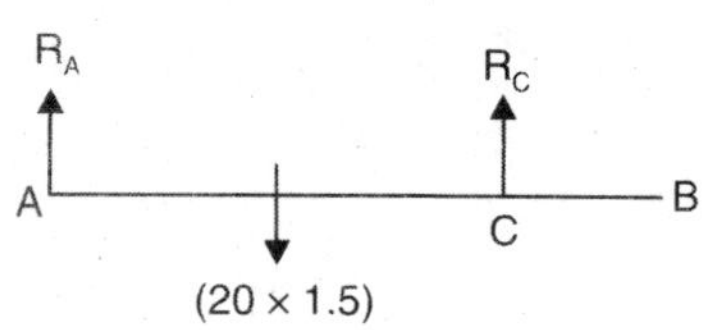

Fig. E4(b)

Now for beam *DE*

$$\Sigma M_D = 0, R_E = 30.2 \text{ KN}$$
$$\Sigma F_x = 0, D_x = 40.1 \text{ KN } (\rightarrow)$$
$$\Sigma F_y = 0, D_y = 44.15 \text{ KN} \uparrow$$

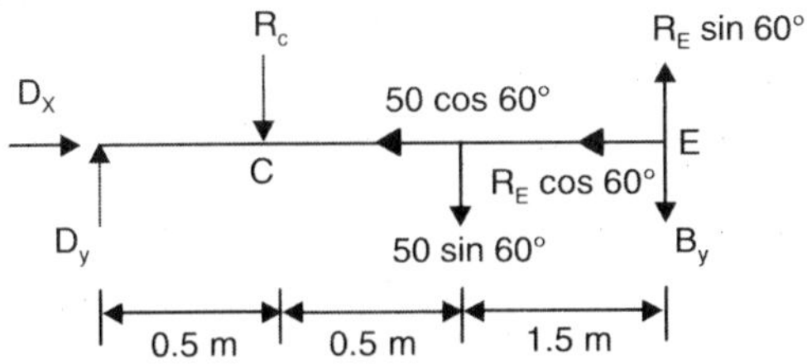

Fig. E4(c)

$$R_D = \sqrt{D_x^2 + D_y^2} = \sqrt{(40.1)^2 + (44.15)^2} = 59.69 \text{ KN}$$

$$\theta = \tan^{-1}\left(\frac{D_y}{D_x}\right) = \tan^{-1}\left(\frac{44.15}{40.1}\right)$$

$$= \mathbf{47.75°}$$ **...Ans.**

Example 5

Solution

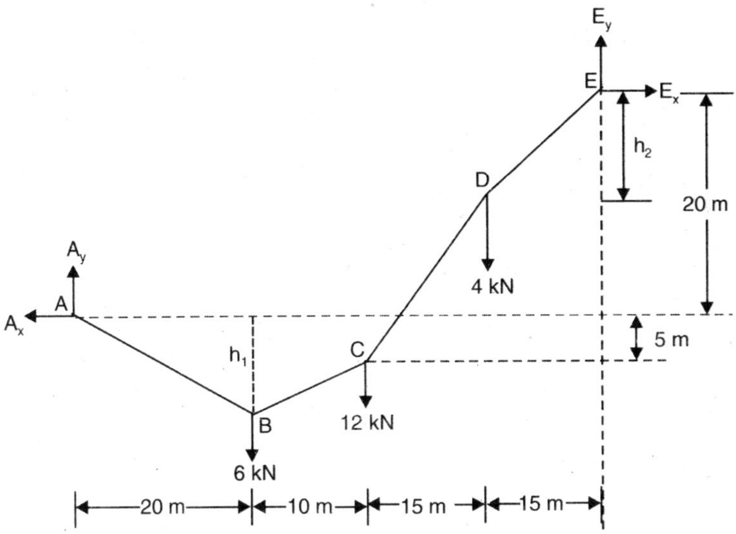

Fig. E5(a)

For entire cable

$$\Sigma M_A = 0$$

$$6 \times 20 + 12 \times 30 + 4 \times 45 + E_x \times 20 - E_y \times 60 = 0 \qquad ...(i)$$

$$\Sigma F_y = 0$$

$$A_y + E_y - 22 = 0 \qquad ...(ii)$$

Taking moment about '*C*' by considering right side part, we get

$$\Sigma M_C = 0$$

$$4 \times 15 + E_x \times 25 - E_y \times 30 = 0 \qquad \ldots(iii)$$

By using equation (*i*) and (*iii*)

$$E_x = 18\text{kN}$$

$$E_y = 17 \text{ kN}$$

$$T_{max} = E_x^2 + E_y^2$$

$$= (18)^2 + (17)^2 = 24.76 \text{ kN} \qquad \ldots\textbf{Ans.}$$

Taking moment about *B*

$$\Sigma M_B = 0$$

$$A_y \times 20 - A_x + h_1 = 0 \qquad \ldots(iv)$$

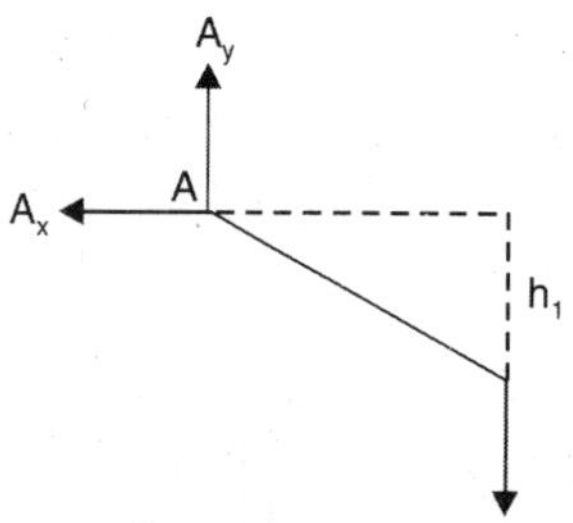

Fig. E5(b)

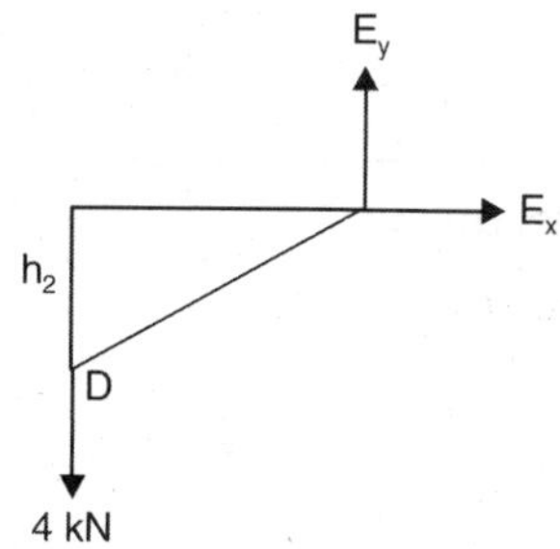

Fig. E5(c)

Taking moment about *D*

$$\Sigma M_D = 0$$

$$-E_x (h_2) + E_y (15) = 0 \qquad \ldots(v)$$

By using these equations we get

$$\textbf{Elevation}\Big|_{D} = \textbf{5.83 m}$$

$$\textbf{Elevation}\Big|_{B} = \textbf{5.56 m} \qquad \ldots\textbf{Ans.}$$

Example 6

Solution

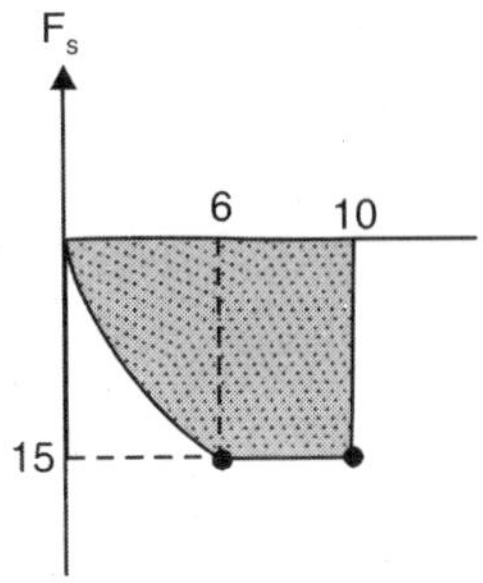

Fig. E6(a)

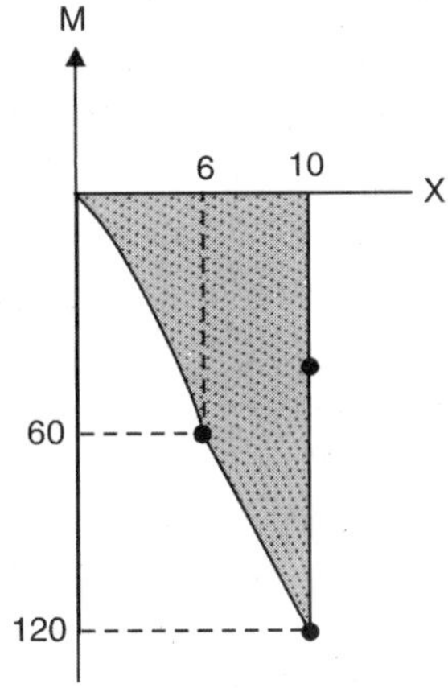

Fig. E6(b)

Example 7

Solution

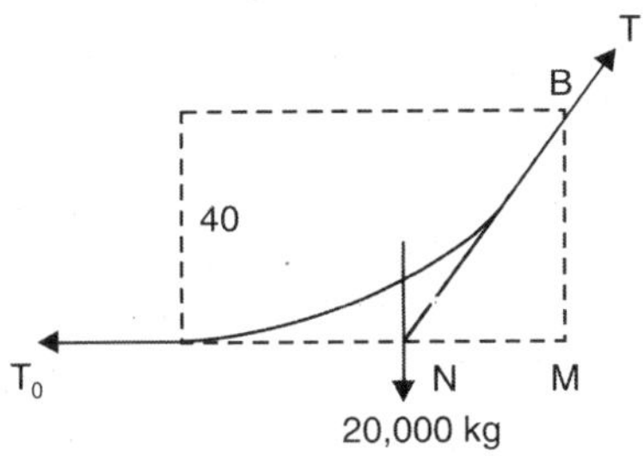

Fig. E7(a)

$$MN = 100 \text{ m}$$
$$BM = 40 \text{ m}$$
$$BN = \sqrt{(100)^2 + (40)^2} = 10\sqrt{116} = 107.7 \text{ m}$$

by force of triangle

$$\frac{T}{BN} = \frac{T_0}{MN} = \frac{20{,}000}{BM}$$
$$T_0 = 50{,}000 \text{ kg}$$
$$\mathbf{T = 53{,}850 \text{ kg}}$$

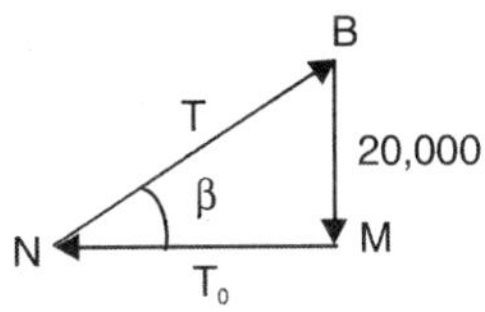

Fig. E7(b)

...Ans.

Example 8

Solution

We are concentrating on only the half part of cable, i.e. from **C** to **B**.

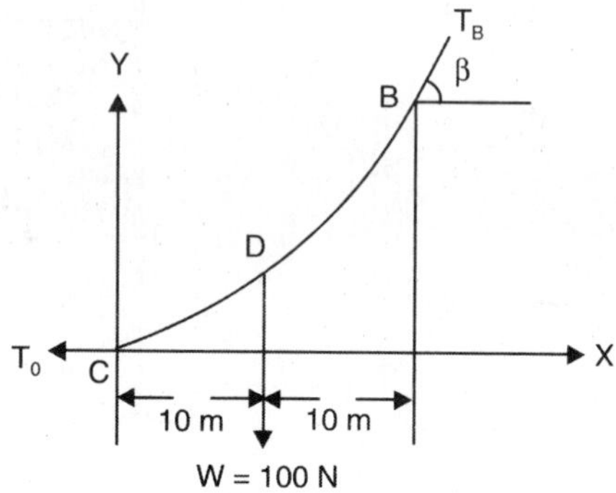

Fig. E8(a)

Taking moment about B

$$\Sigma M_B = 0$$

$$(100)\,(10) - T_0\,(0.5) = 0$$

$$T_0 = 2000 \text{ N}$$

On cable CB there are three forces acting on this so we can construct a force triangle

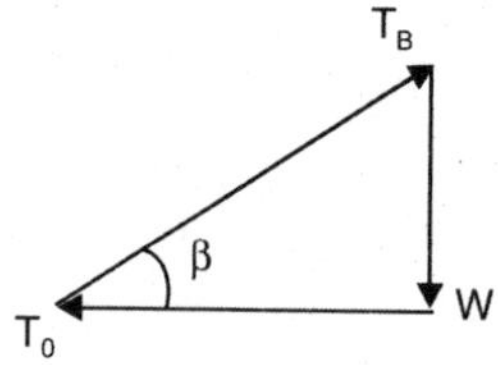

Fig. E8(b)

$$\tan\beta = \frac{W}{T_0} = \frac{100}{2000}$$

$$= \frac{1}{20} \quad \beta = 2.9°$$

$$\approx \mathbf{3°}$$

...Ans.

For sag distance

$$S = x\left[1+\frac{2}{3}\left(\frac{y}{x}\right)^2 + \ldots\right] = 20\left[1+\frac{2}{3}\left(\frac{0.5}{20}\right)^2 + \ldots\right]$$

Note:

$$S = \int_0^x \sqrt{1+\left(\frac{dy}{dx}\right)^2}\,dx$$

Total length = 2S = 2 × 20 = 40 m

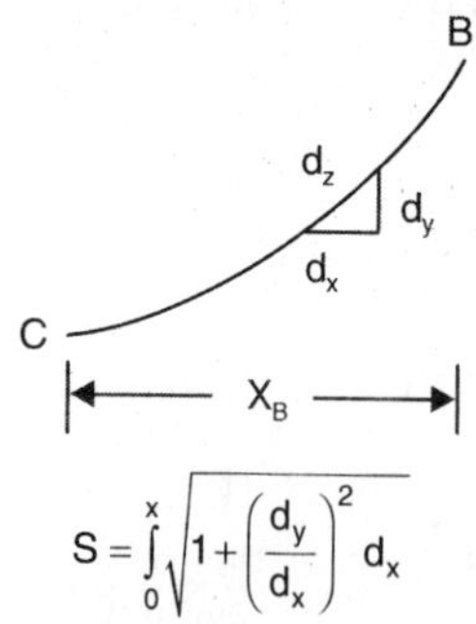

Fig. E8(c)

Example 9

Solution

FBD of the beam AE

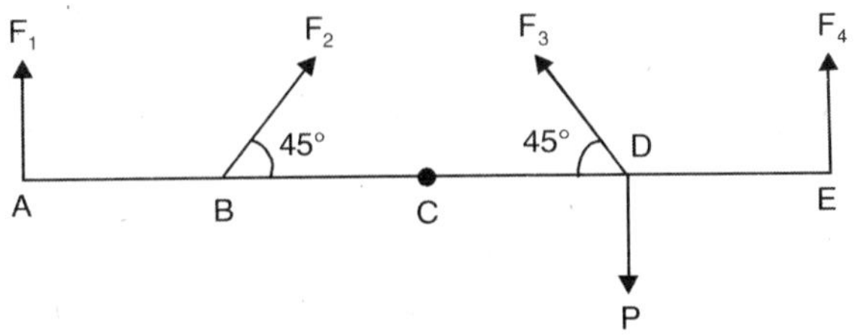

Fig. E9

Here C is a pinned point so we can take AC and CE parts separately.

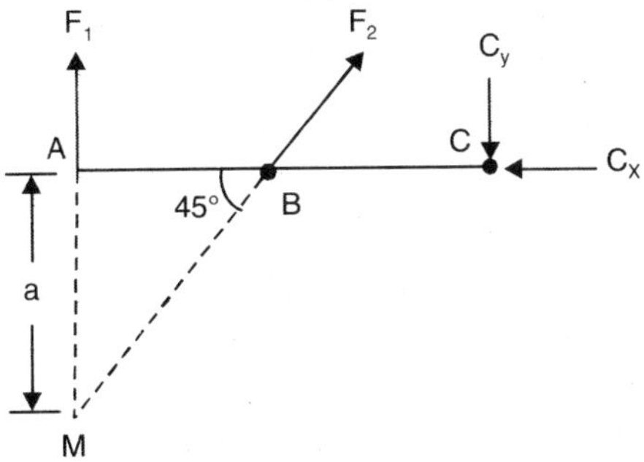

Fig. E9(a)

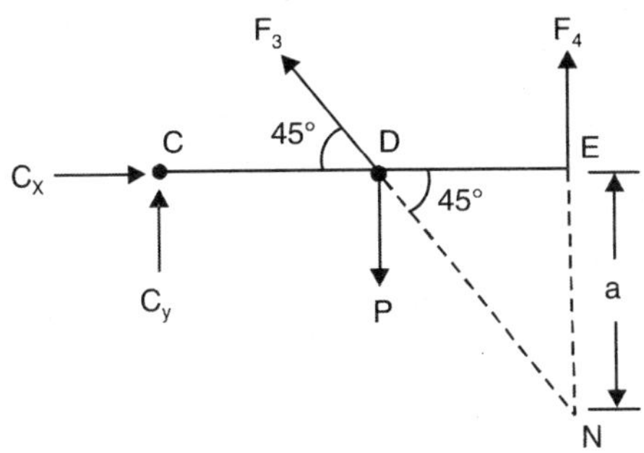

Fig. E9(b)

Concentrate on beam AC

$$\Sigma M_M = 0$$

$$C_x = 2\,C_y \qquad \ldots(i)$$

For beam *CE*

$$\Sigma M_N = 0$$

$$P(a) - C_y(2a) - C_x(a) = 0$$

$$2C_y + C_x = P \qquad \ldots(ii)$$

By equations (*i*) and (*ii*) we get

$$C_x = \frac{P}{2}$$

$$C_y = \frac{P}{4}$$

For beam *AC*

$$\Sigma F_x = 0$$

$$F_2 \cos 45° - C_x = 0$$

$$\frac{F_2}{\sqrt{2}} - \frac{P}{2} = 0$$

$$\therefore F_2 = \frac{P}{\sqrt{2}} \text{ (tensile)}$$

Similarly

$$\Sigma F_y = 0$$

$$T_1 = \frac{P}{2} \text{ (comp)}$$

For beam *CE*

$$\Sigma F_x = 0$$

$$C_x - F_3 \cos 45° = 0$$

$$F_3 = \frac{P}{\sqrt{2}} \text{ (tensile)}$$

$$\Sigma F_y = 0$$

$$\mathbf{F_4} = \frac{\mathbf{P}}{\mathbf{4}} \textbf{ (tensile)} \qquad \textbf{...Ans.}$$

Example 10

Solution

Taking moment at *B*

$$\Sigma M_B = A_y \times 18 - 400 \times 6 = 0$$

$$A_y = 133.33 \text{ N}\uparrow$$

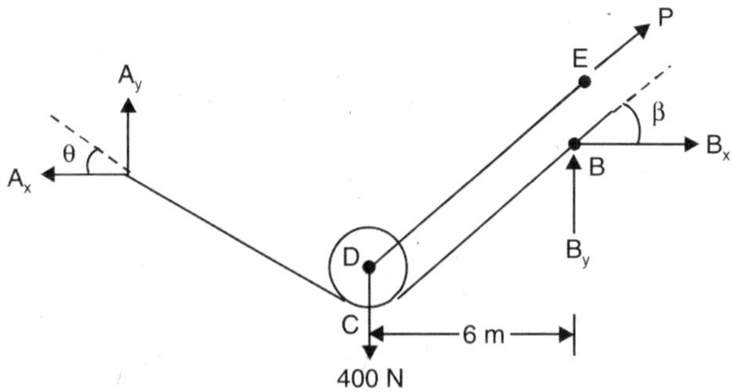

Fig. E10(a)

(P is not considered because of radius of pulley is negligible)

$$\text{Now, } \tan\theta = \frac{4.5}{12} = \frac{A_y}{A_x}$$

$$A_x = 355.5 \text{ N} \leftarrow$$

$$R_A = \sqrt{A_x^2 + A_y^2} = \sqrt{(355.5)^2 + (133.33)^2}$$

$$= \mathbf{379.68 \text{ N}} \quad \text{...Ans.}$$

$$\textbf{Clearly } \mathbf{R_A} = \mathbf{T_{ACB} = 379.68 \text{ N}} \quad \text{...Ans.}$$

Now at point B

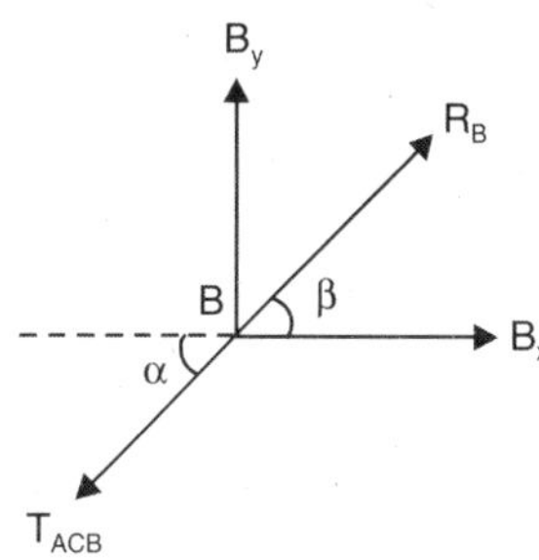

Fig. E10(b)

$$\tan\alpha = \frac{4.5}{6}$$

$$\alpha = 36.87°$$

$$\Sigma F_x = 0$$

$$B_x - 379.68 \cos 36.87° = 0$$

$$B_x = 303.74 \text{ N} \rightarrow$$

$$\Sigma F_y = 0$$

$$B_y + 379.68 \sin 36.87° = 0$$

$$B_y = 227.8 \text{ N} \uparrow$$

$$R_B = \sqrt{B_x^2 + B_y^2} = \sqrt{(303.74)^2 + (227.8)^2}$$

$$= 442.77 \text{ N}$$

$$\beta = \tan^{-1}\left|\frac{B_y}{B_x}\right| = \tan^{-1}\left(\frac{227.8}{379.68}\right)$$

$$= \mathbf{30.96°} \quad \text{...Ans.}$$

For finding the value of P, concentrate on the whole cable ACB,

$$\Sigma F_y = 0$$

$$P \sin 36.87° + 133.33 - 400 + 227.8 = 0$$

$$P = 64.78 \text{ N}$$ **...Ans.**

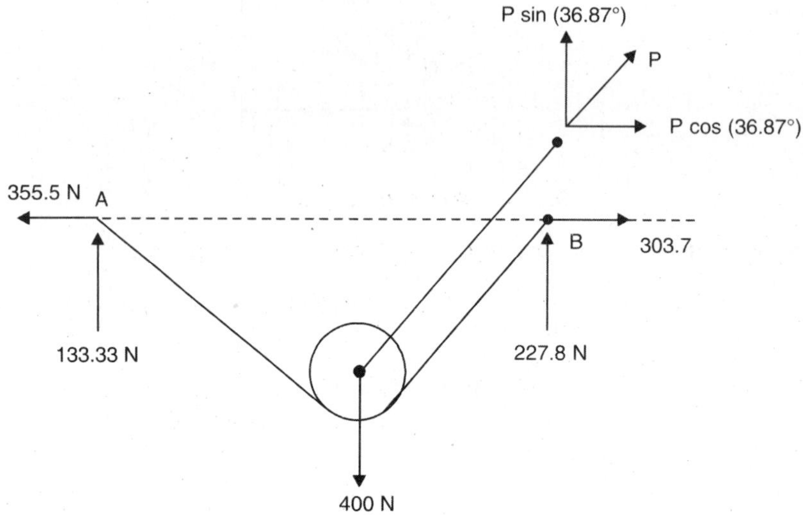

Fig. E10(c)

Virtual Work

Level 1

Example 1

Solution

Take origin at point M or on the roller × co-ordinate of point D because active force P is acting in the horizontal direction

$$x_D = 5\,l\cos\theta$$

$$dx_D = -5\,l\sin\theta\,d\theta$$

$$\Sigma dW = 0$$

$$+Md\theta + P(-5\,l\sin\theta)\,d\theta = 0$$

$$\mathbf{M = 5\,P\mathit{l}\sin\theta} \qquad \textbf{...Ans.}$$

Example 2

Solution

Just give the small vertical displacement of point C, the beam will look like this

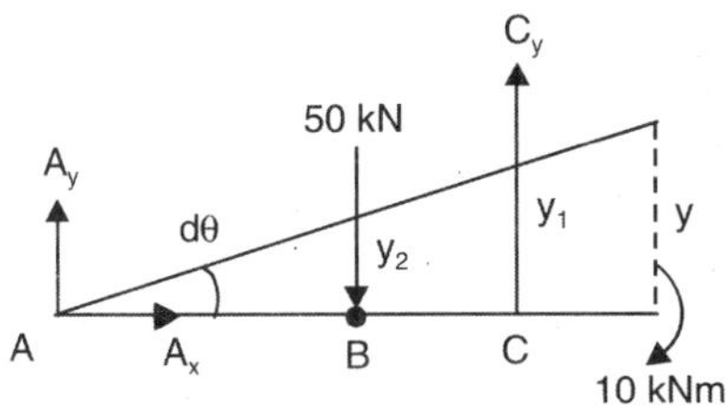

Fig. E2(a)

$$\Sigma dW = 0$$

$$-(50)(y_2) + (C_y)(y_1) - 10d\theta = 0$$

$$-50\left(\frac{3y}{8}\right) + (C_y)\left(\frac{6y}{8}\right) - 10\left(\frac{y}{8}\right) = 0$$

$$-150 + 6\,Cy - 10 = 0$$

$$C_y = 160$$

$$C_y = 26.67\text{ kN} \qquad \textbf{...Ans.}$$

Now consider virtual upward small displacement y at A keeping C in position.
The beam will look like this

$$\Sigma dW = 0$$

$$(A_y\, y) - 50\, y_1 + 10\, d\theta = 0$$

$$(A_y)(y) - 50\left(\frac{y}{2}\right) + 10\left(\frac{y}{6}\right) = 0$$

$$A_y = 25 - \frac{5}{3} = \frac{70}{3}$$

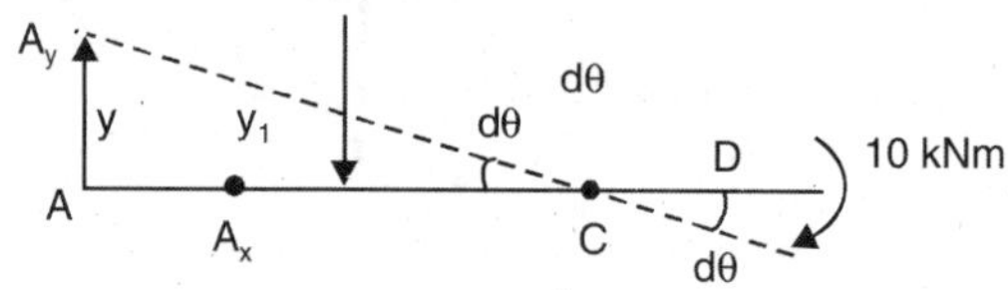

Fig. E2(b)

$$= \mathbf{23.33\ kN}$$ **..Ans.**

and $$A_x = 0$$

Example 3

Solution

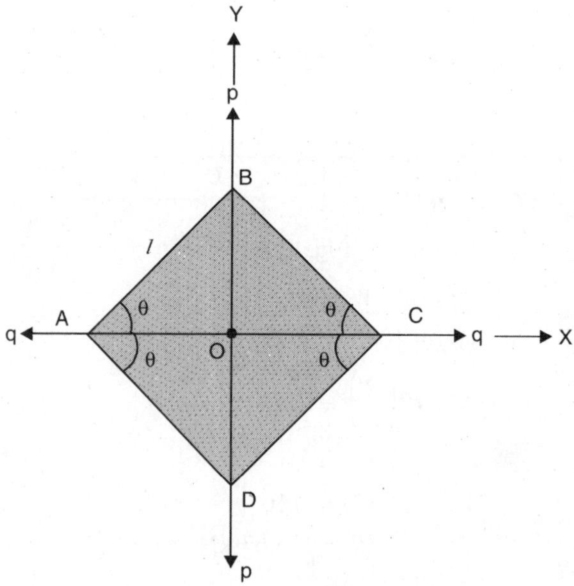

Fig. E3(a)

Taking origin at point 'O'

$$\Sigma dW = 0$$

$$(-q)(+l \sin\theta\, d\theta) + P(l \cos d\theta) + (+q)(-l \sin\theta\, d\theta) - P(-l \cos\theta\, d\theta) = 0$$

$$2q \sin\theta = 2P \cos\theta$$

$$\mathbf{\tan\theta = \frac{P}{q}} \quad \text{...Ans.}$$

Example 4

Solution

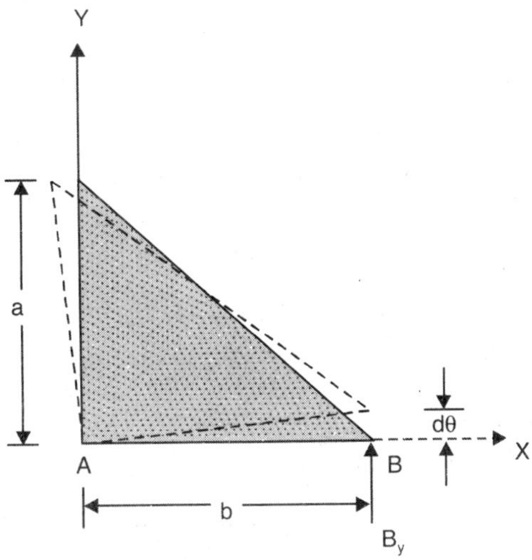

Fig. E4(a)

$$\Sigma dW = 0$$

$$-P(a d\theta) + B_y (b d\theta) = 0$$

$$Pa = B_y \cdot b$$

$$\mathbf{B_y = \frac{Pa}{b}} \quad \text{...Ans.}$$

Example 5

Solution

Let T_1 and T_2 be the tensions in the string of masses m_1 and m_2 respectively.

FBD of m_1 FBD of m_2

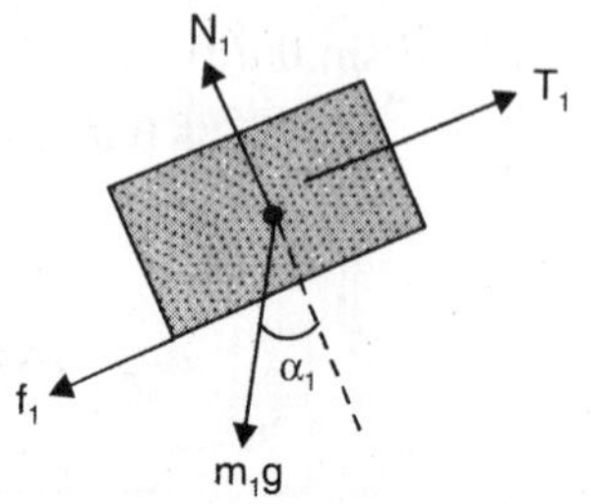

Fig. E5(a)

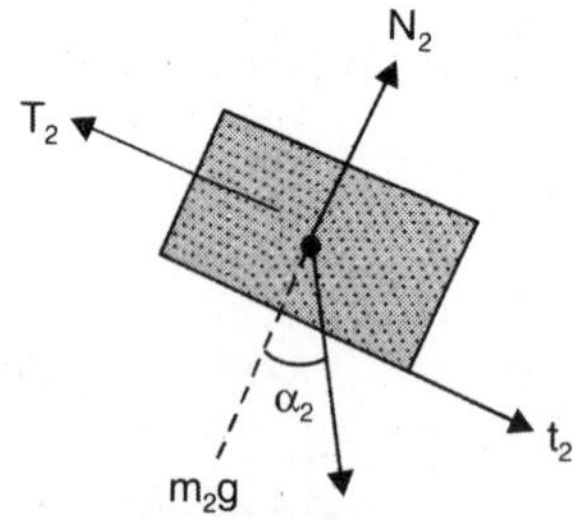

Fig. E5(b)

Give the small displacement dx along the plane up and down for m_1 and m_2 respectively clearly

$$T_1 = T_2$$

so it will not work.

$$\Sigma dW = 0$$

$$-(m_1\, g \sin \alpha_1)(dx) - (\mu_1\, m_1\, g \cos \alpha_1)(dx) + (m_2\, g \sin \alpha_2)(dx) - (\mu_2\, m_2\, g \cos \alpha_2)(dx) = 0$$

$$m_1 (\sin \alpha_1 + \mu_1 \cos \alpha_1) = m_2 (\sin \alpha_2 - \mu_2 \cos \alpha_2)$$

$$\frac{\sin \alpha_1 + \mu_1 \cos \alpha_1}{\sin \alpha_2 \mu_2 \cos \alpha_2} = \frac{m_2}{m_1}$$

$$\boldsymbol{\frac{\sin \alpha_1 + \mu_1 \cos \alpha_1}{\sin \alpha_2 + \mu_2 \cos \alpha_2} = \frac{m_2}{m_1}} \qquad \text{...Ans.}$$

Example 6

Solution

Give the small rotation of rod OA about point O in the clockwise direction then suppose A reaches A' and B reaches B'.

$$\text{Now let } AA' = dy$$

$$BB' = dx$$

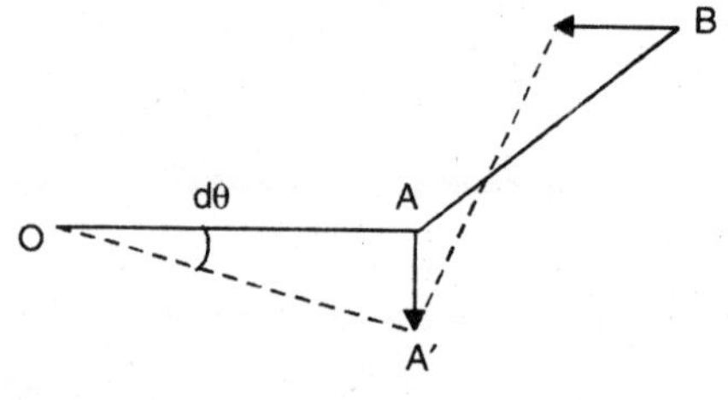

Fig. E6(a)

$$AA' = dy = ad\theta$$

$$BB' = dx$$

$$\Sigma dW = 0$$

$$Md\theta - Pdx = 0 \quad \text{...}(i)$$

Now use the relation

$$x^2 + y^2 = b^2$$

$$dx = -\left(\frac{y}{x}\right)dy = -\left(\frac{y}{x}\right)ad\theta$$

This relation use in the equation (i)

$$\mathbf{P} = \frac{\mathbf{Mx}}{\mathbf{ha}} \quad \textbf{...Ans.}$$

Example 7

Solution

Let dx and dy is the small displacement of B and A respectively.

Now,

By the virtual work principle of equilibrium

$$\Sigma dW = 0$$

$$F_1 \cdot d\,r_1 + F_2 \cdot dr_2 = 0$$

$$-Pdx - W\,dy = 0 \quad \text{...}(i)$$

$$\text{here } x = l \sin\theta$$

$$dx = l \cos\theta\, d\theta \quad \text{...}(ii)$$

$$y = \frac{l}{2}\cos\theta$$

$$dy = -\frac{l}{2}\sin\theta\, d\theta \quad \text{...}(iii)$$

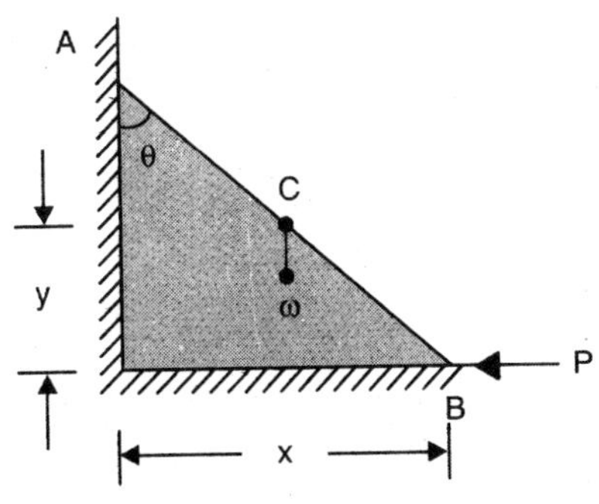

Fig. E7(a)

Using above all equations
We get

$$\mathbf{P} = \frac{1}{2}\mathbf{W} \tan\theta \qquad \text{...Ans.}$$

Example 9

Solution

Let AB be a reference axis for solving this problem.
Depth of G below the AB axis

$$y_G = (l\cos\theta + l_0)$$

$$\Sigma dW = 0$$

$$Md\theta + (-mg)\,dy_G = 0$$

$$Md\theta - mg\,(-l\sin\theta)\,d\theta = 0$$

$$M = mg\,l\sin\theta$$

$$\sin\theta = \frac{M}{mgl}$$

$$\sin\theta\Big|_{max} = 1 = \frac{M}{mgl}$$

$$\mathbf{M} = \mathbf{mg}l \qquad \text{...Ans.}$$

Example 10

Solution

Draw FBD of the rod AB

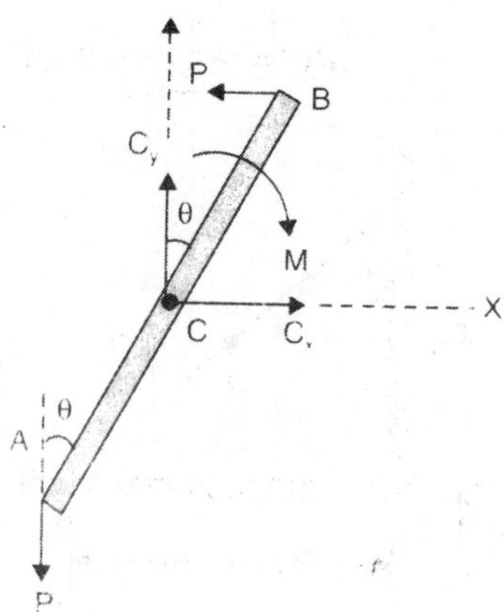

Fig. E10(a)

Let C be the origin

Co-ordinates of points A and B

$$A = (-l\sin\theta, -l\cos\theta)$$
$$B = (l\sin\theta, l\cos\theta)$$
$$\Sigma dW = 0$$
$$(-P)(l\sin\theta\, d\theta) + M\, d\theta$$
$$(-P)(l\cos\theta\, d\theta) = 0$$
$$\mathbf{M = Pl\,(\sin\theta + \cos\theta)} \quad \text{...Ans.}$$

Example 12

Solution

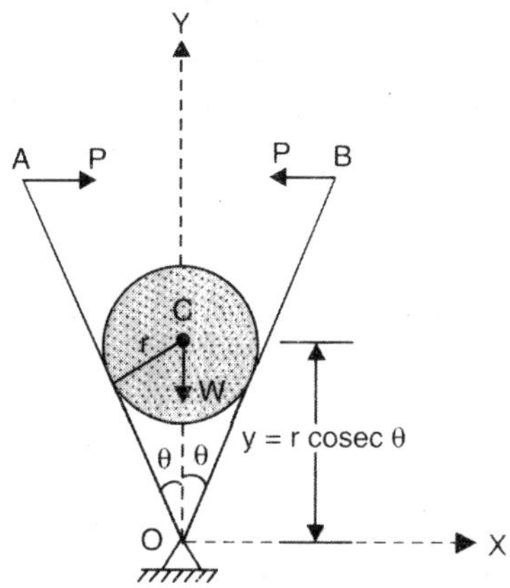

Fig. E12(a)

Using $\Sigma dW = 0$ for the figure drawn above

$$(+P)(-L\cos\theta\, d\theta) + (-P)(L\cos\theta\, d\theta) - W(-r\,\text{cosec}\,\theta\cot\theta\, d\theta) = 0$$

$$-2PL\cos\theta + Wr\frac{\cos\theta}{\sin^2\theta} = 0$$

$$2PL\sin^2\theta = Wr$$

$$\mathbf{P = \frac{Wr}{2L\sin^2\theta}} \quad \text{...Ans.}$$

Example 13

Solution

Let point C be an origin according to that write the y-co-ordinates of the active forces point of A and B.

Depth of A below point C

$$y_A = -a \tan \theta$$

Height of B above point C

$$y_B = L \sin \theta - a \tan \theta$$

$$\Sigma dW = 0$$

$$-P dy_B - q dy_A = 0$$

$$-P (L \cos \theta - a \sec^2 \theta)\, d\theta = q (-a \sec^2 \theta)\, d\theta$$

$$q = P \left(\frac{L \cos \theta}{a \sec^2 \theta} - 1 \right)$$

$$\mathbf{q = P \left(\frac{L \cos^2 \theta}{a} - 1 \right)} \quad \text{...Ans.}$$

Level 2

Example 1

Solution

Draw FBD of whole beams

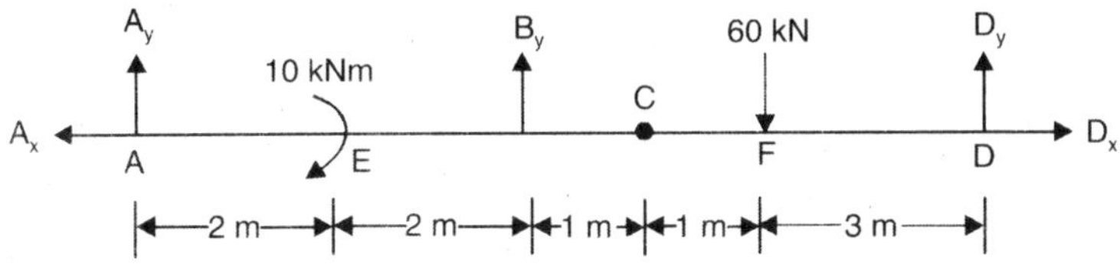

Fig. E1(a)

Step 1:

Give the virtual small displacement of rod *CD*.

$$\Sigma dW = 0$$

$$D_y \times y - 60 \times y_1 = 0$$

$$D_y \times y - 60 \times \frac{y}{4} = 0$$

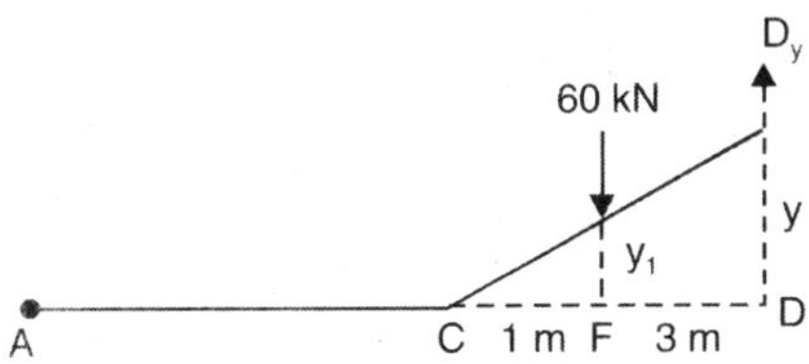

Fig. E1(b)

$$\mathbf{D_y = 15\ kN} \qquad \textbf{...Ans.}$$

Step 2:

$$\Sigma dW = 0$$

$$A_y \times y - R_B \times y_1 + 10\, d\theta = 0$$

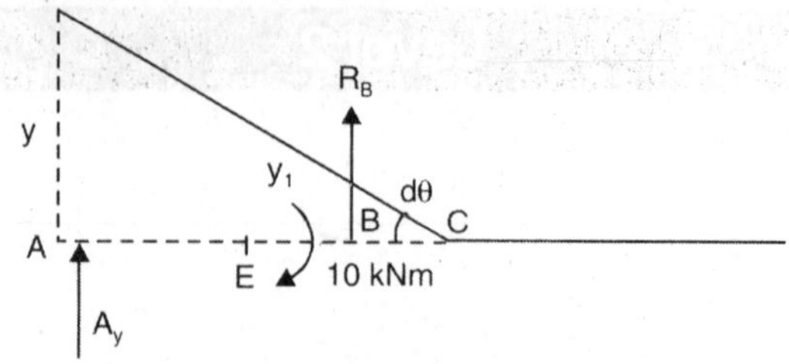

Fig. E1(c)

$$A_y y - R_B \frac{y}{2} + 10\frac{y}{4} = 0$$

$$2A_y - R_B + 5y = 0 \qquad \ldots(i)$$

Step 3:

Now give the vertical displacement y of pinned point C.

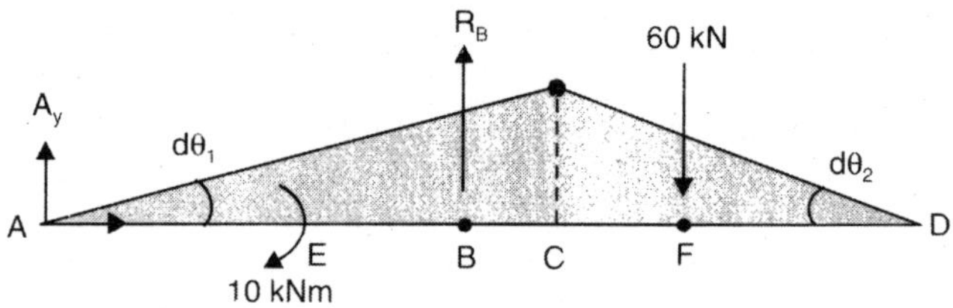

Fig. E1(d)

$$\Sigma dW = 0$$

$$-10d\theta_1 + RB\ y_2 - 60\ y_1 = 0$$

$$-10\left(\frac{y}{5}\right) + R_B\left(\frac{4y}{5}\right) - 60\left(\frac{3y}{4}\right) = 0$$

$$-2 + \left(\frac{4R_B}{5}\right) - 45 = 0$$

$$4R_B = 235$$

$$\mathbf{R_B = 58.75\ kN} \qquad \textbf{...Ans.}$$

Now from (*i*) we get

$$\mathbf{A_y = 13.75\ kN} \qquad \textbf{...Ans.}$$

There is no force in the horizontal direction so,

$$\mathbf{A_x = 0, D_x = 0} \qquad \textbf{...Ans.}$$

Example 2

Solution

Net work done by all active forces on the system is zero.

$$\Sigma dW = 0$$

here in figure

$$L = AB = BC = BD = 0.4 \text{ m}$$

So, by geometry
In ΔBCD

$$\angle BCD = \angle BDC = \alpha \text{ (say)}$$

$$\text{And } \theta + 2\alpha = 90°$$

$$d\theta = -2\, d\alpha \qquad \ldots(i)$$

actual length of the spring

$$L_0 = \sqrt{(0.4)^2 + (0.4)^2} = 0.566 \text{ m}$$

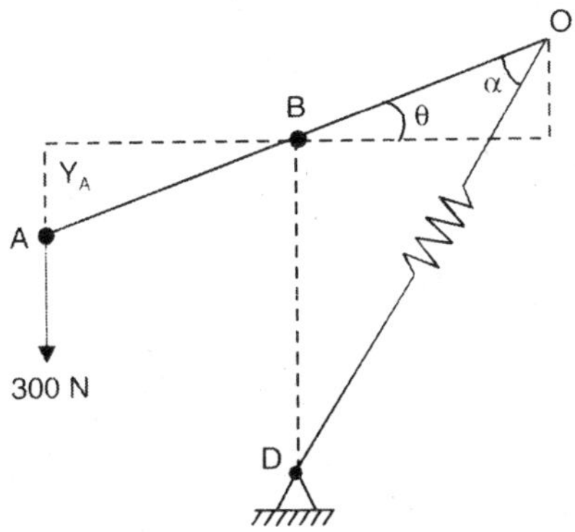

Fig. E2(a)

$$L_1 = DC = 2(0.4) \cos \alpha$$

Elongation in the spring

$$= L_1 - L_0$$

$$\Sigma dW = 0$$

Active forces are acting at point A and C that gives the small displacement. So,

$$\vec{F}_A \cdot d\vec{r}_A + \vec{F}_C \cdot d\vec{r}_C = 0$$

$$(300)\,(L \cos \theta\, d\theta) - (5000)\,(L_1 - L_0)\,(dL_1) = 0$$

$$L_1 = 0.8 \cos \alpha$$

$$dL_1 = 0.8 \sin \alpha \left(\frac{d\theta}{2}\right) \text{ by } (i) = +0.4 \sin \alpha\, d\theta$$

$$(300)\,(L \cos \theta\, d\theta) - (5000)\,(L_1 - L_0)\,(0.4 \sin \alpha\, d\theta) = 0 \quad (ii)$$

Using sine rule
In ΔBCD

$$\frac{L}{\sin\alpha} = \frac{L_1}{\cos\theta}$$

$$L\cos\theta = L_1 \sin\alpha \qquad \ldots(iii)$$

Using (*iii*) in (*ii*) we get

$$L_1 = 0.666 \text{ m}, L_0 = 0.566 \text{ m}$$

$$L_1 = 2L\cos\alpha$$

$$\text{We get } \alpha = \mathbf{33.6°}$$

$$\text{Sought angle } \theta = \mathbf{22.7°}$$

...Ans.

Example 3

Solution

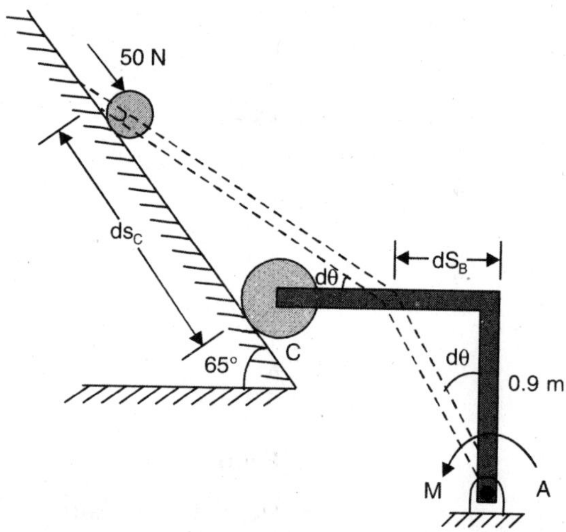

Fig. E3(a)

$$\text{Clearly } dS_B = dS_C \cos 65°$$

$$\text{And } 0.9\, d\theta = dS_C \cos 65° \qquad \ldots(i)$$

$$\Sigma dW = 0$$

$$\vec{F}_A \cdot d\vec{r}_A + \vec{F}_B \cdot d\vec{r}_B = 0$$

$$+ M\, d\theta - (50)\,(dS_C) = 0$$

By using (*i*)

$$dS_C = 2.13\, d\theta$$

$$M = 50(2.13)$$

$$\boldsymbol{M} = \mathbf{106.48 \text{ Nm}}$$

...Ans.

Example 4

Solution

Let O be the origin

$$y_A = L\cos\theta$$

$$x_B = r\sec\theta$$

$$\Sigma dW = 0$$

$$(-100)(dx_B) - 80(dy_B) = 0$$

$$(100)(L\sin\theta)\,d\theta = 80\,(r\sec\theta\tan\theta)\,d\theta$$

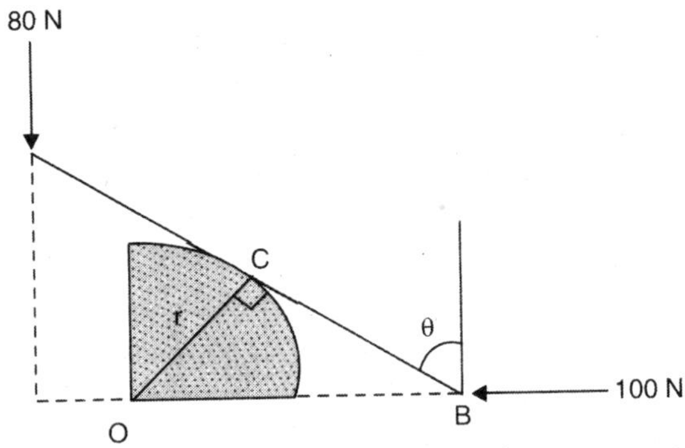

Fig. E4

$$\cos^2\theta = \frac{(100)(0.04)}{(80)(0.2)}$$

we get $\theta = \mathbf{60°}$ **...Ans.**

Example 5

Solution

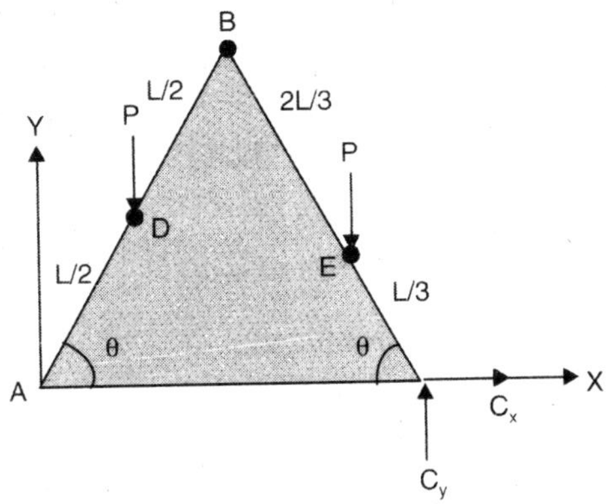

Fig. E5(a)

Assume hinge point A as the origin.
Use our lovely formula

$$\Sigma dW = 0$$

$$C \times d(2L\cos\theta\, d\theta) - Pd\left(\frac{L}{2}\sin\theta\, d\theta\right) - Pd\left(\frac{L}{3}\sin\theta\, d\theta\right) = 0$$

$$\mathbf{Cx = \frac{5}{12} P \cot\theta}$$ **...Ans.**

Example 6

Solution

Clearly take the origin about point E.
Now use

$$\begin{aligned} \Sigma dW &= 0 \\ Pd\,(3l\cos\theta) + q\,d(2l\sin\theta) &= 0 \\ -3P\sin\theta\, d\theta + 2q\cos\theta\, d\theta &= 0 \\ 3P &= 2q\cot\theta \\ \mathbf{\frac{2q}{3P}} &= \mathbf{\tan\theta} \end{aligned}$$ **...Ans.**

Example 7

Solution

FBD by beam

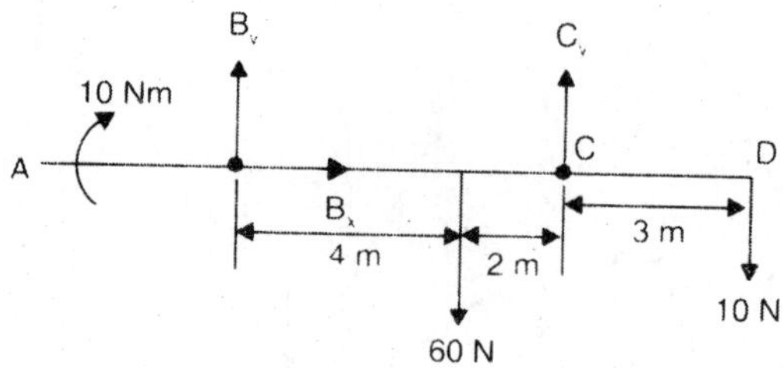

Fig. E7(a)

Give a small vertical displacement C, keeping B in position
The beam will look like this

$$\begin{aligned} \Sigma dW &= 0 \\ -10y + C_y y_1 - 60\, y_2 - 10\, d\theta &= 0 \end{aligned}$$

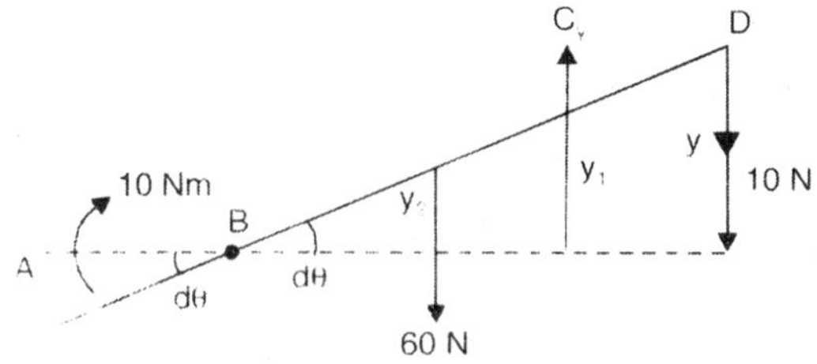

Fig. E7(b)

$$-10y + C_y\left(\frac{6y}{9}\right) - 60\left(\frac{4y}{9}\right) - 10\left(\frac{7}{9}\right) = 0$$

$$90 + 6\,C_y - 240 - 10 = 0$$

$$6\,C_y = 340$$

$$\mathbf{C_y = 56.66\ N}$$

Now for point B

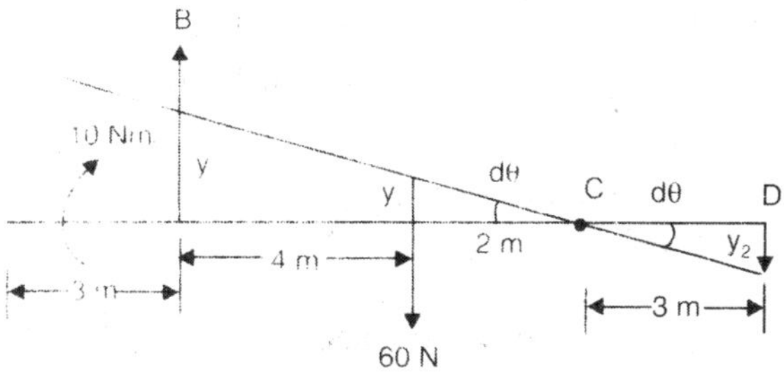

Fig. E7(c)

$$\Sigma dW = 0$$

$$(10\,x\,y_2) - (60\,y_1) + (B_y\,y) + 10\,d\theta = 0$$

$$10\,(3d\theta) - 60\,(2d\theta) + (6d\theta)\,B_y + 10\,(d\theta) = 0$$

$$30 - 120 + 6\,B_y - 10 = 0$$

$$6\,B_y = 100$$

$$\mathbf{B_y = 16.66\ N}$$ **...Ans.**

Example 10

Solution

Take the origin at point at A and draw the FBD of the whole frame.

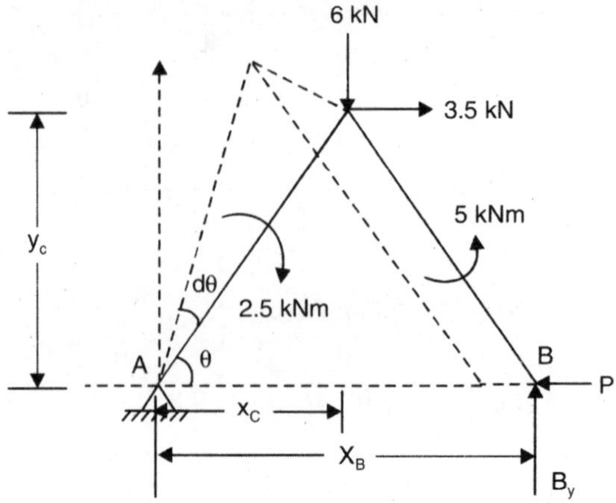

Fig. E10(a)

Give the virtual rotation of the frame

$$(-P)\ dx_B + 3.5\ dy_C - 6(dx_C) - 2.5\ d\theta - 5d\theta = 0$$

$$(-P)\ (-2L \sin\theta\ d\theta) + (3.5)\ (-L \sin\theta\ d\theta) - 6(L \cos\theta\ d\theta) - 2.5\ d\theta - 5\ d\theta = 0$$

$$2\ PL \sin\theta - 3.5\ L \sin\theta - 6\ L \cos\theta = 7.5$$

substitute $L = 0.6$, $\theta = 60°$

we get,

P = 10.7 kN **...Ans.**

Level 3

Example 1

Solution

Let AB be the rod of length $2l$ and G its center of gravity at a height y above the point C.

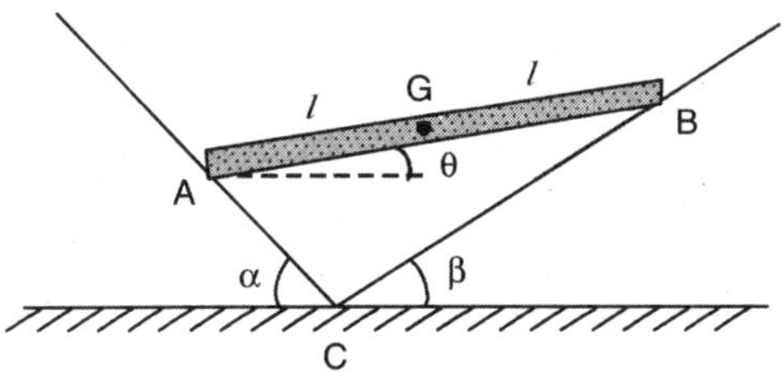

Fig. E2(a)

Clearly

$$y = l \sin \theta + AC \sin \alpha$$

$$= l \sin \theta + 2l \sin \alpha \frac{\sin(\beta+\theta)}{\sin(\alpha+\beta)}$$

$$dy = l \cos \theta \, d\theta - 2l \sin \alpha \frac{\cos(\beta+\theta)}{\sin(\alpha+\beta)} d\theta$$

But for equilibrium

$$\text{We get } \tan^{-1} \frac{\sin(\beta+\alpha)}{2\sin\alpha \sin\beta} \qquad \text{...Ans.}$$

Example 2

Solution

By geometry

$$x = R\cos\theta + l\cos\phi \qquad ...(i)$$

$$R\sin\theta = l\sin\phi \qquad ...(ii)$$

$$\Sigma\, dW = 0$$

$$-Md\theta - Fdx = 0 \qquad ...(iii)$$

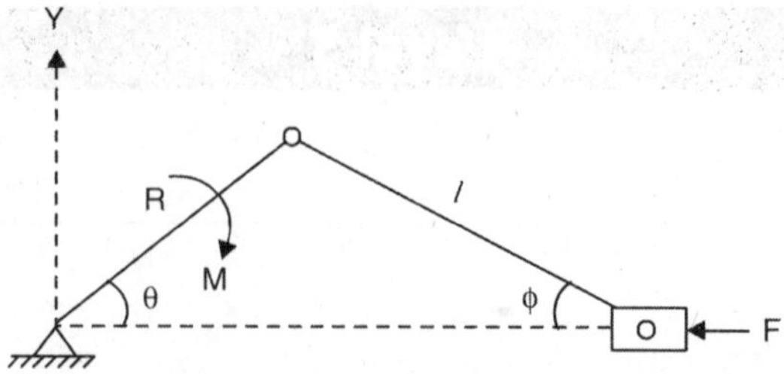

Fig. E3(a)

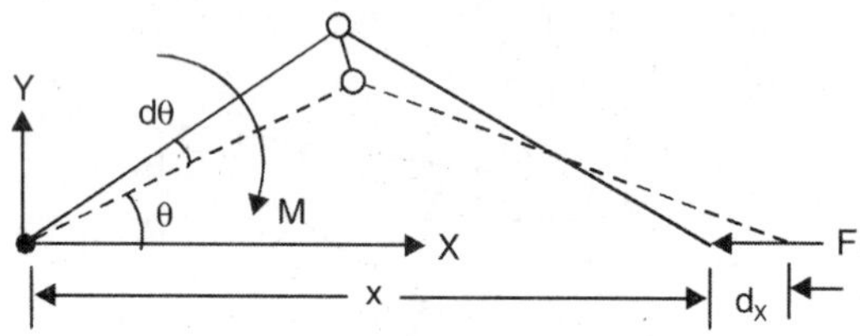

Fig. E3(b)

by (*i*)

$$dx = -R \sin d\theta - l \sin \phi \, d\phi \qquad ...(iv)$$

by (*ii*)

$$\cos^2 \phi = 1 - \left(\frac{R}{l}\right)^2 \sin^2 \theta \qquad ...(v)$$

using all above equations
we get,

$$\mathbf{M = FR\sin\theta\left[1 + \frac{R\cos\theta}{l\sqrt{1\left(\frac{R}{l}\right)^2 \sin^2\theta}}\right]} \qquad \textbf{...Ans.}$$

Example 3

Solution

Clearly here active forces are weight of the rod *AB*, *BC* and a external force acting at point *C* in horizontal direction
Let *A* be the origin
For *AB* rod

$$AE = EB = L$$

For *BC* rod

$$BD = DC = \frac{L}{2}$$

$$E = (L\cos\theta_1 - L\sin\theta_1)$$

$$D = \left(2L\cos\theta_1 + \frac{L}{2}\cos\theta_2, (2L\sin\theta_1 + L/2\sin\theta_2)\right)$$

$$C = (2\text{ L}\cos\theta_1 + \text{L}\cos\theta_2, -(2\text{ L}\sin\theta_1 + \text{L}\sin\theta_2))$$

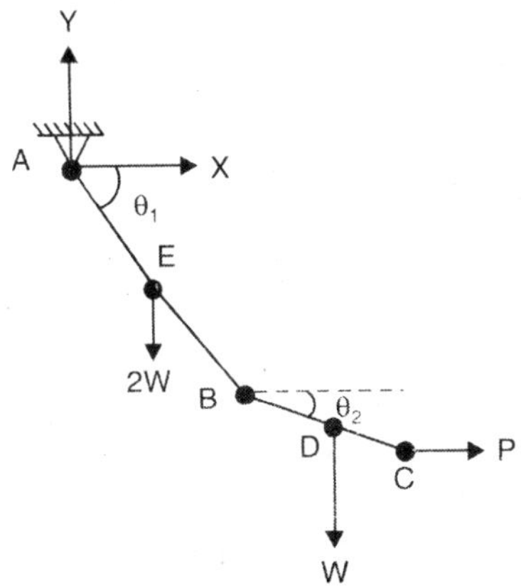

Fig. E4(a)

This is a type of two degree of freedom

Let θ_1 be constant

Now from virtual work

$$\Sigma dW = 0$$

$$0 + W\left(0 + \frac{L}{2}\cos\theta_2 d\theta_2\right) + P(0 - L\sin\theta_2 d\theta_2) = 0$$

we get

$$\theta_2 = \tan^{-1}\left(\frac{\mathbf{W}}{\mathbf{2P}}\right)$$

again let θ_2 be constant

$$\text{We get } \theta_1 = \tan^{-1}\left(\frac{\mathbf{2W}}{\mathbf{P}}\right) \qquad \textbf{...Ans.}$$

Example 5

Solution

$$OA = l,\ CA = CB = R$$

$$CG = \frac{3}{8}R$$

Let depth of G below the fixed point O is y.

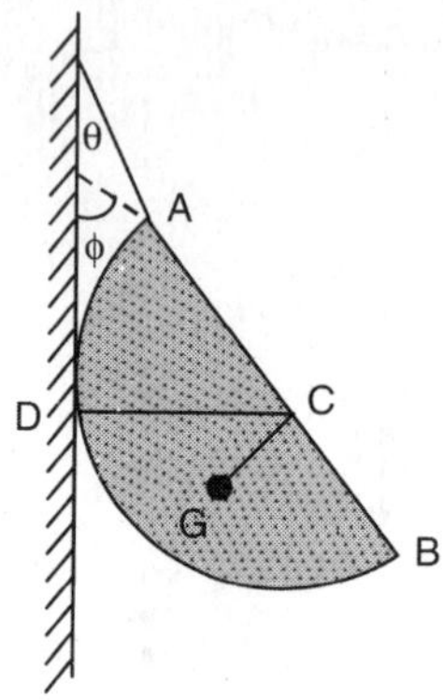

Fig. E5(a)

So,

$$y = l\cos\theta + R\cos\phi + \frac{3}{8}R\sin\phi$$

$$dy = -l\sin\theta\, d\theta + R\left(\frac{3}{8}\cos\phi\sin\phi\right)d\phi$$

But for equilibrium $dy = 0$

$$-l\sin\theta\, d\theta + R\left(\frac{3}{8}\cos\phi\sin\phi\right)d\phi = 0 \qquad \ldots(i)$$

for relation between $d\theta$ and $d\phi$

$$CD = R\sin\phi + l\sin\theta$$

$$\text{But } CD = R$$

So,

$$R = R\sin\phi + l\sin\theta$$

$$0 = R\cos\phi\, d\phi + l\cos\theta\, d\theta$$

$$d\phi = \frac{l\cos\theta d\theta}{R\cos\phi} \qquad \ldots(ii)$$

by using (*i*) and (*ii*)
we get,

$$\phi = \tan^{-1}\left(\frac{3}{8}+\tan\theta\right) \qquad \textbf{...Ans.}$$

Example 6

Solution

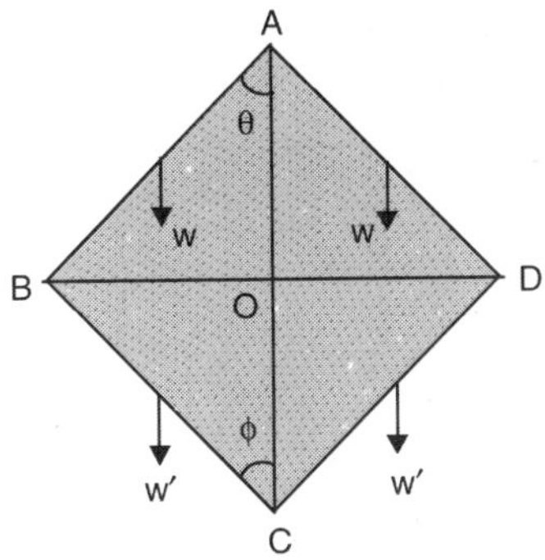

Fig. E6(a)

Clearly all forces are in downward vertical direction so we concentrate only on the *y*-coordinates of active force points.
Let T be the tension in the string.

$$\text{Let } AB = AD = 2a$$

$$BC = DC = 2b$$

$$\text{And } \angle BCA = \phi$$

$$\Sigma dW = 0$$

$$2W\, d(a\cos\theta) + 2W'\, d\,(2a\cos\theta + b\cos\phi) - T\, d\,(2a\cos\theta + 2b\cos\phi) = 0$$

$$W\, a\sin\theta\, d\theta + W'\,(2a\sin\theta\, d\theta + b\sin\phi\, d\phi) = T\,(a\sin\theta\, d\theta + b\sin\phi\, d\phi) \qquad \ldots(i)$$

By diagram

$$a\sin\theta = b\sin\phi$$

So,

$$a\cos\theta\, d\theta = b\cos\phi\, d\phi$$

$$\therefore d\phi = \frac{a\cos\theta}{b\cos\phi}d\theta \qquad \ldots(ii)$$

Using value of $d\phi$ from (*ii*) in (*i*)
We get

$$\mathbf{T = W' + (W + W')\left(\frac{\tan\theta}{\tan\theta + \cot\theta}\right)} \qquad \text{...Ans.}$$

Example 7

Solution

$$\text{Let } AB = l_0$$
$$BC = l_1$$
$$AC = l_2$$

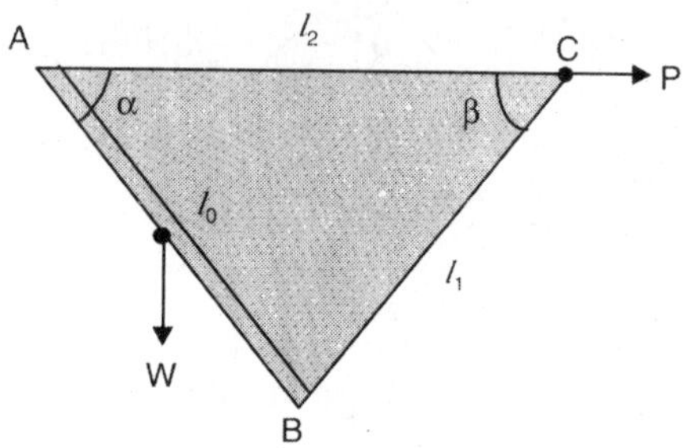

Fig. E7(a)

$$\Sigma dW = 0$$

$$P\, dl_2 + W\left(d\left(\frac{l_0}{2}\sin\alpha\right)\right) = 0 \qquad \text{...}(i)$$

$$\text{Now, } l_2 = l_0 \cos\alpha + l_1 \cos\beta$$

$$dl_2 = -\, l_0 \sin\alpha\, d\alpha - l_1 \sin\beta\, d\beta \qquad \text{...}(ii)$$

Here $d\alpha$ and $d\beta$ are unknowns but can be connect by this equation

$$l_0 \sin\alpha = l_1 \sin\beta \qquad \text{...}(iii)$$

Using above all equations we get

$$\text{P} = \frac{\mathbf{W\cos\alpha\cos\beta}}{\mathbf{2\sin(\alpha+\beta)}} \qquad \text{...Ans.}$$

Space Forces

Level 1

Example 1

Solution

Top view of the plate

Let D be the mid point of AC where 1.2 kN load acts. Weight of plate is 2.4 kN will act vertically at G.

Suppose T_A, T_B and T_C are the tensions in the string passing through point A, B and C respectively.

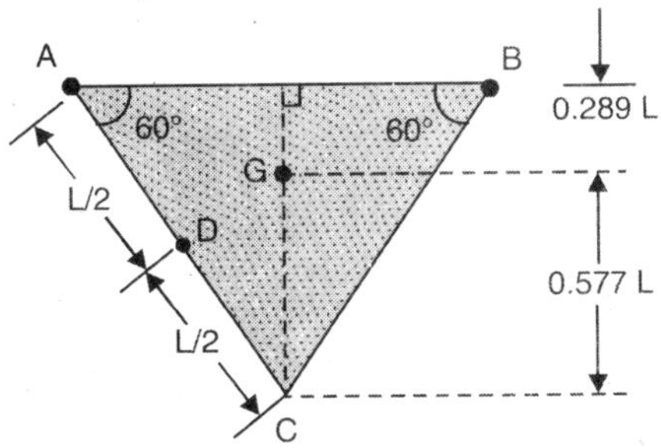

Fig. E1(a)

$$\Sigma M_{AB} = 0$$

$$L\,(0.866)\,T_C - (1.2)\left(\frac{L}{2}\right)\sin 60° - (2.4)\,(0.289)\,L = 0$$

$$\mathbf{T_C = 1.4\ kN} \qquad \text{...Ans.}$$

Similarly

$$\Sigma M_{AC} = 0$$

$$L\,(0.866)\,T_B - (0.289\,L)\,(2.4) = 0$$

$$\mathbf{T_B = 0.8\ kN} \qquad \text{...Ans.}$$

$$\Sigma M_{BC} = 0$$

$$(0.866\,L)\,T_A - (1.2)\left(\frac{L}{2}\right)\sin 60° - (2.4)\,(0.289\,L) = 0$$

$$\mathbf{T_A = 1.4\ kN} \qquad \text{...Ans.}$$

Example 2

Solution

Top-view of the plate

Suppose T_A, T_B and T_C are the tensions in the string A, B and C.

$$T_A + T_B + T_C = 1.5 \qquad \text{...}(i)$$

$$\Sigma M_{x\text{-axis}} = 0$$

$$T_A\,(3 \sin 25°) + T_B\,(3 \sin 30°) - T_C\,(3 \sin 60°) = 0$$

$$0.845\,T_A + T_B - 1.732\,T_C = 0 \qquad \text{...}(ii)$$

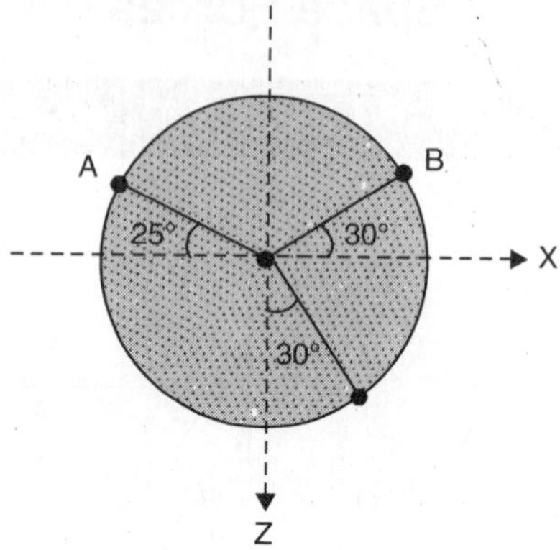

Fig. E2(a)

similarly

$$\Sigma M_{z\text{-axis}} = 0$$

$$T_A\,(3 \cos 25°) - T_B\,(3 \cos 30°) - T_C\,(3 \sin 30°) = 0$$

$$T_C - 1.813\ T_A + 1.732\ T_B = 0 \qquad \ldots(iii)$$

Using all above equations

We get $\mathbf{T_A = 0.627\ kN}$

$\mathbf{T_B = 0.36\ kN}$

$\mathbf{T_C = 0.514\ kN}$ **...Ans.**

Example 3

Solution

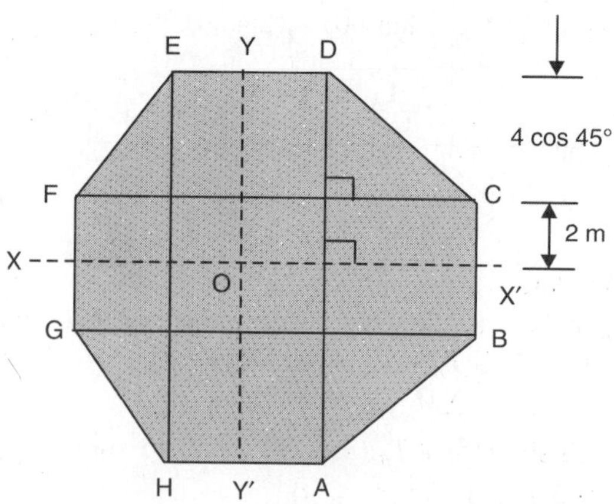

Fig. E3(a)

Clearly $R = 2600$ kN

Now just tilt the original problem figure in the way that shown in Fig. E3(a)
Assume resultant R acts at a point M on the xy plane

$$M = (\bar{x}, \bar{y})$$

$\Sigma M_{xx'}$

$$R\bar{y} = (500)(4.83) + (300)(4.83) + (300)(2) - (600)(2) - (400)(4.83) - (500)(2)$$

$$\bar{y} = 0.13 \text{ m}$$

$\Sigma M_{yy'}$

$$R\bar{x} = 300(2) + 300(4.83) + 600(4.83) - 500(2) - (500)(4.83) - 400(2)$$

$$\bar{x} = 0.28 \text{ m}$$

So resultant force will pass through the point $(\bar{\mathbf{x}}, \bar{\mathbf{y}})$ **= (0.28 m, 0.13 m)** on the xy plane.

Example 4

Solution

Let $\vec{R}$ be the resultant of the forces

$$\vec{R} = -40j - 12j - 8j - 20j$$
$$= -80j \text{ kN}$$

Let resultant vector $\vec{R}$ intersects at the point $M(x, z)$ on the XZ plane
Apply Varigenon theorem about origin

Moment of resultant about origin = Algebraic sum of moments of all forces.

$$\begin{vmatrix} \hat{i} & \hat{j} & \hat{k} \\ 0 & 80 & 0 \\ x & 0 & z \end{vmatrix}_R = \begin{vmatrix} \hat{i} & \hat{j} & \hat{k} \\ 0 & 12 & 0 \\ 10 & 0 & 0 \end{vmatrix}_{P_2} + \begin{vmatrix} \hat{i} & \hat{j} & \hat{k} \\ 0 & 8 & 0 \\ 10 & 0 & 5 \end{vmatrix}_C + \begin{vmatrix} \hat{i} & \hat{j} & \hat{k} \\ 0 & 20 & 0 \\ 4 & 0 & 10 \end{vmatrix}_D$$

$$(-80\bar{z})\hat{i} + (80\bar{x})\hat{k} = -240\hat{i} + 280\hat{k}$$

equating coefficient of $\hat{i}$ and $\hat{k}$.
we get

$$\bar{x} = 3.5$$
$$\bar{z} = 3$$

Finally
Resultant acts at the point **(3.5, 0,3)** **...Ans.**

Example 5

Solution

Calculate

$$M_{BH} = \lambda_{BH} \cdot M_H$$
$$H \equiv (0, a, a)\ ,\ B = (a, 0, 0)$$

$$\lambda_{BH} = \frac{(0-a)\hat{i} + (a-0)\hat{j} + (a-0)\hat{k}}{\sqrt{a^2 + a^2 + a^2}}$$

$$= \frac{-a\hat{i} + a\hat{j} + a\hat{k}}{\sqrt{3}\,a}$$

$$\lambda_{BH} = \frac{-\hat{i} + \hat{j} + \hat{k}}{\sqrt{3}}$$

$$M_H = \begin{vmatrix} \hat{i} & \hat{j} & \hat{k} \\ (a-0) & (0-a) & (a-a) \\ 0 & P/\sqrt{2} & -P/\sqrt{2} \end{vmatrix}$$

$$M_H = \frac{Pa}{\sqrt{2}}\left(\hat{i} + \hat{j} + \hat{k}\right)$$

$$M_{BH} = \frac{(\hat{i} + \hat{j} + \hat{k})}{\sqrt{3}} \cdot \frac{Pa}{\sqrt{2}}\left(\hat{i} + \hat{j} + \hat{k}\right)$$

$$= \frac{Pa}{\sqrt{6}}(-1+1+1)$$

$$= \frac{\mathbf{Pa}}{\sqrt{\mathbf{6}}} \qquad \textbf{...Ans.}$$

Level 2

Example 1

Solution

Write vector expression for each force

$$\vec{F}_1 = \frac{F_1(8\hat{i} + 4\hat{j} + 8\hat{k})}{\sqrt{8^2 + 4^2 + 8^2}} = (-0.667F_1)\hat{i} + (0.333F_1)\hat{j} + (0.667F_1)\hat{k}$$

$$\vec{F}_2 = \frac{F_2(-5\hat{i} - 6\hat{j} + 4\hat{k})}{\sqrt{5^2 + 6^2 + 4^2}} = (-0.57F_2)\hat{i} + (-0.684F_2)\hat{j} + (0.456F_2)\hat{k}$$

$$\vec{F}_3 = \frac{F_3(-8\hat{i} + 4\hat{j} - 2\hat{k})}{\sqrt{8^2 + 4^2 + 2^2}} = (-0.873F_3)\hat{i} + (-0.436F_2)\hat{j} + (-0.218F_2)\hat{k}$$

$$W = W(-\hat{j})$$

now apply equations of equilibrium

$$\Sigma F_x = 0,\ \Sigma F_y = 0,\ \Sigma F_z = 0$$

we get as

$$\mathbf{F_2} = \mathbf{1.032\ W\ (comp)}$$

$$\mathbf{F_1} = \mathbf{0.108\ W\ (tensile)}$$

$$\mathbf{F_3} = \mathbf{0.741\ W\ (tensile)}$$

Example 2

Solution

FBD of the bar

$$\text{Clearly } BG = AG = 3.5 \text{ m}$$

$$\text{And } 7 = \sqrt{2^2 + 6^2 + h^2}$$

$$h = 3 \text{ m}$$

Apply condition of equilibrium

$$\Sigma F_x = 0$$

$$B_x - A_x = 0 \quad \text{...}(i)$$

$$\Sigma F_y = 0$$

$$B_y - A_y = 0 \quad \text{...}(ii)$$

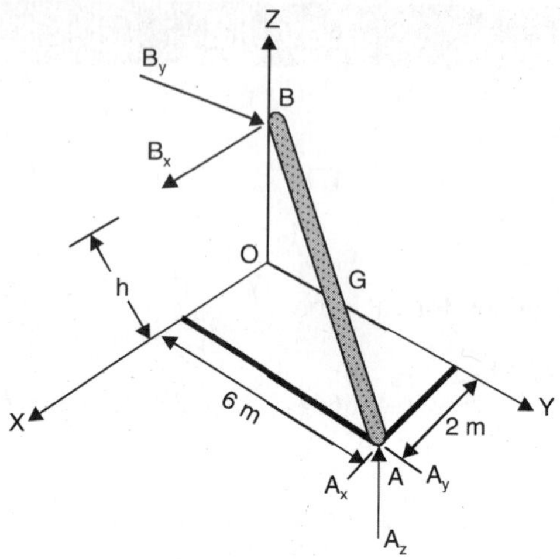

Fig. E2(a)

$$\Sigma F_z = 0$$

$$A_z - 981 = 0 \qquad \text{...(iii)}$$

Point A is socket joint having maximum number of unknown forces passing through it

$$\Sigma M_A = 0$$

$$\begin{vmatrix} \hat{i} & \hat{j} & \hat{k} \\ (02) & (06) & (30) \\ B_x & 0 & 0 \end{vmatrix}_{B_x} + \begin{vmatrix} \hat{i} & \hat{j} & \hat{k} \\ (02) & (06) & (30) \\ 0 & B_y & 0 \end{vmatrix}_{B_y} + \begin{vmatrix} \hat{i} & \hat{j} & \hat{k} \\ (12) & (36) & (1.50) \\ 0 & 0 & 981 \end{vmatrix}_{mg} = 0$$

After solving this determinants
We get

$$B_y = 981 \text{ N}$$
$$B_x = 327 \text{ N}$$

Using equations (*i*) and (*ii*)
We get

$$A_x = 327 \text{ N}$$
$$A_y = 981 \text{ N}$$

Reaction offered by socket joint A

$$\mathbf{R_A} = \sqrt{\mathbf{A_x^2 + A_y^2 + A_z^2}}$$

$$= \sqrt{\mathbf{(327)^2 + (981)^2 + (981)^2}}$$

$$\mathbf{R_A} = \mathbf{1425.36\ N} \qquad \textbf{...Ans.}$$

Basics of Dynamics

Level 1

Example 3

Solution

Method 1

By using datum level method in which total length of string is taken and then differentiation is done by proper variables.

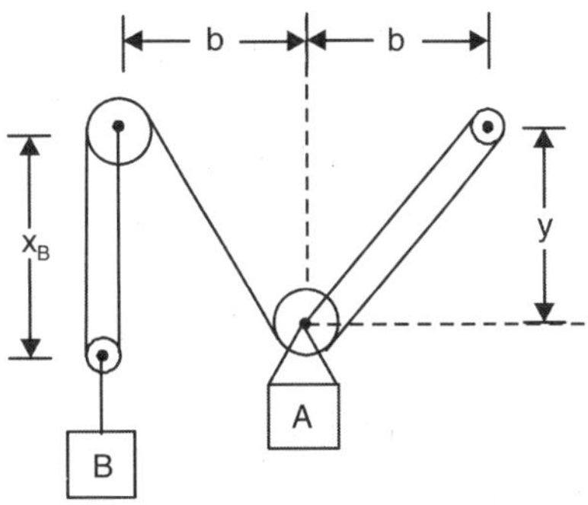

Fig. E3

Total length of cable

$$L = 2X_B + 2\sqrt{b^2 + y^2} + \sqrt{b^2 + y^2} + C$$

$$\frac{dL}{dt} = 2V_B + \left(\frac{3}{2}\right)\frac{2}{\sqrt{b^2 + y^2}} \times (2y) \times V_A$$

$$0 = 2V_B + \frac{3y(V_A)}{\sqrt{b^2 + y^2}}$$

$$\mathbf{V_B = \frac{3y}{2\sqrt{b^2 + y^2}} \cdot V_A} \qquad \textbf{...Ans.}$$

Method 2

Internal work done by cable on the system is zero

$$\Sigma dW_i = 0 \qquad ...(i)$$

Using (*i*)

$$3\bar{T}_A \cdot d\bar{r}_A + 3\bar{T}_B \cdot d\bar{r}_B = 0$$

$$-3R_A dS_A \cos\alpha + 2T_B \cdot dS_B = 0$$

$$(T_A = T_B)$$

$$-3dS_A \cos\alpha + 2dS_B = 0$$

$$-3V_A \cos\alpha + 2V_B = 0$$

$$-\frac{3}{2}V_A \cos\alpha = V_B$$

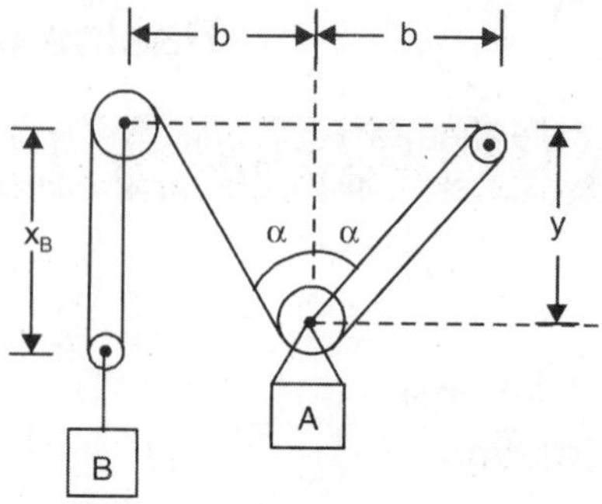

Fig. E4

$$\mathbf{V_B} = \left|\frac{\mathbf{3}}{\mathbf{2}}\left(\frac{\mathbf{y}}{\sqrt{\mathbf{b}^2+\mathbf{y}^2}}\right)\mathbf{V_A}\right| \qquad \textbf{...Ans.}$$

Example 7

Solution

Clearly

$$\tan\theta = \frac{3}{4} = 0.75$$

$$\theta = 37°$$

$$R = \sqrt{3^2+3^2} = 5$$

$$V_x = -3 \text{ m/s}$$

$$V_y = 4 \text{ m/s}$$

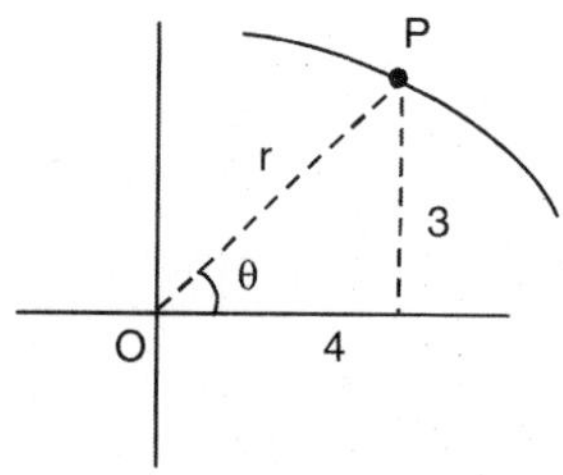

Fig. E7

$$V = \sqrt{V_x^2+V_y^2} = \sqrt{(-3)^2+(4)^2}$$

$$\tan\alpha = \left|\frac{V_y}{V_x}\right| = \frac{4}{3}$$

$$\alpha = \tan^{-1}\left(\frac{4}{3}\right) = 53.13°$$

$$a_r = 10 \text{ m/s}^2$$

$$a_\theta = \mathbf{15\ m/s^2}$$

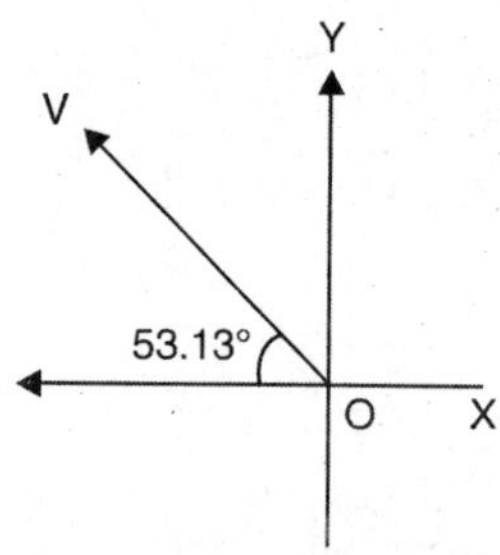

Fig. E7(a)

$$a = \sqrt{a_r^2 + a_\theta^2} = \sqrt{(10)^2 + (15)^2}$$

$$= 18.02 \text{ m/s}^2$$

$$\tan\beta = \left|\frac{a_\theta}{a_r}\right| = \frac{15}{10}$$

$$\beta = \tan^{-1}\left(\frac{3}{2}\right) = 56.31°$$

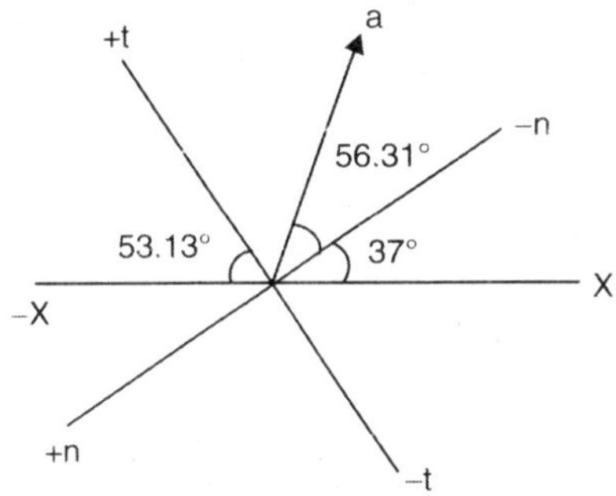

Fig. E7(b)

$$V_r = 0$$

$$\dot{r} = 0$$

$$V_\theta = 5$$

$$\dot{\theta} = \frac{5}{5} = \mathbf{1 \text{ rad/sec}}$$

$$a_r = \ddot{r} - r\dot{\theta}^2$$

$$10 = \ddot{r} - 5 \times 1 \times 1$$

$$\ddot{r} = \mathbf{15 \text{ m/s}^2}$$

$$a_\theta = 2\dot{r}\dot{\theta} + r\ddot{\theta}$$

$$15 = 2\,(0)\,(1) + 5\,(\ddot{\theta})$$

$$\ddot{\theta} = \mathbf{3 \text{ rad/s}^2}$$ ...**Ans.**

$$\rho = \frac{V^2}{a_n} = \frac{V^2}{10} = \frac{5 \times 5}{10}$$

$$\rho = \mathbf{2.5 \text{ m}}$$ ...**Ans.**

Example 9

Solution

The problem given here is like the Doppler effect for wave motion.

V_s = Velocity of plane from which the shots are fired.

V_t = Velocity of target plane

V = Muzzle velocity

r = rate of fire of the gun

For (*i*) Problem statement

$$\mathbf{r}' = \left(\frac{\mathbf{V} + \mathbf{V_s} + \mathbf{V_t}}{\mathbf{V}}\right)\mathbf{r}$$

for (*ii*) Problem statement

$$\mathbf{r}' = \left(\frac{\mathbf{V} - \mathbf{V_s} - \mathbf{V_t}}{\mathbf{V}}\right)\mathbf{r} \qquad \text{...Ans.}$$

Example 11

Solution

Clearly

According to the problem

$$V_A = 800 \text{ km/h}$$

$$\bar{V}_A = 800\,\hat{i}$$

$$\bar{V}_B = V_B \cos 45\hat{i} + V_B \sin 45\hat{j}$$

$$\bar{V}_{B/A} = \bar{V}_B - \bar{V}_A$$

$$= \left(V_B \cos 45°\hat{i} + V_B \sin 45°\hat{j}\right) - \left(800\hat{i}\right)$$

$$\bar{V}_{B/A} = \left(V_B \cos 4\hat{5} - 800\right)\hat{i} + \left(V_B \sin 45°\right)\hat{j} \qquad \text{...}(i)$$

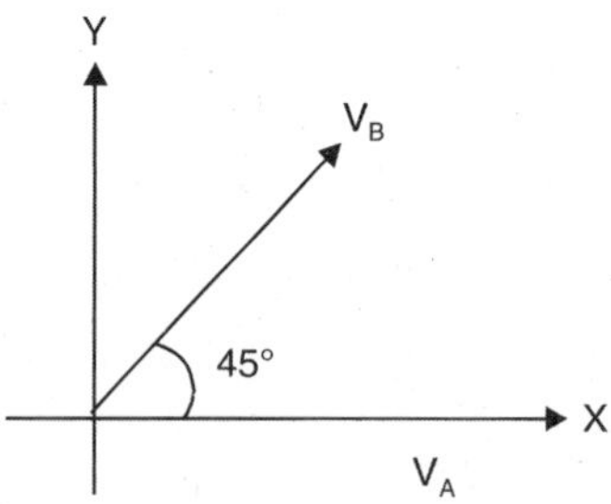

Fig. E11(a)

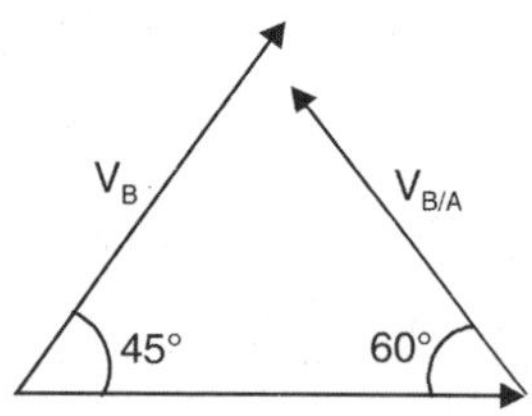

Fig. E11(b)

$$\left(V_{B/A}\cos 60°\left(-\hat{i}\right) + V_{B/A}\sin 60°\left(\hat{j}\right)\right) \qquad \text{...}(ii)$$

equating both above equation we get,

$$\mathbf{V_B} = \mathbf{717 \text{ km/h}}$$

$$\mathbf{V_{B/A}} = \mathbf{586 \text{ km/h}} \qquad \text{...Ans.}$$

Method 2

$$\frac{V_A}{\sin 75°} = \frac{V_B}{\sin 60°}$$

$$V_B = \frac{800}{\sin 75°}(\sin 60°)$$

$$\mathbf{V_B} = \mathbf{717 \text{ km/h}} \qquad \text{...Ans.}$$

Example 17

Solution

In triangle AOB

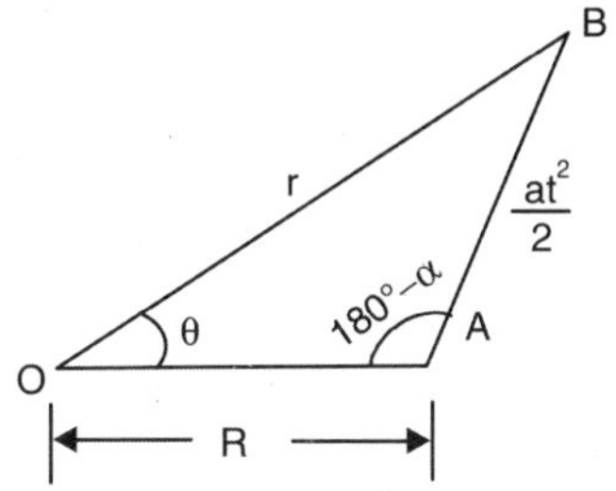

Fig. E17(a)

$$\frac{r}{\sin(180°\alpha)} = \frac{R}{\sin(\alpha\theta)} = \frac{(at^2)/2}{\sin\theta}$$

$$\frac{r}{\sin\alpha} = \frac{R}{\sin(\alpha\theta)}$$

Eliminate θ and then differentiate r with respect to time.

Level 2

Example 1

For the 1st train
By *V-t* curve
We can write here

$$40 \times (OM) = 7 \qquad \ldots(i)$$

$$\frac{1}{2} \times (LM) \times (MN) = 1 \qquad (\Delta LM = 40)$$

$$20\ (MN) = 1 \qquad \ldots(ii)$$

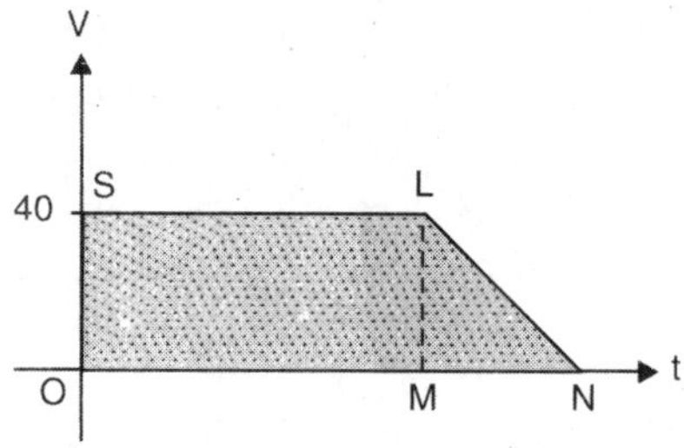

Fig. E1

For 2nd train
By *V-t* curve

$$8 = \frac{1}{2}\ (CD)\ (OE) \qquad \ldots(iii)$$

by using above all equations
we get,

$$CD = V_{max} = \mathbf{71.5\ km/hr.} \qquad \ldots\textbf{Ans.}$$

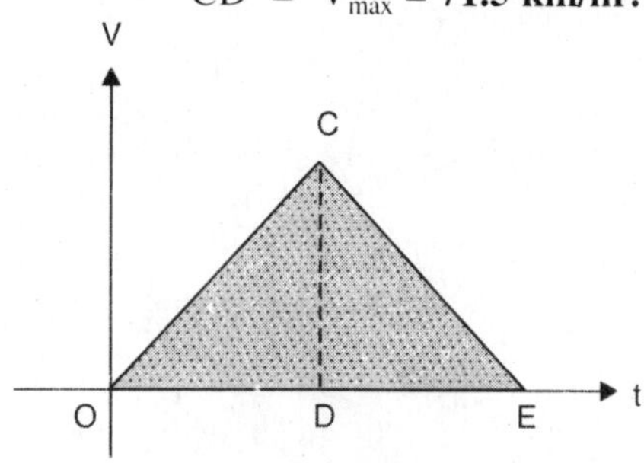

Fig. E1(a)

Example 2

Solution

As we know that

$$a = V\frac{dv}{ds}$$

$$\int_{S_1}^{S_2} ads = \int_{S_1}^{S_2} ads$$

$$\therefore a = \frac{V_2^2 - V_1^2}{2(S_2 - S_1)}$$

$$a\Big|_{0 \le S \le 3} = \frac{3600 - 2500}{2(0.3 - 0)}$$

$$= \mathbf{1833.33\ km/hr^2}$$ **...Ans.**

Example 3

Solution

By sine rule in ΔABC

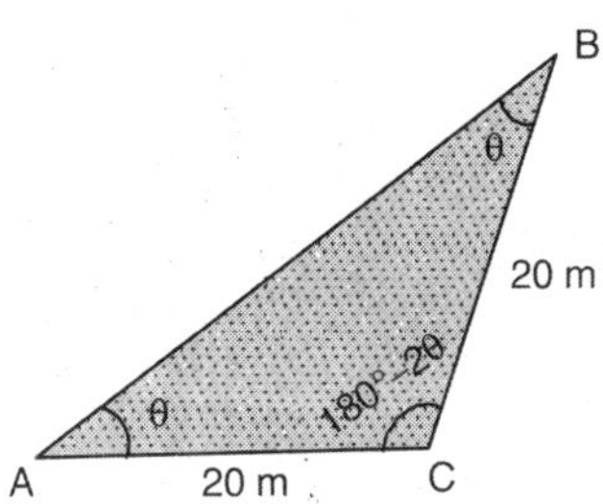

Fig. E3(a)

$$\frac{AB}{\sin(180° - 2\theta)} = \frac{20}{\sin\theta}$$

$$AB = 40\cos\theta$$

$$AB = \dot{r} = 40\cos\theta$$

$$\dot{r} = -40\sin\theta\ddot{\theta}$$

$$\text{for } \theta = 45°$$

$$r = 40\cos(45°) = 20\sqrt{2}\text{ m}$$

$$\dot{r} = -40\sin(45°)\,\dot{\theta}$$

$$= -20\sqrt{2}\,\dot{\theta}$$

$$\bar{V} = \dot{r}\hat{e}_r + r\dot{\theta}\hat{e}_\theta$$

$$\bar{V} = -20\sqrt{2}\,\dot{\theta}\,\hat{e}_r + (40\cos 45°)\,\dot{\theta}\,\hat{e}_\theta$$

$$|\bar{V}| = \sqrt{V_r^2 + V_\theta^2}$$

$$25 = \sqrt{\left(-20\sqrt{2}\,\dot{\theta}\right)^2 + \left(\left(20\sqrt{2}\right)\dot{\theta}\right)^2}$$

$$25 = \dot{\theta}\sqrt{800+800}$$

$$\dot{\theta} = \frac{25}{400} = \frac{1}{16} = \mathbf{0.06\ rad/s}$$

Example 4

Solution

$$S = 2t^3 - 24t + 6$$

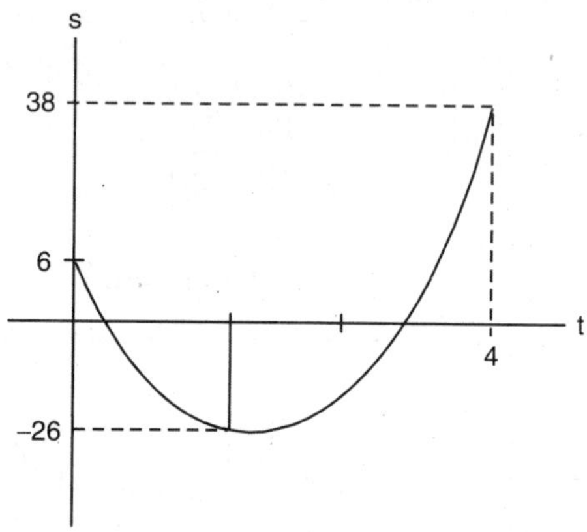

Fig. E4

$$\frac{ds}{dt} = 6t^2 - 24$$

$$0 = 6t^2 - 24$$

According to the problem

$$u = 72 \text{ m/s}$$

$$72 = 6t^2 - 24$$

$$t = 4S$$

Now $\quad u = 6t^2 - 24$

$$\frac{dv}{dt} = 12\,t$$

When $V = 30$

$$30 = 6t^2 - 24$$

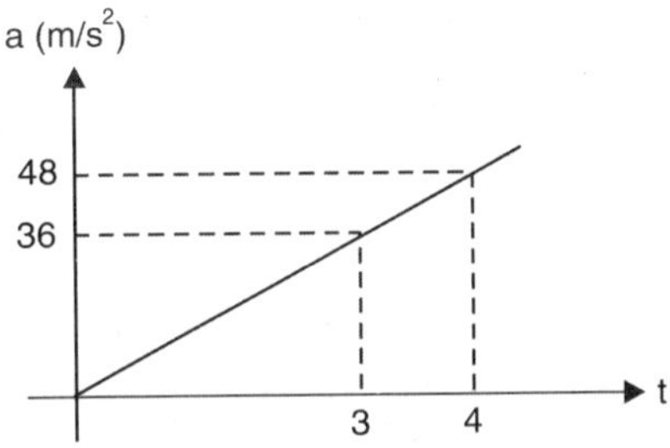

Fig. E4(a)

$$t = \sqrt{54/6} = 3 \text{ sec}$$

$$\text{so, } a = \frac{dv}{dt} = 12\,(t) = 12 \times 3 = \mathbf{36\ m/s^2}$$

Example 5

Solution

Method 1

$$L = \text{Total length of cable}$$

$$L = 3y_B + 2y_A + C \text{ (see figure)}$$

Differentiate with respect to time

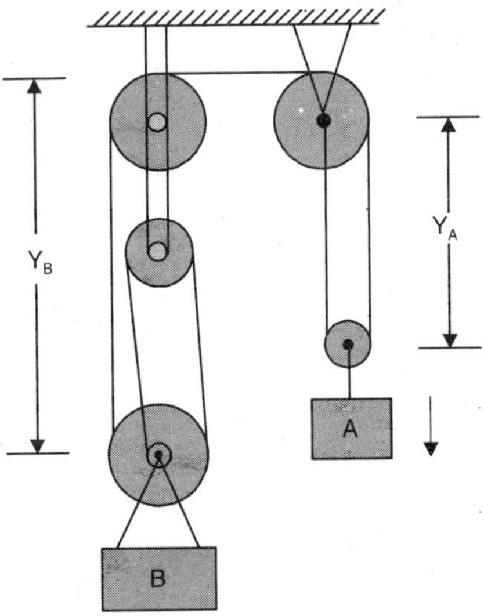

Fig. E5(a)

$$0 = 3V_B + 2V_A$$
$$V_B = 0.2 \text{ m/s} \uparrow$$

Method 2

$$\Sigma dW_i = 0$$
$$3\bar{T}\, d\bar{r}_B + 2\bar{T}\, d\bar{r}_A = 0$$
$$3T\, d\, S_B - 2T\, dS_A = 0$$

Differentiate it w.r.t time

$$3V_B - 2V_A = 0$$
$$\mathbf{V_B = 0.2 \text{ m/s} \uparrow}$$

...Ans.

Example 6

Solution

By datum level method or $\Sigma dW_i = 0$ we can calculate the constraint relation among *A*, *B* and *D*.

$$V_A + 2V_D + V_B = 0 \quad \text{...(i)}$$
$$a_A + 2a_D + a_B = 0 \quad \text{...(ii)}$$
$$V_D = 75 \text{ mm/s} \downarrow$$
$$V_A = 300 \text{ mm/s}$$
$$\text{At } \Delta x = 200 \text{ mm}$$

For *A*

$$V2A = u_A^2 + 2a_A\, S_A$$
$$(300)^2 = 0 + 2\, a_A\, \Delta x_A$$
$$= 2a_A\, (200)$$
$$a_A = 225 \text{ m/s}^2$$

$$\Delta t = \frac{3000}{225} = 1.33 \text{ sec}$$

For *D*

$$\Delta x_D = 75 \times 1.33$$
$$= 100 \text{ mm}$$
$$V_B = -300 - 75(2)$$
$$= 450 \text{ mm/s} \uparrow$$
$$a_B = 255 \text{ mm/s}^2$$
$$V_A + 2\, V_D + V_B = 0$$
$$V_B = 150 \text{ mm/s} \uparrow$$

$$V_B^2 = u_B^2 + 2\, a_B\, \Delta x_B$$
$$(450)^2 = (150)^2 + 2 \times 255 \times \Delta x_B$$
$$\mathbf{\Delta x_B = 400 \text{ mm} \uparrow}$$

...Ans.

Example 7

Solution

There are two constraint relations for different motor arrangement.

$$V_D + 2V_B = 0 \qquad \ldots(i)$$

$$a_D + 2a_B = 0 \qquad \ldots(ii)$$

$$a_D = 8 \text{ m/s}^2$$

$$a_B = -4 \text{ ms}^2$$

$$V_C + 2V_A = 0 \qquad \ldots(iii)$$

$$4 + 2V_A = 0$$

$$V_A = 2 \text{ m/s}$$

For A

$$\Delta x_A = 2t$$

For B

$$\Delta x_B = 0 + \frac{1}{2} \times 4 \times t^2$$

$$= 2t^2$$

$$\Delta x_A = \Delta x_B - 3$$

$$2t = 2t^2 - 3$$

$$2t^2 - 2t - 3 = 0$$

$$t = 1.82 \text{ sec.}$$

...Ans.

Example 8

Solution

Trajectory or path of the milk can be assumed as a projectile.

For $A \rightarrow B$

$$y = x\tan\alpha - \frac{g}{2}\left(\frac{x^2}{u^2\cos 2\alpha}\right)$$

$$\alpha = 40°, \; y = h - 140, \; x = 80$$

$$u = 1200 \text{ mm/s}$$

$$g = 9810 \text{ mm/s}^2$$

$$h - 140 = 80 \tan 40° + 4905 \left(\frac{80}{120\cos 40°}\right)^2$$

$$h = 244.3 \text{ mm}$$

For $A \rightarrow C$

$$x = 80 + 66 = 146 \text{ mm}$$

$$h - 140 = 146 \tan 40° + (4905)\left(\frac{146}{1200 \cos 40°}\right)^2$$

$$h = 386.2 \text{ mm}$$

Range of h so that milk will enter in the glass.

$$244.3 \text{ mm} < h < 386.2 \text{ mm}$$ **...Ans.**

Example 9

Solution

$$a = \frac{dV}{dt}$$

$$\int_4^t a\,dt = \int_{50}^V dV$$

$$-10\int_4^t dt = \int_{50}^V dV$$

$$(90 - 10t) = v \qquad ...(i)$$

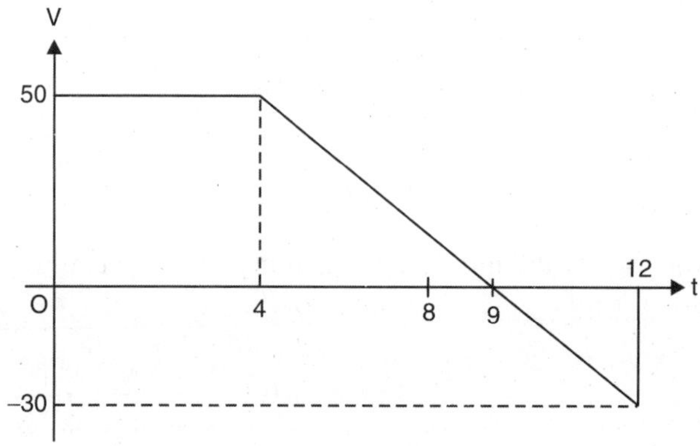

Fig. E9

Now for distance

$$V = \frac{dx}{dt}$$

$$\int v\,dt = \int dx$$

$$\int_4^t (90 - 10t)\,dt = \int_0^x dx$$

$$(90t - 5t^2) - 280 = x$$

$$x_{net} = x + x_o = 90t - 5t^2 - 280 + (50) \times 4$$

$$x_{net} = -5t^2 + 90t - 80$$

Calculate $x_{max}\Big|_{t=9}$ = **325 m**

Here t = 9 (think how)

Example 13

Solution

For radius of curvature

$$\rho = \frac{\left[1+\left(\frac{dx}{dy}\right)^2\right]^{3/2}}{\left(\frac{d^2x}{dy^2}\right)}$$

For $\frac{dx}{dy}$

$$\frac{x^2}{a^2} - \frac{y^2}{b^2} = 1$$

We will find $\left.\frac{dx}{dy}\right|_{(a,\,0)} = \frac{a^2}{b^2}\cdot\frac{y}{x} = 0$

$$\left.\frac{dx}{dy}\right|_{(a,\,0)} = \left.\frac{a^2}{b^2x^2}\left(xy\frac{dx}{dy}\right)\right|_{(a,\,0)}$$

$$= \frac{a}{b^2} \quad \text{...(ii)}$$

Using (i) and (ii) in above formula for p
We get

$$\rho = \frac{b^2}{a}$$

$$a_N = \frac{V_0^2}{\rho} = \frac{V_0^2}{(b^2/a)}$$

$$\mathbf{a_A} = \left(\frac{\mathbf{a}}{\mathbf{b^2}}\right)\mathbf{V_0^2} \quad \text{...Ans.}$$

Example 16

Solution

According to arrangement of figure total length of cable = 10 m

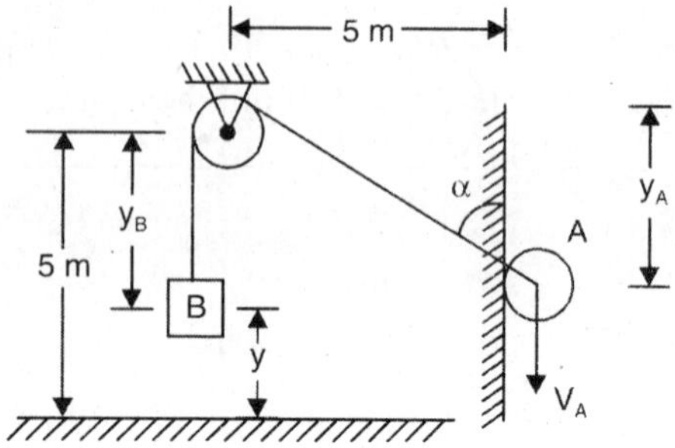

Fig. E16(a)

$$y_B = (5 - y)$$

$$10 = (5 - y) + \sqrt{(y_A^2 + 25)}$$

$$\sqrt{(y_A^2 + 25)} = y + 5 \qquad \ldots(i)$$

$$\text{Now use } \Sigma dW_i = 0$$

$$\bar{T}_A d\bar{r}_A + \bar{T}_B \cdot d\bar{r}_B = 0$$

$$-(TdS_A \cos\alpha) + TdS_B = 0$$

$$V_B = V_A \cos\alpha = V_A \frac{y_A}{\sqrt{y_A^2 + 25}} \qquad \ldots(ii)$$

By equation (*i*) and (*ii*)
We get

$$\mathbf{V_B = \frac{3\sqrt{(y+5)^2 - 25}}{y+5}} \qquad \textbf{...Ans.}$$

Example 17

Solution

$$S(t) = \begin{cases} t^2 & 0 \le t \le 10 \\ 20t - 100 & 10 \le t \le 15 \end{cases}$$

For *V*-*t* graph
$0 \le t \le 10$

$$S = t^2$$

$$V = \frac{ds}{dt} = 2t$$

$20 \le t \le 15$

$$S = 20\,t - 100$$

$$V = \frac{ds}{dt} = 20$$

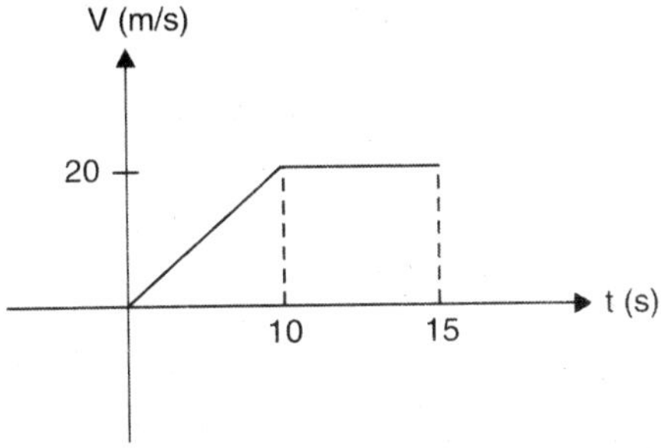

Fig. E17(a)

For *a-t* graph

$$a\,(f) = \frac{dv}{dt} \quad 0 \le t \le 10$$

$$= \mathbf{2\ m/s^2} \qquad \textbf{...Ans.}$$

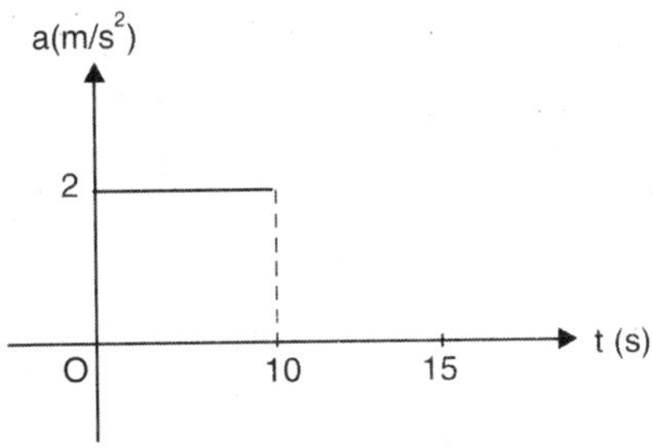

Fig. E17(b)

Example 18

Solution

$$\omega' t = \omega t + a\,\theta/b + \theta \text{ (think how)}$$

we get

$$\theta = \frac{b(\omega' - \omega)t}{a+b}$$

so,
$$\dot{\theta} = \frac{b(\omega' - \omega)t}{a+b}$$

(*i*) $t_0 = \dfrac{2\pi(a+b)}{b(\omega' - \omega)}$

(*ii*) Time taken by *P* to make a circuit of circle

$$t_0 = \frac{2\pi}{\theta + \omega} = \frac{2\pi(a+b)}{a\omega + b\omega'}$$

Example 19

Solution

Clearly time of flight of the shell

$$= \text{time of } \left(n - \frac{1}{2}\right) \text{ revolutions of the particle}$$

$$\frac{2u\sin\theta}{g} = \left(\frac{2n-1}{2}\right) \quad \left(\frac{2\pi R}{v}\right) \qquad \ldots(i)$$

$$\frac{u^2 \sin 2\theta}{g} = 20\sqrt{3} \qquad \ldots(ii)$$

using (*i*) and (*ii*)
we get

$$\boldsymbol{\theta_{min} = 30°} \qquad \textbf{...Ans.}$$

$$\mathbf{u = 20\ m/s} \qquad \textbf{...Ans.}$$

Example 20

Solution

$$t = \frac{x}{v_0 \cos(\theta + \alpha)} \qquad \ldots(i)$$

By equation of trajectory

$$y = x \tan \alpha_0 - \frac{gx^2}{2u^2 \cos^2 \alpha_0} \qquad \ldots(ii)$$

$$\text{here } y = x_R \tan \alpha$$

$$\alpha_0 = \theta + \alpha$$

Then we get by equation (*ii*)

$$x_R = \left(\frac{2}{g}\right) v_0^2 \cos^2 (\theta + \alpha) [\tan (\theta + \alpha) - \tan \alpha]$$

range of the inclined plane

$$R = \frac{x_R}{\cos\alpha}$$

$$= \left(\frac{2}{g\cos\alpha}\right) v_0^2 \cos^2 (\theta + \alpha) [\tan (\theta + \alpha) - \tan \alpha] \quad ...(iii)$$

Now let $M = \tan (\theta + \alpha)$

Then we can write equation (*iii*) as

$$R = \frac{2V_0^2 (M \tan\alpha)}{g(M^2 + 1)\cos\alpha}$$

For maximum range

$$\frac{dR}{dM} = 0$$

we get $$\mathbf{R} = \frac{\mathbf{v_0^2}}{\mathbf{g(\sin\alpha + 1)}} \quad \textbf{...Ans.}$$

$$\theta_{max} = \tan^{-1}\left(\frac{1}{\tan\alpha + \sec\alpha}\right) \quad \textbf{...Ans.}$$

Example 28

Solution

Relative angular velocity vanishes when $\bar{V}_{ABy} = 0$

$$V_1 \sin \alpha_1 = V_2 \sin \alpha_2$$

According to the problem figure

$$u \cos \alpha + v \cos \beta = 0$$

$$\frac{\cos\alpha}{\sqrt{R_1}} + \frac{\cos\beta}{\sqrt{R_2}} = 0$$

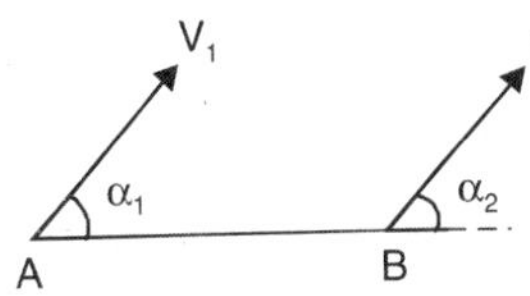

Fig. E.29(a)

$$\frac{\cos\alpha}{\sqrt{R_1}} + \frac{\cos(\beta+\alpha)}{\sqrt{R_2}} = 0 \quad ...(i)$$

Now,

$$R_1 \sin \alpha = R_2 \sin \beta \quad ...(ii)$$

By using above two equations
We get,

$$\cos\theta = \frac{\sqrt{R_1}\sqrt{R_2}}{\left(R_1\sqrt{R_1R_2}+R_2\right)} \quad \text{..Ans.}$$

Example 29

Solution

This is like a mathematical problem.

$$a = -kv^2$$

$$\frac{dv}{dt} = -kV^2$$

$$\int_8^V \frac{dv}{V^2} = -k\int_0^t dt$$

$$v = \frac{8}{1+8kt}$$

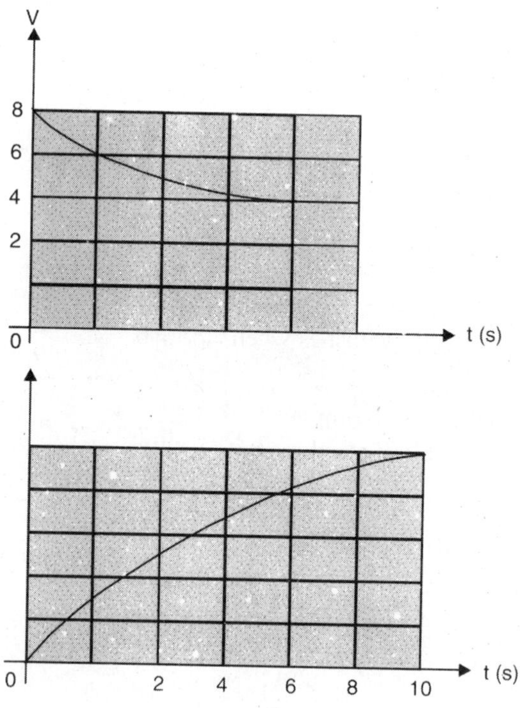

Fig. E29

now for K

$$\text{put } V = 4 \text{ knots}$$

$$t = \frac{1}{6} \text{ hr}$$

$$k = \frac{3}{4}$$

$$v = \mathbf{\frac{8}{1+6t}} \qquad \textbf{...Ans.}$$

$$V = \frac{ds}{dt}$$

$$\int_0^t v di = \int_0^s ds$$

$$\mathbf{s = \frac{4}{3}\ln(1+6t)} \qquad \textbf{...Ans.}$$

Example 30

Solution

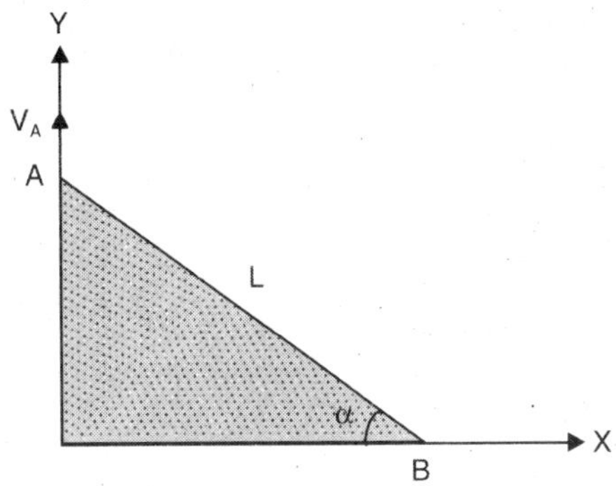

Fig. E30(a)

Along the line AB relative velocity will be same because rod length is constant.

$$V_B \cos\alpha = V_A \sin\alpha$$

$$V_B = V_A \tan\alpha$$

$$V_B = V_A \cdot \frac{y}{\sqrt{1^2 - y^2}}$$

Again differentiate it we get

$$\mathbf{a_B} = -\frac{\mathbf{L^2 V_A^2}}{\mathbf{(L^2 - y^2)^{3/2}}} \qquad \textbf{...Ans.}$$

Example 31

Solution

$$y = \frac{10}{\theta} = 10(\theta^{-1})$$

$$\dot{y} = -10\,(\theta^{-2})\,\dot{\theta}$$

$$\overline{V} = \dot{r}\hat{e}_r + r\dot{\theta}\hat{e}_\theta$$

$$\dot{r} = \dot{y} = -10\,(\dot{\theta}^2)\,\dot{\theta}$$

$$r\dot{\theta} = y\dot{\theta} = 10\,(\theta^{-1})\,\dot{\theta}$$

Now $\theta = 1$ radian

$$V^2 = V_r^2 + V_\theta^2$$

$$6^2 = 100\,(\dot{\theta})^2 + 100\,(\dot{\theta})^2$$

$$\dot{\theta} = 0.424 \text{ rad/s}$$

$$V_r = -4.24 \text{ m/s}$$

$$\mathbf{V_\theta = 4.24\ m/s} \qquad \textbf{...Ans.}$$

Level 3

Example 1

Solution

$$\Sigma dW_i = 0$$

We get,

$$V_A = (3/2)\ V_B$$

Let $\quad V_B = V$

$$V_A = (1.5)\ V$$

$$\bar{V}_A = \left(-V_A \cos 30°\right)\hat{i} - \left(V_A \sin 20°\right)\hat{j}$$

$$= \left(-1.5V \cos 30°\right)\hat{i} - \left(1.5V \sin 20°\right)\hat{j}$$

$$= \left(-1.3V\right)\hat{i} - \left(0.75\right)\hat{j}$$

For $\bar{V}_B = -V\cos\theta\hat{i} + V\sin\theta\hat{j}$

$$\bar{V}_{B/A} = \bar{V}_B - \bar{V}_A$$

We get

$$\bar{V}_{B/A} = \left(-V\cos\theta + 1.3V\right)\hat{i} + \left(V\sin\theta + 0.75V\right)\hat{j} \qquad \ldots(i)$$

But according to problem

$$\left|\bar{V}_{B/A}\right| = 5.6\,\text{m/s at } 70°\ \left(\text{1st quard}\right)$$

$$\bar{V}_{BA} = \left(5.6\cos 70°\right)\hat{i} + \left(5.6\sin 70°\right)\hat{j}$$

$$= \left(1.915\right)\hat{i} + \left(5.262\right)\hat{j} \qquad \ldots(ii)$$

Equating (*i*) and (*ii*)

We get,

$$\theta = 35.2°,\ 64.6°$$

$$\mathbf{V_A = 5.94\ m/s\ 9.47\ m/s}$$

$$\mathbf{V_B = 3.96\ m/s,\ 6.31\ m/s} \qquad \textbf{...Ans.}$$

Example 11

Solution

As the speeds of the both particle are constant.

So,

$$S_A = 0.7 \times 2 = 1.4 \text{ m}$$
$$S_B = 1.5 \times 2 = 3.0 \text{ m}$$

$$\theta_A = \frac{S_A}{r} = \frac{1.4}{5}$$
$$= 0.28 \text{ rad}$$
$$= 16.04°$$

$$\theta_B = \frac{S_B}{r} = \frac{3.0}{5}$$
$$= 0.6 \text{ rad}$$
$$= 34.38°$$

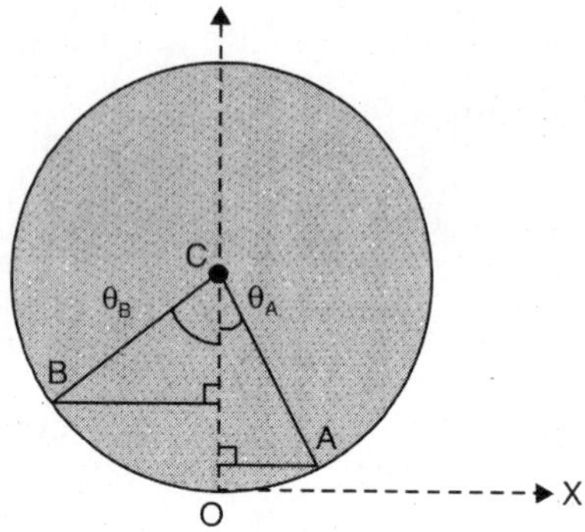

Fig. E11(a)

$$\bar{r}_A = 5\sin\theta_A\hat{i} - (5 - 5\cos\theta_A)\hat{j}$$
$$= 5\sin(16.04°)\hat{i} + (5 - 5\cos(16.04))\hat{j}$$
$$\bar{r}_A = 1.382\hat{i} + 0.195\hat{j}$$
$$\bar{r}_B = -5\sin\theta_B\hat{i} + (5 - 5\cos\theta_B)\hat{j}$$
$$r_B = -2.823\hat{i} + 0.873\hat{j}$$

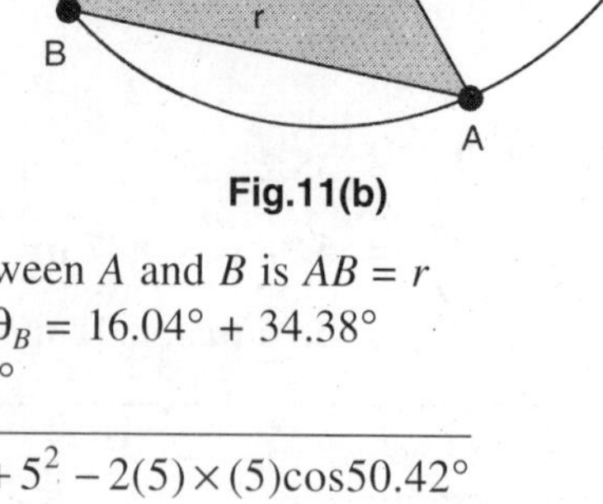

Fig.11(b)

Shortest distance between A and B is $AB = r$

$$\theta = \theta_A + \theta_B = 16.04° + 34.38°$$
$$= 50.42°$$

$$\text{Hence } r = \sqrt{5^2 + 5^2 - 2(5)\times(5)\cos 50.42°}$$

$$\mathbf{r = 4.259 \text{ m}}$$ ...Ans.

Newton's Laws

Level 1

Example 3

Solution

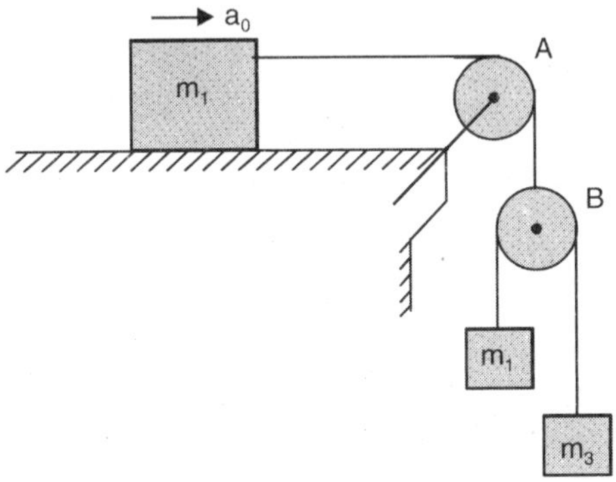

Fig. E3(a)

For m_1 :

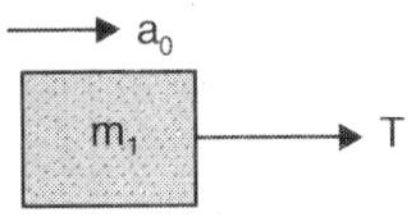

Fig. E3(b)

$$T = m_1 a_0 \qquad \ldots(i)$$

For m_2 :

Frame of reference of pulley B.

$$T_1 + m_2 a_0 - m_2 g = m_2 a$$

$$m_2 g - T_1 = m_2 (a_0 - a)$$

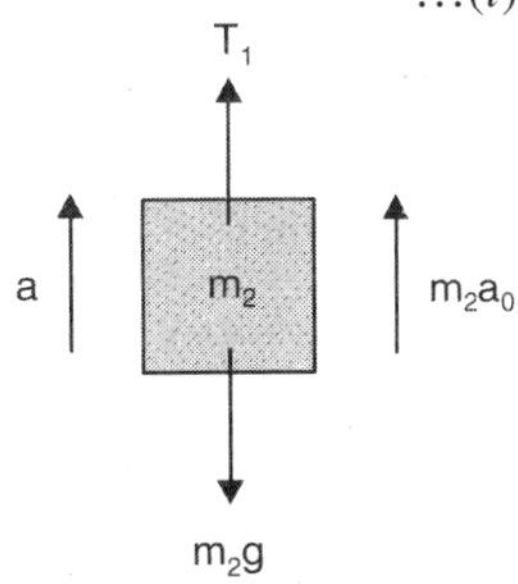

Fig. E3(c)

For m_3 :
Frame of reference of pulley B.

$$m_3 g - T_1 - m_3 a_0 = m_3 a$$

$$m_3 g - T_1 = m_3(a_0 + a) \quad \ldots(iii)$$

$$T = 2T_1 \quad ..(iv)$$

Using above all we get

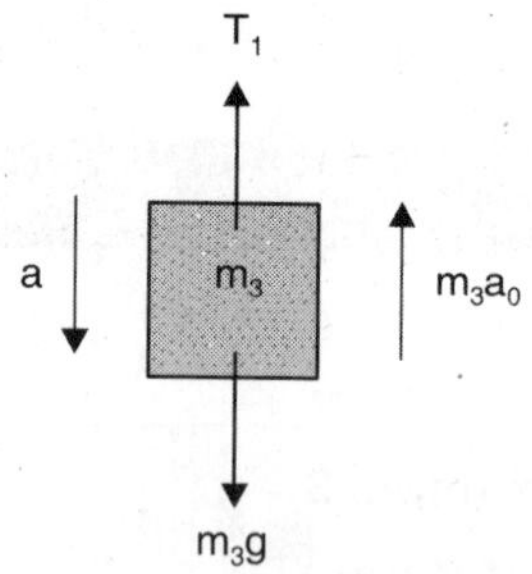

Fig. E3(d)

$$\mathbf{a_0 = \frac{g}{1 + \frac{m_1}{4}\left(\frac{1}{m_2} + \frac{1}{m_3}\right)}} \quad \textbf{...Ans.}$$

Example 4

Solution

At highest point

$$a_t = 0$$

$$\text{And } a_n = \frac{V^2}{\rho}$$

$$a_n = \frac{V^2}{\rho} = \frac{(50 \times 5/18)^2}{21}$$

$$a_n = 9.816 \text{ m/s}^2$$

$$\mathbf{a = a_n = 9.816 \text{ m/s}^2} \quad \textbf{...Ans.}$$

Example 5

Solution

FBD for each block

For m_1

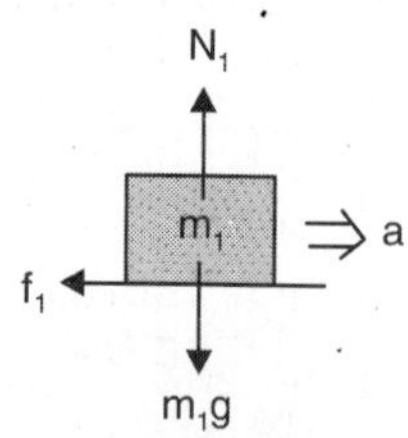

Fig. E5(a)

$$N_1 = m_1 g$$

$$f_1 = m_1 a$$

$$\mu_s N_1 = m_1 a$$

$$0.4 \times 2 \times \times 9.8 = 2 \times a$$

$$a = 3.9 \approx 4 \text{ m/s}^2$$

$$a_{\max}\big|_{m_1} = 4 \text{ m/s}^2$$

For m_2

$$T - f_2 = (m_1 + m_2)\, a$$
$$T - \mu N_2 = (2 + 4) \times 4$$
$$T - 0.2 \times (2 + 4) \times 9.8 = (2 + 4) \times 4$$
$$T = 35.77 \text{ N}$$

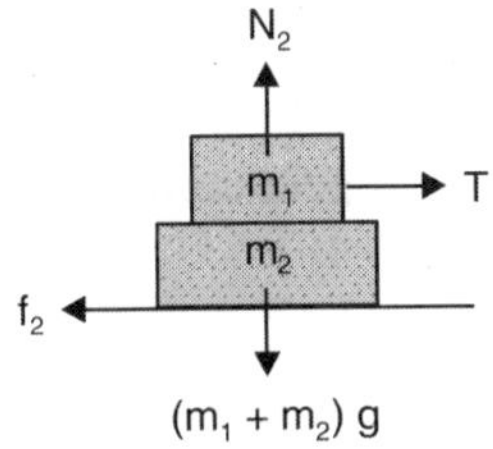

Fig. E5(b)

For m_3

$$m_3\, g - T = m_3\, a$$
$$m_3\,(9.8 - 4) = T$$
$$m_3 = \frac{35.77}{(9.84)}$$

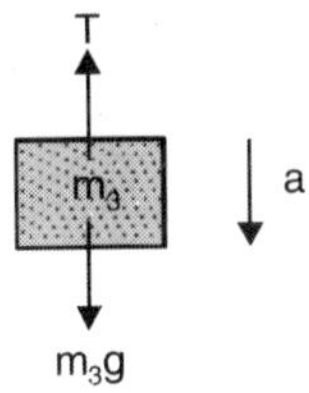

Fig. E5(c)

$$= \mathbf{6.15\ kg}$$ **...Ans.**

Example 6

Solution

$$\mathbf{a_A = 0}$$
$$\mathbf{a_B = 4.114\ m/s^2}$$
$$\mathbf{a_C = 0.162\ m/s^2}$$

Example 7

Solution

We are interested in writing the equation of motion of the fluid particle in the frame of reference of container, as resultant of these two forces should be perpendicular to the water level surface so along x, the resultant is zero.

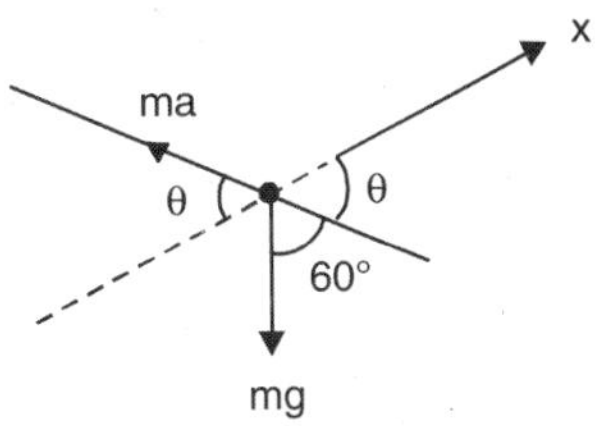

Fig. 7(a)

So, $ma \cos\theta = mg \cos(60° + \theta)$

$3 \cos\theta = 10\,(\cos 60° \cos\theta - \sin 60° \sin\theta)$

we get

$$\tan\theta = \frac{2}{5\sqrt{3}} = 0.23$$
$$\theta = \mathbf{\tan^{-1}(0.23)}$$ **...Ans.**

Example 8

Solution

For impending motion to the left, i.e. force of friction will act in opposite direction as shown

FBD of A

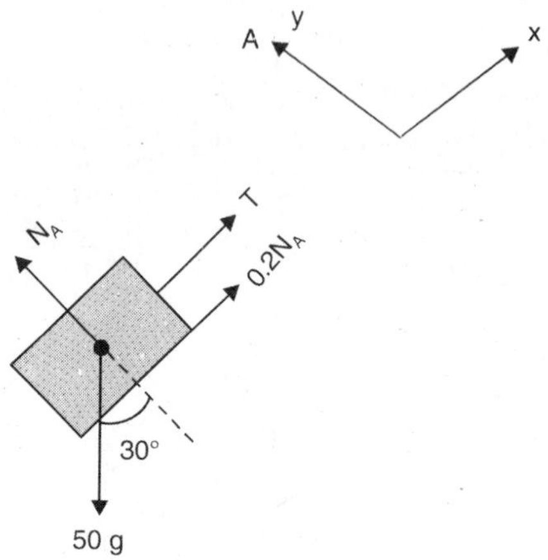

Fig. E8(a)

Along x-axis

$$T + 0.2\ N_A - 50\ g \sin 30° = 0 \qquad \ldots(i)$$

Along y-axis

$$N_A - 50\ g \cos 30° = 0$$
$$N_A = 424.79 \text{ N}$$

Now use N_A in equation (i) we get

$$T = 160.29 \text{ N}$$

For block B along y-axis $N_B = 100$ g

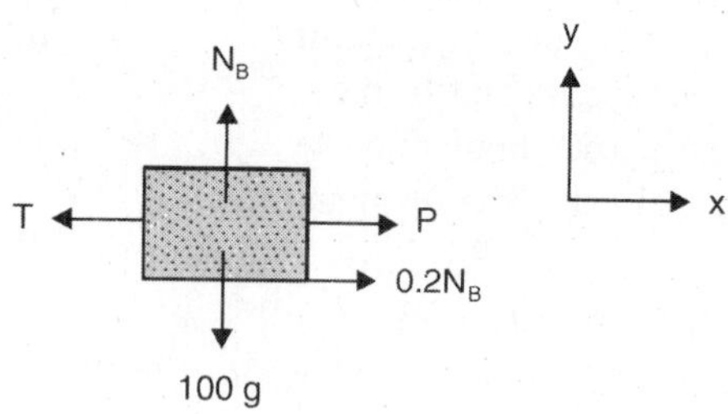

Fig. E8(b)

Along X-axis

$$P + 0.2N_B - T = 0$$
$$P + 0.2 \times 100 \times 9.8 - 160.29 = 0$$
$$P = -35.91 \text{ N}$$

(Negative value of P is not possible then we conclude here that the $P_{min} = 0$, system cannot move left)
For impending motion of block to right, only direction of friction forces will be opposite in sense.
We get after simplifying according to above

$$\textbf{Sought } P = \textbf{526.41 N} \qquad \textbf{...Ans.}$$

Example 9

Solution

At point A

$$a_n = \frac{V^2}{\rho} = \frac{(50\times(5/18))^2}{(40+0.6)}$$

$$= 4.751 \text{ m/s}^2$$

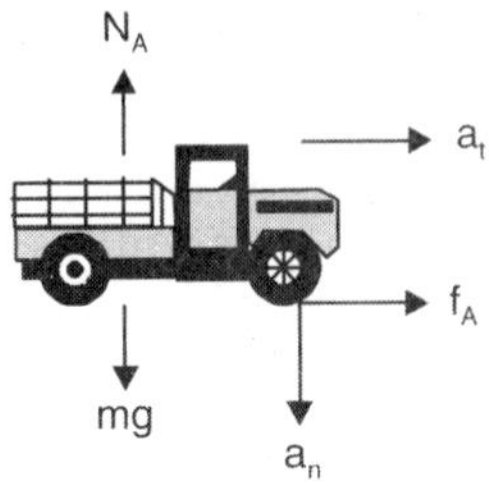

Fig. E9(a)

so,

$$mg - N_A = m\, a_n$$

$$N_A = mg - ma_n$$

$$= 250 \times 9.8 - 250 \times 4.751$$

$$N_A = \textbf{1264.685 N} \qquad \textbf{...Ans.}$$

Now, according to the problem speed changes from 50 km/h to 100 km/h in 10 seconds.

$$V = u + a_t\, t$$

$$\left(\frac{100\times 5}{18}\right) = (13.889) + a_t \times (10)$$

$$a_t \doteq 1.389 \text{ m/s}^2$$

According to the problem

$$a_A = a_B$$

$$a_n\big|_A = a_n\big|_B$$

$$4.751 = \frac{V_B^2}{\rho_B} = \frac{(27.778)^2}{\rho_B}$$

$$\text{Then } \rho_B = 162.41 \text{ m}$$

$$R_B = \text{Radius of curvature at } B$$

$$= 162.41 + 0.6$$

$$= \textbf{163 m} \qquad \textbf{...Ans.}$$

At B

$$\Sigma F_y = m a_y$$

$$N_B - mg = \frac{mV_B^2}{\rho_B}$$

$$N_B = (250 \times 9.8) + (250)(4.751)$$

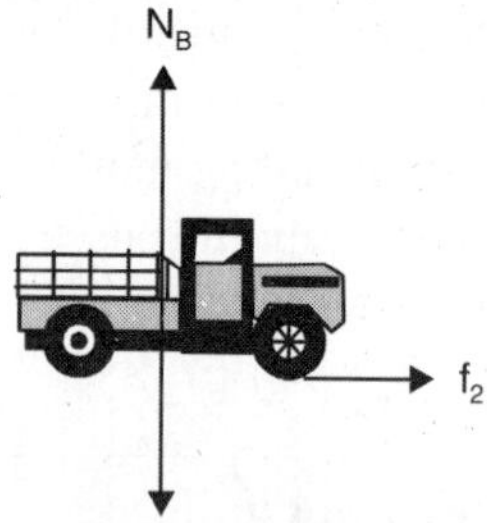

Fig. E9(b)

$$= \mathbf{3640.25\ N}$$ **...Ans.**

Example 11

Solution

$$\rho = \text{Density of ball.}$$

$$2\rho = \text{Density of water.}$$

Now velocity of sphere when it comes with the contact of water level

$$v = \sqrt{2 \times 9.8 \times 19.6}$$

$$= 19.6 \text{ m/s}$$

Since sphere feels retardation inside the water due to upthrust by water.
Mass of sphere = v ρg

$$\text{Upthrust} = v\,(2\rho g)$$

$$\text{Weight} = v\rho g$$

$$\Sigma F_y = m\,a_y\,v(2\rho)\,g$$

$$v(2\rho)g - v\rho g = v\rho a$$

$$a = g$$

$$= 9.8 \text{ m/s}^2$$

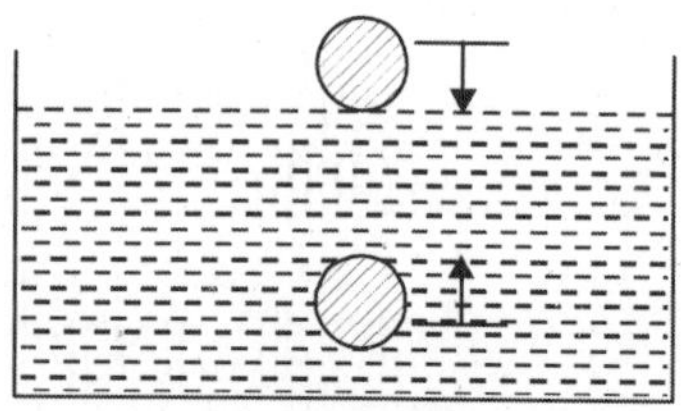

Fig. E11(a)

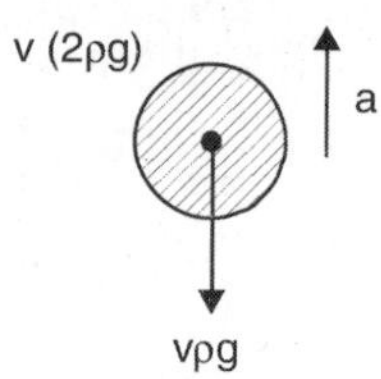

Fig. E11(b)

$$\text{So sought time} = 2\left(\frac{v}{a}\right)$$

$$= 2\left(\frac{19.6}{9.8}\right)$$

$$= \mathbf{2 \times 2 = 4\ sec.}$$ **...Ans.**

Example 12

Solution

Free body diagram of the particle

Along X-axis

$$T \cos 30° - N \cos 60° = mr\omega^2$$

$$(r = 0.2 \cos 30°) \qquad \ldots(i)$$

Along Y-axis

$$T \sin 30° + N \sin 60° = mg \qquad \ldots(ii)$$

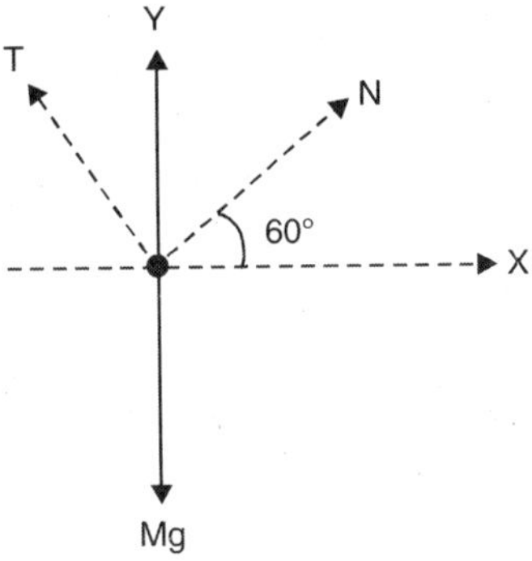

Fig. E12(a)

When particle leaves constant $N = 0$

And by equation (*i*) and (*ii*) we get

$$\omega = 9.9 \text{ rad/s}$$

By kinematical relationship

$$\omega = \omega_0 + \alpha t$$

$$\omega_0 = 0$$

$$9.9 = (0.5)\, t$$

$$\therefore \quad t = \frac{9.9}{0.5} = \mathbf{19.8 \text{ second}}$$

And $\mathbf{T = 0.39\ N}$ **...Ans.**

Example 13

Solution

In the frame of reference of bracket the FBD of the drum is

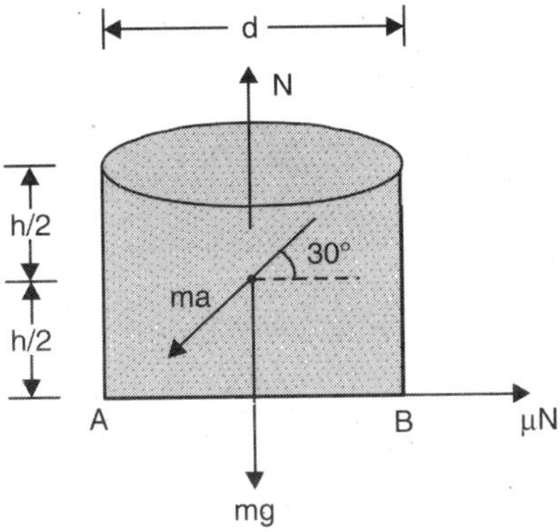

Fig. E13(a)

For equilibrium, for the observer is sitted on the bracket, Algebraic sum of all forces along each axis should be zero.

Along x-axis

$$\mu N - ma \cos 30° = 0 \qquad \ldots(i)$$

Along y-axis

$$N - mg - ma \sin 30° = 0 \quad \text{...(ii)}$$

For overturning

$$\Sigma M_A = 0$$

$$ma \cos 30° \times h/2 - m\, a \sin 30° \times d/2 - mg \times d/2 = 0$$

$$\boldsymbol{h/d = 4} \quad \textbf{...Ans.}$$

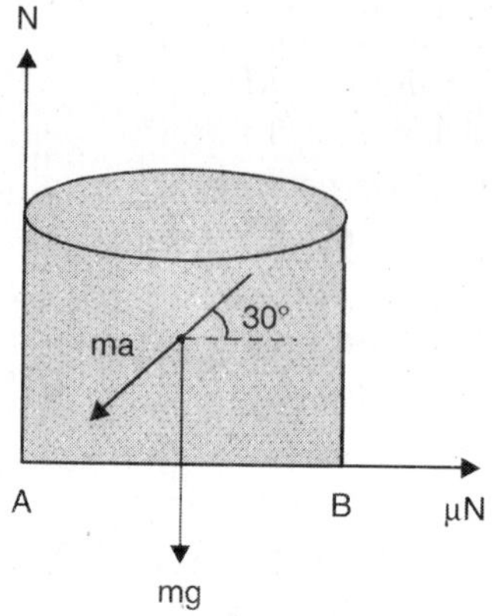

Fig. E13(b)

Example 14

Solution

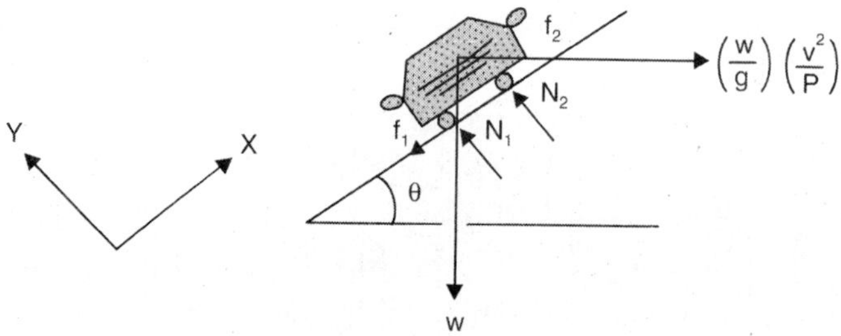

Fig. E14(a)

For dynamic equilibrium

$$N_1 + N_2 - W \cos \theta = \left(\frac{W}{g}\right)\left(\frac{V^2}{\rho}\right) \sin \theta$$

$$N_1 + N_2 = W \cos \theta + \left(\frac{W}{g}\right)\left(\frac{V^2}{\rho}\right) \sin \theta \quad \text{...(i)}$$

Again

$$\sum F_x = 0$$

$$f_1 + f_2 = \left(\frac{W}{g}\right)\left(\frac{V^2}{\rho}\right)\cos\theta - W\sin\theta$$

$$\mu\,(N_1 + N_2) = \left(\frac{W}{g}\right)\left(\frac{V^2}{\rho}\right)\cos\theta - W\sin\theta \qquad \ldots(ii)$$

Using (*i*) and (*ii*)
We get

$$V = \sqrt{\rho g \tan(\phi + \theta)}$$

$$\theta = 20°,\ \rho = 100 \text{ m},$$

$$g = 9.81 \text{ m/s}^2,\ \phi = \tan^{-1}(0.2) = 11.31°$$

$$V_{max} = \sqrt{(100)(9.81)\tan(31.31°)}$$

$$= 24.427 \text{ m/s}$$

$$\approx \mathbf{88\ km/h}$$ **...Ans.**

Example 15

Solution

FBD of the block.

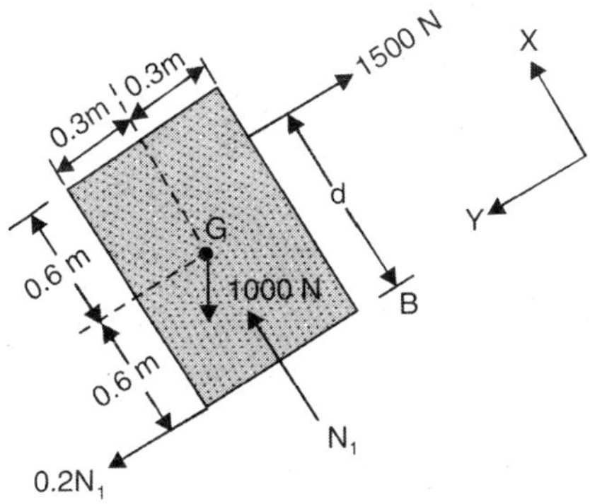

Fig. E15(a)

$$\Sigma F_y = ma_y$$

$$N_1 = 1000\cos 37° = 800 \text{ N}$$

$$\Sigma F_x = ma_x$$

$$1500 - (0.2)(800) - 1000 \sin 37° = \left(\frac{1000}{9.81}\right) a_x$$

$$a_x = 7.26 \text{ m/s}^2$$

$$\text{Now} \quad \Sigma M_A = 0, \Sigma M_B = 0$$

$$\mathbf{d = 0.696 \text{ m}}$$

$$\text{and} \quad \mathbf{d = 0.376 \text{ m}}$$ **...Ans.**

Example 16

Solution

FBD of beam

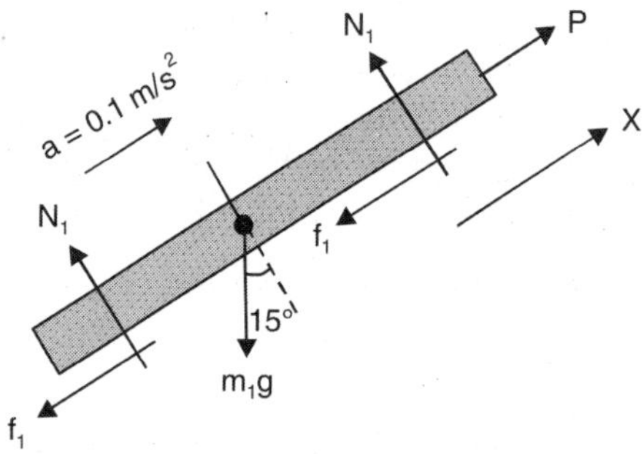

Fig. E16(a)

$$\Sigma F_x = m\, a_x$$

$$P - 2 f_1 - m_1 g \sin \alpha = m_1 a$$

$$P - 2f_1 = (100)(0.1) + 100 \times 9.81 \times \sin 15°$$

$$P - 2f_1 = 263.902 \text{ N}$$

FBD of cylinder

$$\Sigma \tau_0 = I_0\, \alpha$$

$$f_1 \times 2R - (m_2\, g \sin a) R = \left(\frac{3}{2} m_2 R^2\right) \times a$$

$$\text{Now } \alpha = \frac{a}{R}$$

Applying all above equations, we get

$$\mathbf{P = 317.682 \text{ N}}$$ **...Ans.**

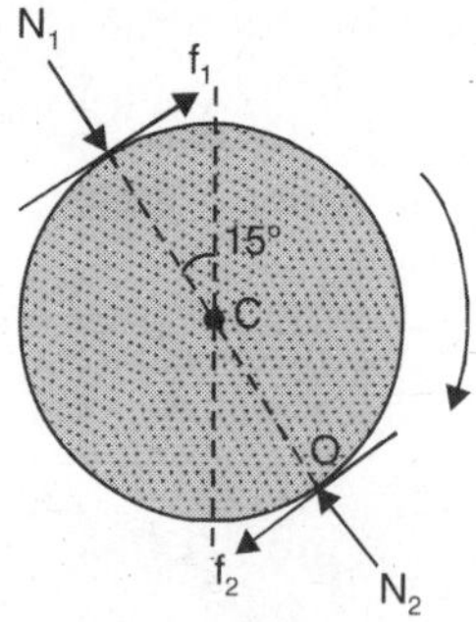

Fig. E16(b)

Example 18

Solution

Let a_0 = acceleration of prism

FBD of prism

$$\Sigma F_x = m\, a_x$$

$$N \sin\theta = M\, a_0 \qquad \ldots(i)$$

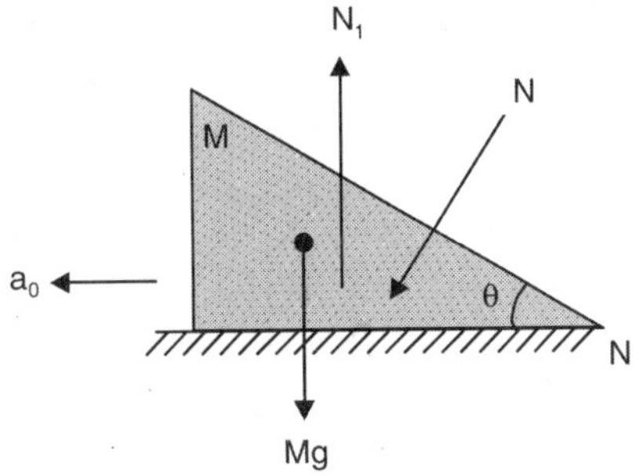

Fig. E18(a)

For block of mass m

Frame of reference of wedge M.

Along the plane

$$ma_0 \cos\theta + mg \sin\theta = ma \quad \ldots(ii)$$

Perpendicular to the plane

$$N + m\, a_0 \sin\theta = mg \cos\theta \quad \ldots(iii)$$

Using above all equation we get,

$$\mathbf{a} = \frac{(\mathbf{M}+\mathbf{m})\mathbf{g}\sin\theta}{\mathbf{M}+\mathbf{m}\sin^2\theta} \qquad \ldots\textbf{Ans.}$$

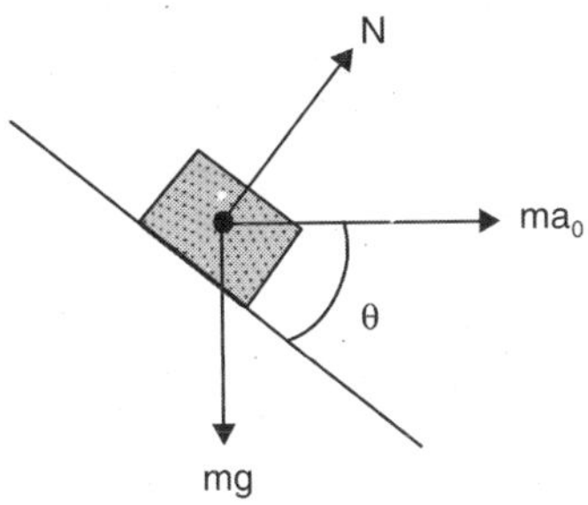

Fig. E18(b)

Example 22

Solution

In the frame of reference of elevator.

FBD of particle

Along the plane

$$mg \sin\theta + ma \sin\theta = ma_0$$

$$a_0 = (g + a) \sin\theta$$

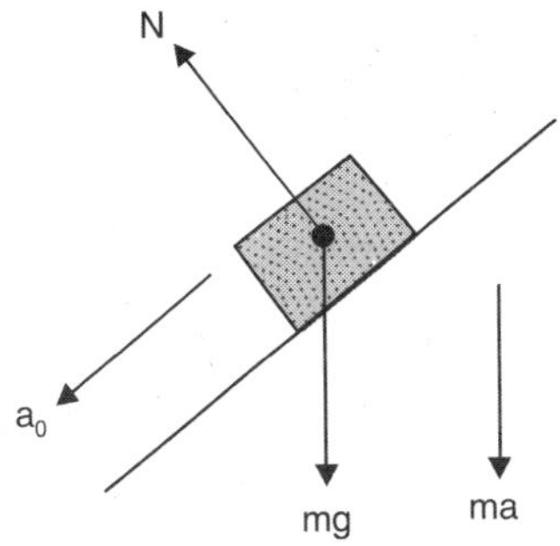

Fig. E22(a)

Now

S = distance traveled by particle with respect to elavator = $L/\cos\theta$

$$\frac{L}{\cos\theta} = \frac{1}{2}(g + a) \sin\theta \cdot t^2$$

$$\text{Sought time} = \left[\frac{\mathbf{2L}}{(\mathbf{g}+\mathbf{a})\ \mathbf{sin\theta\ cos\theta}}\right]^{1/2} \qquad \ldots\textbf{Ans.}$$

Example 23

Solution

Clearly concentrate on FBD of the boy and cage.

Cage:

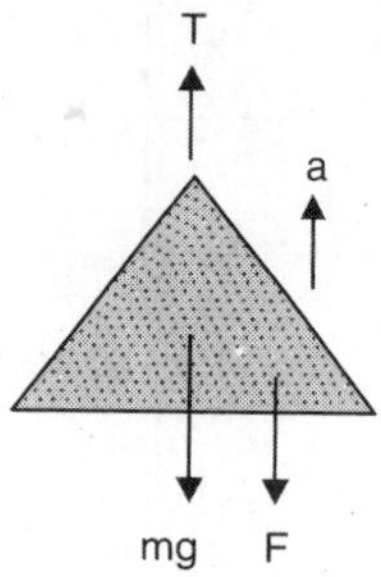

Fig. E23(a)

Boy:

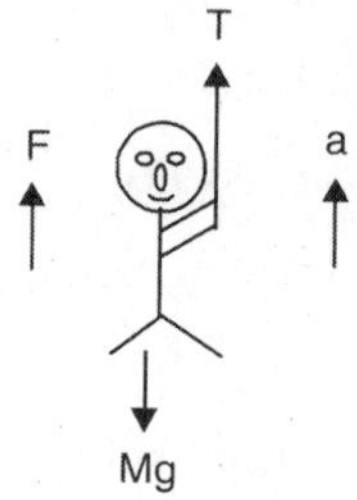

Fig. E23(b)

$$T\text{-}F\text{-}mg = ma \quad \ldots(i)$$

$$T + F - Mg = Ma \quad \ldots(ii)$$

By these two equations, we get

$$\mathbf{a} = \left(\frac{\mathbf{2}}{\mathbf{Mm}}\right)\mathbf{F} - \mathbf{g} \quad \textbf{...Ans.}$$

$$\mathbf{T = F\ (M + m) - 2\ m\ Mg} \quad \textbf{...Ans.}$$

Level 2

Example 1

Solution

Let m be the mass of the particle P of the liquid having co-ordinate (x, y).

In the rotating frame of reference the forces acting on the liquid particle is as shown in figure

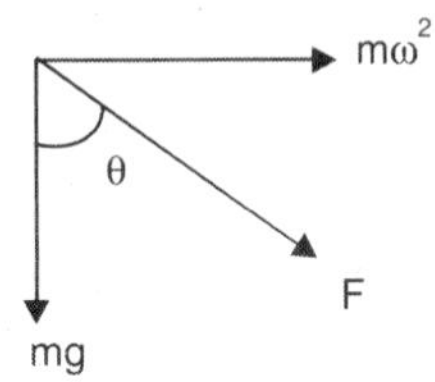

Fig. E1(a)

$$\tan\theta = \frac{mx\omega^2}{mg}$$

$$\frac{dy}{dx} = \frac{xw^2}{g}$$

$$\int_0^y dy = \frac{w^2}{g}\int_0^x x\,dx \Rightarrow y = \frac{w^2x^2}{2g}$$

Here take $y = h, x = R$

Now, $\omega = 2\pi \times 2 = 4\pi$ rad/s

$R = 0.05$ m

$$h = \frac{(4\pi)^2 \times (0.05)^2}{2 \times 9.8}$$

$= \mathbf{2\ cm}$ **...Ans.**

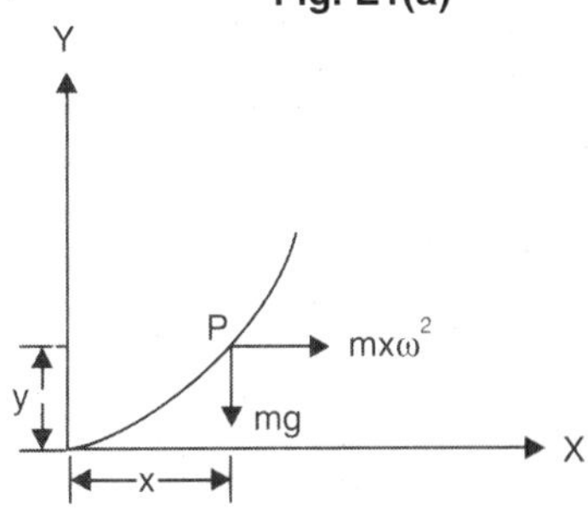

Fig. E1(b)

Example 2

Solution

For largest allowable acceleration bundle may just slide to left.

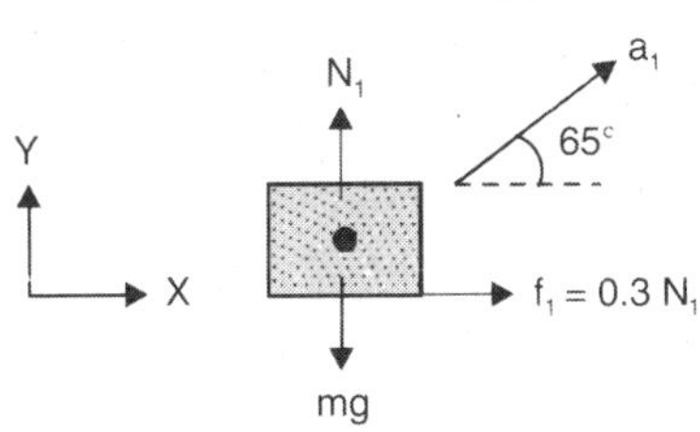

Fig. E2(a)

$$\Sigma F_y = ma_y$$
$$N_1 - mg = m\, a_1 \sin 65° \quad ...(i)$$
$$\Sigma F_x = ma_x$$
$$0.3\ N_1 = m\, a_1 \cos 65° \quad ...(ii)$$

By (*i*) and (*ii*)
We get

$$\mathbf{a_1 = 19.526\ m/s^2} \quad ...\text{Ans.}$$

For largest allowable deceleration

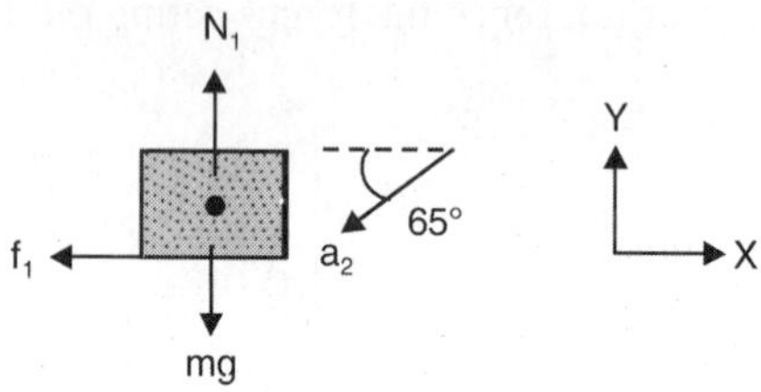

Fig. E2(b)

$$\Sigma F_x = m\, a_x$$
$$0.3\ N_1 = m\, a_2 \cos 65° \quad ...(iii)$$
$$\Sigma F_y = m\, a_y$$
$$mg - N_1 = m\, a_2 \sin 65° \quad ...(iv)$$

By (*iii*) and (*iv*)
We get

$$a_2 = 4.238\ m/s^2 \quad \textbf{...Ans.}$$

Example 3

Solution

FBD of slider at any general position.
For dynamic equilibrium

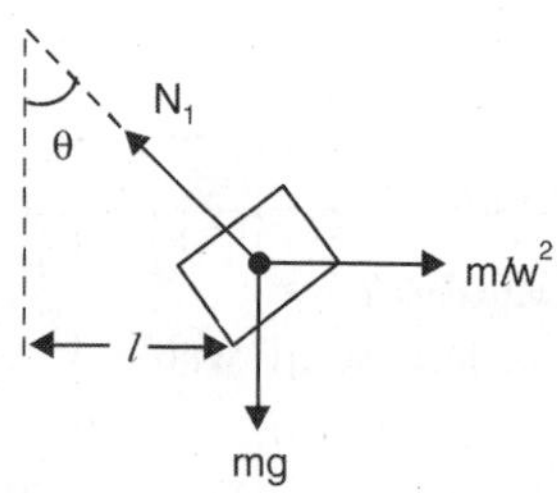

Fig. E3(a)

$$\Sigma F_x = ma_x = 0$$
$$mg \cos \theta - N_1 + ml\omega^2 \sin \theta = 0$$
$$\text{For } \theta = 70°$$
$$N_1 = 11.214\ N$$

$$\Sigma F_y = ml\omega^2 \cos \theta - mg \sin \theta$$
$$= (0.5)\ (0.6)\ (\sin 70°)\ (6 \times 6)\ (\cos 70°) - (0.5)\ (10)\ (\sin 70°)$$
$$= -\ 1.138\ N$$

Friction force acts upward (or forward) on the rod because impending motion of slider is downward (or backward).

$$\mu N_1 = F$$

$$\mu = \frac{1.138}{11.214}$$

$$= 0.102 \qquad \text{...Ans.}$$

$$\text{For } \theta = 30°$$

Impending motion of slider is upward (or forward)

And we get $\mu = \mathbf{0.32}$ **...Ans.**

Example 7

Solution

FBD of block

$$\Sigma F_y = m\,a_y$$

$$F - T - W = ma \qquad ...(1)$$

$$F = (\text{volume of block}) \times (\text{density of water}) \times (g + a)$$

$$= \left(\frac{1}{800}\right) \times (1000) \times (10 + 1)$$

$$= 13.75 \text{ N}$$

$$W = 10 \text{ N}$$

$$a = 1 \text{ m/s}^2$$

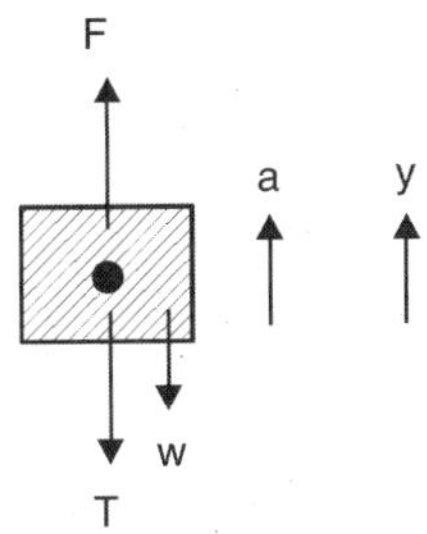

Fig. E7(a)

Putting all in equation (i)

We get

$$\mathbf{T = 2.75\ N} \qquad \text{...Ans.}$$

$$\text{When } T = 0$$

$$F - W = ma$$

Now, a = 3.75 m/s² $\qquad$ **...Ans.**

Example 8

Solution

Here according to the problem in first case block of weight 40 N is attached and allowed to fall freely and in the second case block is removed and only 40 N force is applied on the string.

For case I

For block

$$mg - T = ma$$

$$T = mg - ma$$
$$= m(g - a)$$
$$T = \frac{w}{g}(g - a) \qquad \ldots(i)$$

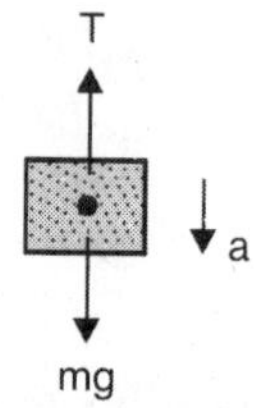

Fig. E8(a)

For drum

Newton's law in rotation

$$\Sigma\tau_G = I_G\,\alpha$$
$$\left(\frac{m_1 r 2D}{2}\right)\alpha \qquad \ldots(ii)$$
$$w_1 = m_1 g$$

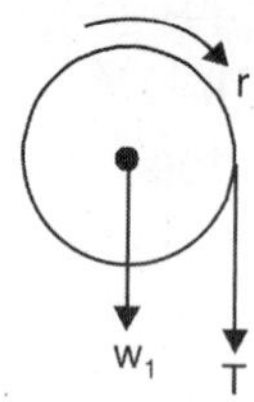

Fig. E8(b)

By (*i*) and (*ii*)

We get

$$\alpha = \mathbf{9.689\ rad/s} \qquad \textbf{...Ans.}$$

Case II:

$$40 \times r_D = I_G\,\alpha$$
$$\alpha = \mathbf{17.44\ rad/s^2} \qquad \textbf{...Ans.}$$

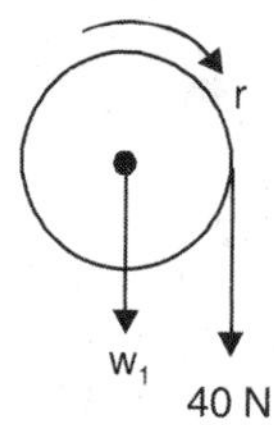

Fig. E8(c)

Example 9

Solution

Applying Newton's law of rotation

$$\Sigma\tau_A = I_A\,\alpha$$
$$mg \times \left(\frac{l}{2}\right) = \left(\frac{ml^2}{3}\right)\alpha$$
$$mg \times (0.8) = \frac{m}{3}(1.6)^2\left(\frac{a}{0.8}\right)$$
$$a = 7.358\ \text{m/s}^2$$

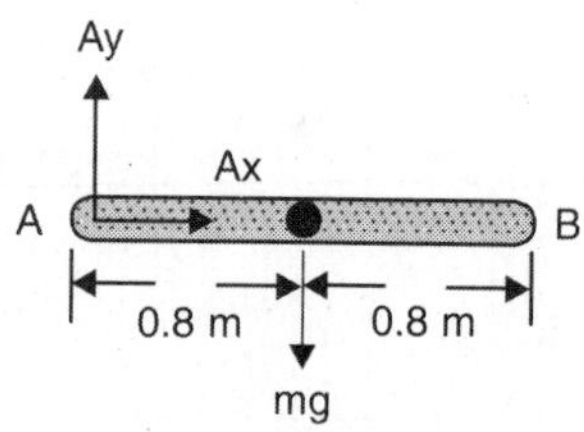

Fig. E9(a)

Now,

$$\Sigma F_y = ma_y$$
$$mg - A_y = ma$$

$$20 \times (9.8) - A_y = 20\ (7.35)$$

$$A_y = 49.05 \text{ N} \uparrow$$

There is no horizontal force acting on the rod so

$$A_x = 0$$

$$\mathbf{R_A = 49.05\ N} \qquad \textbf{...Ans.}$$

Example 10

Solution

For block

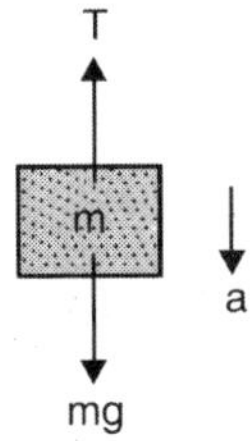

Fig. E10(a)

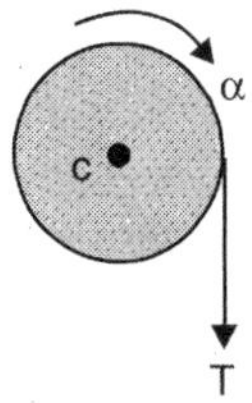

Fig. E10(b)

For disk

$$\Sigma F_y = ma_y$$

$$mg - T = ma \qquad ...(i)$$

Newton's law of rotation

$$\Sigma \tau_C = I_C\, \alpha$$

$$T \times R = \frac{MR^2}{2} \times \left(\frac{a}{R}\right) \qquad ...(ii)$$

By using (*i*) and (*ii*) we get

$$\alpha = \frac{\mathbf{2\ mg}}{\mathbf{R(2m + M)}} \qquad \textbf{...Ans.}$$

$$\text{And } a = R\alpha = \frac{2mg}{(2m + M)}$$

Now

$$S = 2\,R$$

$$u = 0$$

$$V^2 = u^2 + 2as$$

$$V^2 = 2 \times \frac{2mg}{(2m + M)} \times 2\text{R}$$

$$V = \mathbf{2\sqrt{\frac{2\ mgR}{2\ m + M}}} \qquad \textbf{...Ans.}$$

Example 11

Solution

Concept behind this problem is that lateral thrust on outer rail must be equal to centrifugal force

$$F = \text{lateral thrust}$$

Then

$$F = \frac{mv^2}{\rho}$$

$$= \left(\frac{w}{g}\right)\left(\frac{v^2}{\rho}\right) \qquad \ldots(i)$$

locomotive starts from rest at A, velocity v after covering $(R–S)$ distance is

$$v^2 = u^2 + 2a_t\,\Delta s$$

$$= 2\,(g/10)\,(R - S)$$

$$\frac{v^2}{p} = \left(\frac{g}{5}\right)\left(\frac{RS}{\rho}\right) \qquad \ldots(ii)$$

Now put value of $\frac{v^2}{\rho}$ from equation (*ii*) to (*i*)

We get

$$F = \left(\frac{W}{5}\right)\frac{(RS)S}{R^2}$$

For maximum thrust

$$\frac{dF}{dS} = 0$$

$$\text{We get } S = \frac{R}{2}$$

$$\text{Sought } \mathbf{F_{max}} = \mathbf{30\ kN}$$ **...Ans.**

Example 12

Solution

For rod AB

$$\Sigma F_y = m\,a_y$$

$$2w - T = \left(\frac{2w}{g}\right) a_G \qquad \ldots(i)$$

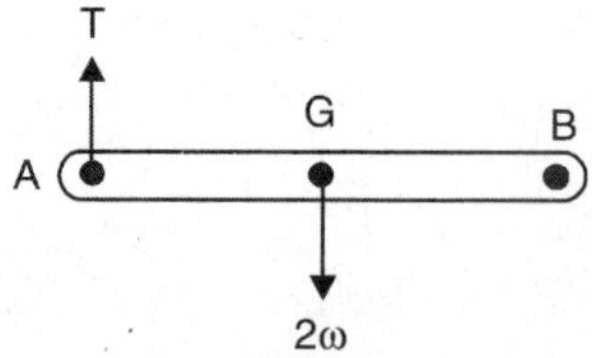

Fig. E12(a)

$$\Sigma\tau_G = I_G\,\alpha$$

$$T \times \left(\frac{L}{2}\right) = \left(\frac{2w}{g}\right)\left(\frac{L^2}{12}\right)\alpha$$

$$\left(\frac{TL}{2}\right) = \left(\frac{2w}{g}\right)\left(\frac{L^2}{12}\right)\left(\frac{a}{L/2}\right) \qquad \ldots(ii)$$

From equation (*i*) and (*ii*), we get,

$$a = \frac{3g}{4}$$

$$\alpha = \frac{3g}{2L}$$

Now,

$$\bar{a}_B = \bar{a}_{B/G} + \bar{a}_G$$

$$= \left(\frac{3g}{2L} \times \frac{L}{2}\right)(-\hat{j}) + \frac{3g}{4}(-\hat{j})$$

$$= \frac{3g}{4}(-\hat{j}) + \frac{3g}{4}(-\hat{j})$$

$$|\bar{\mathbf{a}}_{\mathbf{B}}| = \frac{\mathbf{3g}}{\mathbf{2}} \downarrow \qquad \textbf{...Ans.}$$

Example 13

Solution

We are interested to solve this problem in the frame of reference of car.

Now FBD of bob

Applying Newton's law

$$\Sigma F_t = m\,a_t$$

$$m\,a_0 \sin\theta + mg\cos\theta = m\,a_t$$

$$m\,a_0 \sin\theta + mg\cos\theta = m\frac{vdv}{Rd\theta}$$

$$\int_0^\theta Ra_0 \sin\theta d\theta + \int_0^\theta Rg\cos\theta\, d\theta = \int_0^v vdv$$

$$2\,a_0\,R(1 - \cos\theta) + 2\,gR\sin\theta = V^2 \qquad \ldots(i)$$

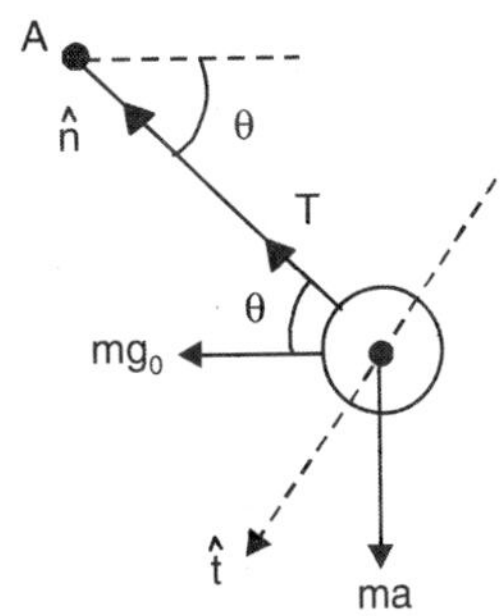

Fig. E13(a)

$$\Sigma F_n = m a_n$$

$$T - mg \sin\theta = m\frac{V^2}{R} \qquad \ldots(ii)$$

Using equation (*i*) in equation (*ii*)
We get

$$\mathbf{T = m\,[3g \sin\theta + a_0\,(2 - 3\cos\theta)]} \qquad \ldots\textbf{Ans.}$$

$$\mathbf{T =}\Big|_{\theta=\frac{\pi}{2}} = \mathbf{m\left[3g + 2a_0\right]} \qquad \ldots\textbf{Ans.}$$

Example 14

Solution

Let at any instant of motion of chain, length of the chain inside the tube be x.

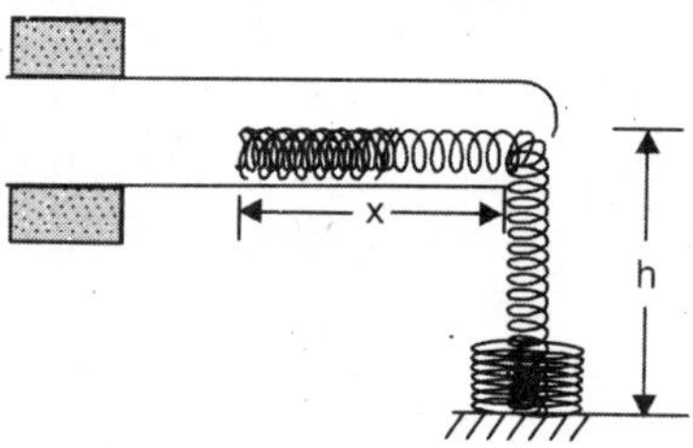

Fig. E14 (a)

$$m\vec{a} = \vec{F} + \vec{V}_{\text{rel}}\frac{dm}{dt}$$

$$\text{here } \vec{V}_{\text{rel}} = 0$$

$$\lambda x a = T \qquad \ldots(i)$$

For hanging part

$$\lambda h a = \lambda\, hg - T \qquad \ldots(ii)$$

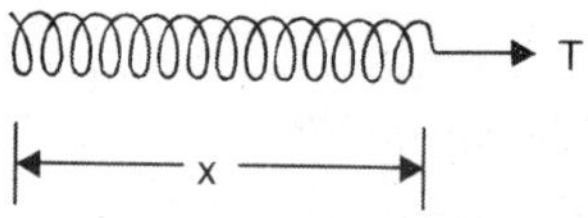

Fig. E14(b)

By using (*i*) and (*ii*)

$$(x + h)\,V\frac{dv}{ds} = gh$$

$$-(x + h)\,V\frac{dv}{ds} = gh$$

$$\int_0^V V\,dV = -gh\int_{(l-h)}^{0}\frac{dx}{x+h}$$

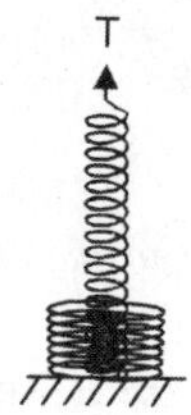

Fig. E14(c)

$$\text{We get } \mathbf{v} = \sqrt{\mathbf{2gh\ln}\left(\frac{\mathbf{l}}{\mathbf{h}}\right)} \qquad \ldots\textbf{Ans.}$$

Example 16

Solution

Since, the board is accelerating upwards and man is on the board, so man is also accelerating upward with acceleration a_2.

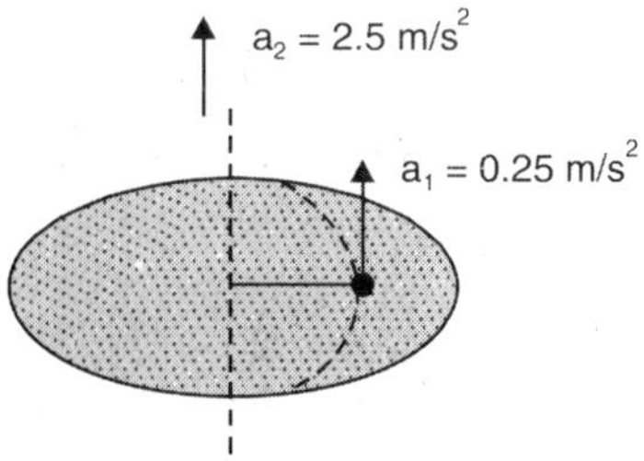

Fig. 16

For man

$$N - mg = ma_2$$

$$N = m(g + a_2) \quad ...(i)$$

Suppose man covers distance x on the board

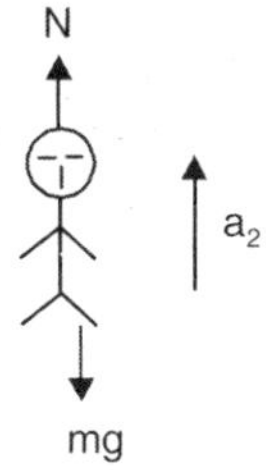

Fig. 16(a)

So,

$$V^2 = u^2 + 2\,a_t\,x$$

$$V^2 = 0 + 2 \times \left(\frac{25}{100}\right)x$$

$$V^2 = \frac{x}{2} \quad ...(ii)$$

Now,

Resultant acceleration of man on the board due to tangential and normal acceleration is vector sum of both and this is provided by frictional force. Hence,

$$m\left[\left(\frac{V^2}{R}\right)^2 + a2t\right]^{1/2} = mN \quad ...(iii)$$

Use (ii) and (i) in (iii)

$$m\left[\left(\frac{x}{2R}\right)^2 + a2t\right]^{1/2} = \text{m}m\,(g + a_2)$$

We get,

Sought distance $\mathbf{x = 2\sqrt{6}\ m}$ **...Ans.**

Example 17

Solution

Situation before rope cut for m_2.

$$T_1 = m_2\,g \quad ...(i)$$

T_1 = Stretched force in left spring to hold the block m_2 in equilibrium

For m_3

$$m_3 g + T_2 - T_0 = 0 \quad \text{...(ii)}$$

T_2 = Stretched force in right spring to hold the block m_3 in equilibrium

$$T_0 = \text{Tension in the rope}$$

$$T_0 = (m_1 + m_2) g \quad \text{...(iii)}$$

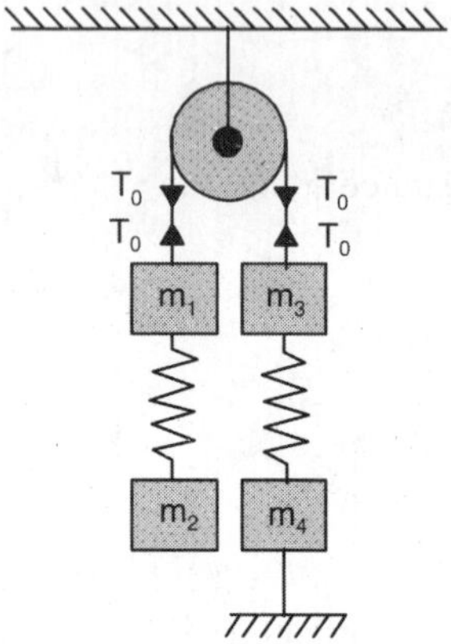

Fig. E17(a)

Use value of T_0 in equation (*ii*)

We get,

$$T_2 = (m_1 + m_2 - m_3) g \quad \text{...(iv)}$$

Situation after rope cut for m_1

$$m_1 g + T_1 - T_0 = m_1 a_1 \quad \text{...(v)}$$

For m_2

$$m_2 g - T_1 = m_2 a_2 \quad \text{...(vi)}$$

For m_3

$$T_2 + m_2 g - T_0 = m_2 a_3 \quad \text{...(vii)}$$

For m_4

$$m_4 g - T_2 = m_4 a_4 \quad \text{...(viii)}$$

Using above all we can get our desire result as

$$A_1 = 0$$

$$A_2 = 0$$

$$A_3 = 0$$

$$\mathbf{A_4} = \frac{(\mathbf{m_3 + m_4 m_1 m_2})}{\mathbf{m_4}}\mathbf{g} \quad \textbf{...Ans.}$$

Example 18

Solution

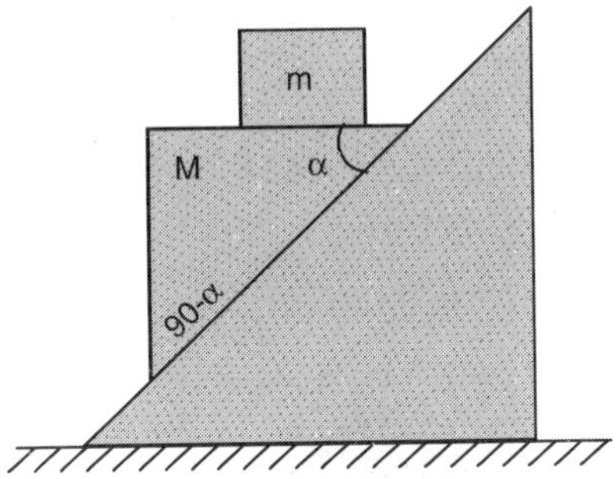

Fig. E18(a)

Let

$$a_m = \text{Acceleration of mass } m$$

$$a_M = \text{Acceleration of mass } M$$

FBD for block m

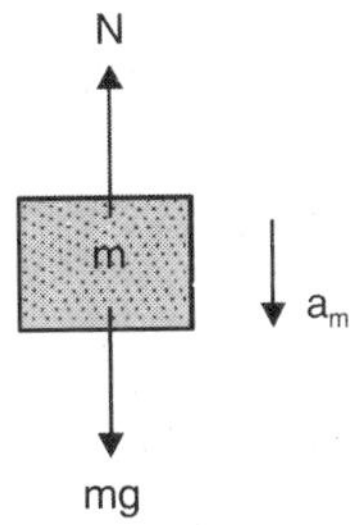

Fig. E18(b)

$$mg - N = m\,a_m \qquad \ldots(i)$$

F_{BD} for block M

$$\sum F_x = m\,a_x$$

$$Mg \sin\alpha + N \sin\alpha = M\,a_M \qquad \ldots(ii)$$

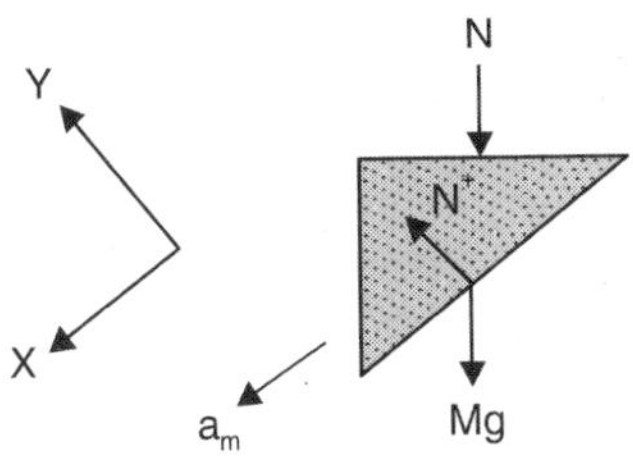

Fig. E18(c)

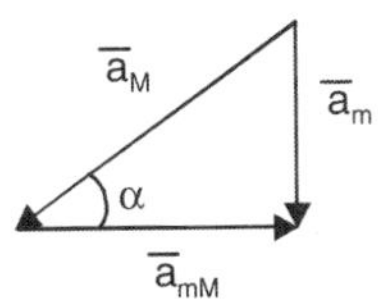

Fig. E18(d)

Now relation between a_m and a_M

$$\sin\alpha = \frac{a_m}{a_M} \qquad \ldots(iii)$$

Using above all equations we get

$$a_m = \frac{(\mathbf{M}+\mathbf{m})\mathbf{g}\sin^2\alpha}{(\mathbf{M}+\mathbf{m}\sin^2\alpha)} \qquad \ldots\textbf{Ans.}$$

Example 19

Solution

$$m\vec{a} = \vec{F} + \vec{V}_{rel}\frac{dm}{dt}$$

$$\lambda = \text{Mass per unit length}$$

$$mV_x = \left(\frac{dV_x}{dx}\right) = F_x - V_x\frac{dm}{dt}$$

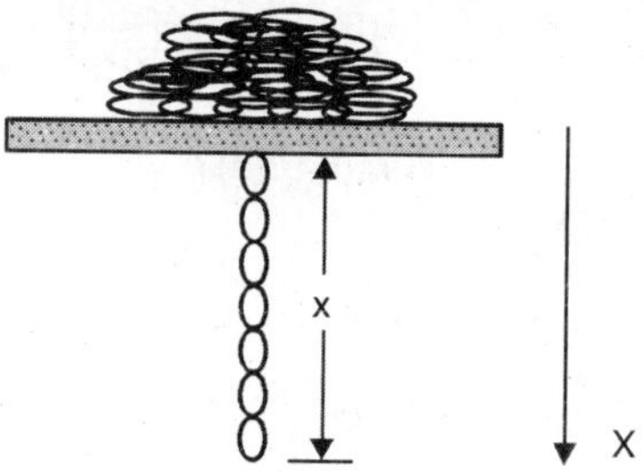

Fig.19(a)

$$\lambda x V\frac{dv}{dx} = \lambda\, xg - V\,(\lambda V)$$

$$xvdv + v^2\,dx = gx\,dx$$

$$\therefore \quad \frac{d(v^2x^2)}{2x} = gx\,dx$$

$$d\,(v^2\,x^2) = 2gx^2\,dx$$

$$v^2\,x^2 = 2g\frac{x^3}{3}$$

$$\mathbf{v} = \sqrt{\frac{\mathbf{2gx}}{\mathbf{3}}}$$ **...Ans.**

Example 20

Solution

Consider the situation, snapping of the right string. The *FBD* of rod is

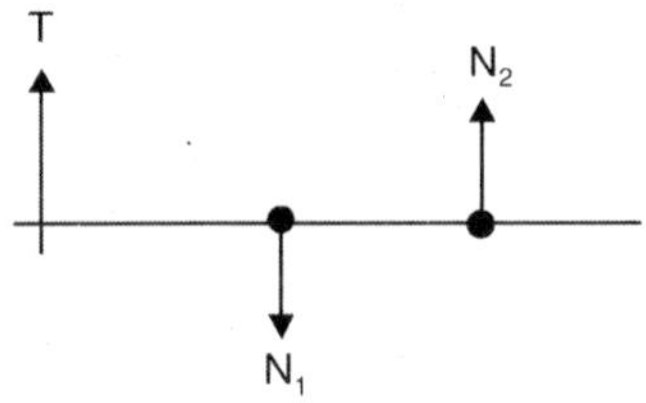

Fig. E20(a)

$$-N_1 + N_2 + T = 0 \quad \ldots(i)$$

Taking moment about '*O*'

$$N_1 l - 2N_2 l = 0$$

$$N_1 = 2\,N_2 \quad \ldots(ii)$$

By equations (*i*) and (*ii*) we get,

$$T = N_2 \quad \ldots(iii)$$

Now, let write the equations for loads

For m_1

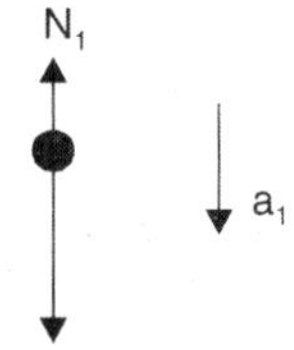

Fig. E20(b)

$$m_1 g - N_1 = m_1 a_1 \quad \ldots(iv)$$

For m_2

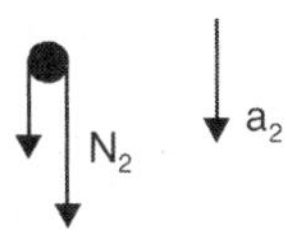

Fig. E20(c)

$$N_2 + m_2 g = m_2 a_2 \quad \ldots(v)$$

By constraint relation

$$a_1 = 2a_2 \quad \ldots(vi)$$

Using all above we get

$$\mathbf{T} = \left(\frac{\mathbf{m_1 m_2}}{\mathbf{4m_2 + m_1}}\right)\mathbf{g} \quad \textbf{...Ans.}$$

Example 21

Solution

Clearly, block B is attached with A and C so both remain stationary relative to the trolley of mass M. It means all blocks move with same accelerations with respect to ground observer. Let it be a_0.

FBD for block A

$$\Sigma F_y = m\,a_y$$

$$T_1 \cos\theta = m\,a_0 \quad \ldots(i)$$

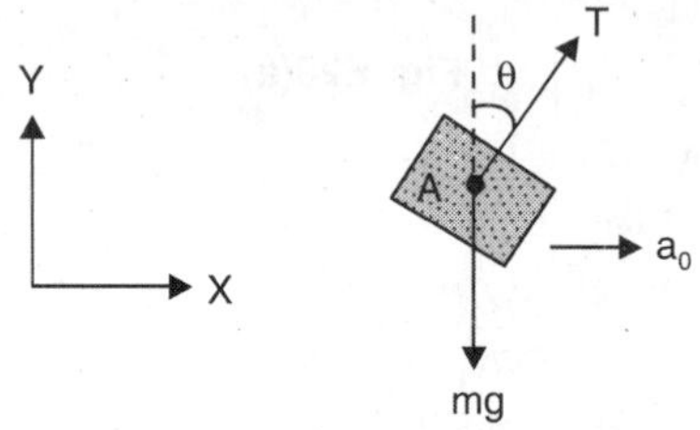

Fig. E21(a)

$$\Sigma F_x = m\,a_x$$

$$T_1 \sin\theta = m\,a_0 \quad \ldots(ii)$$

For block B

$$\Sigma F_y = m\,a_y$$

$$N_1 = mg \quad \ldots(iii)$$

$$\Sigma F_x = m\,a_x$$

$$T_2 + f_1 - T_1 = m\,a_0$$

$$T_2 + \mu N_1 - T_1 = m\,a_0 \quad \ldots(iv)$$

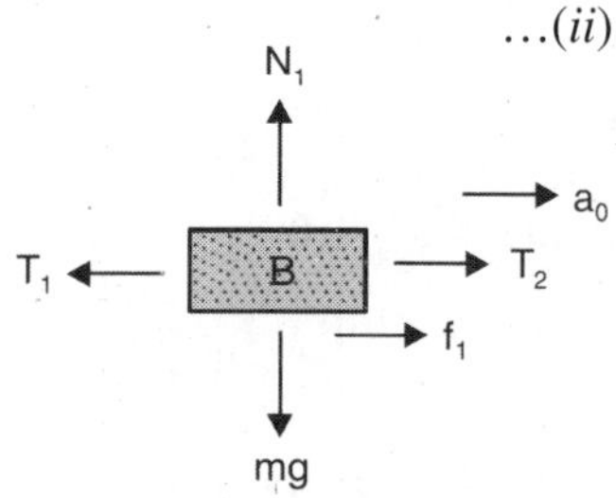

Fig. E21(b)

For block C

$$\Sigma F_y = m\,a_y$$

$$T_2 = f_2 + mg$$

$$T_2 = \mu N_2 + mg \quad \ldots(v)$$

$$\Sigma F_x = m\,a_x$$

$$N_2 = m\,a_0 \quad \ldots(vi)$$

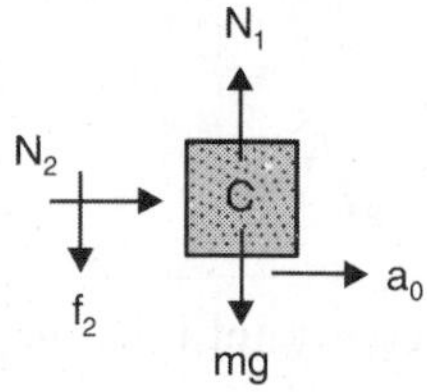

Fig. 21(c)

Using all above equations
We get

$$\theta = 37° \quad \ldots \textbf{Ans.}$$

F_{BD} of trolley

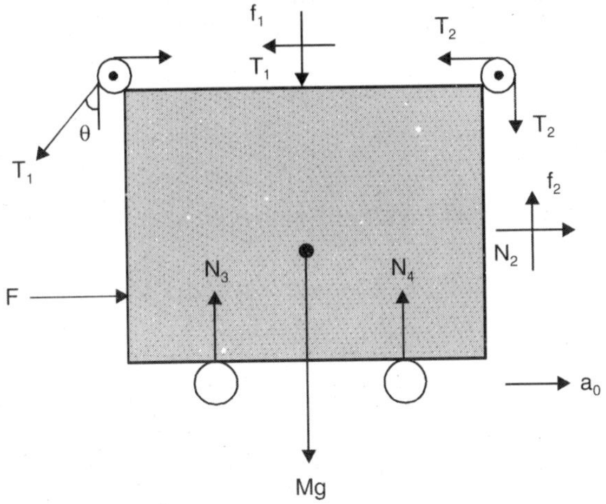

Fig. E21(d)

Simply use

$$\Sigma F_x = m a_x$$

$$F - T_1 \sin \theta + T_1 - f_1 - T_2 - N_2 = M a_0$$

$$\mathbf{F = 315}$$ **...Ans.**

Level 3

Example 1

Solution

Let a_1, a_2, a_3 denote the accelerations with which the strings are slipping over the pulleys.

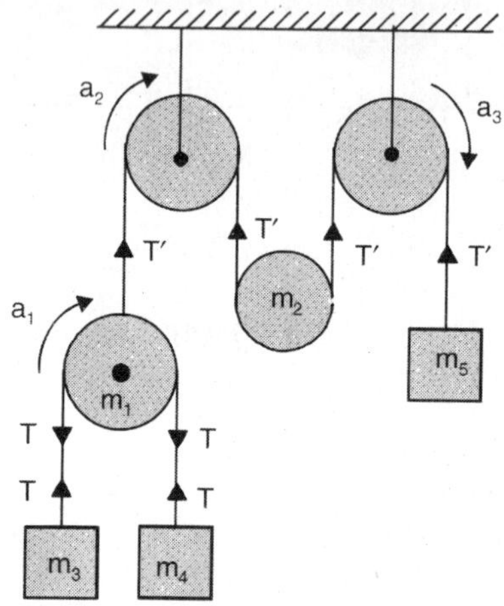

Fig. E1(a)

$$a_{m1} = \text{Acceleration of pulley of mass } m_1 = a_2$$

$$a_{m2} = \text{Acceleration of pulley of mass } m_2 = \frac{(a_3 + a_2)}{2}$$

$$a_{m3} = \text{Acceleration of block of mass } m_3 = a_2 + a_1$$

$$a_{m4} = \text{Acceleration of block of mass } m_4 = a_2 - a_1$$

$$a_{m5} = \text{Acceleration of block of mass } m_5 = -a_3$$

There are five unknowns as a_1, a_2, a_3, T and T'

Equations for each masses are

For m_1

$$T' - 2T - m_1g = m_1 a_2 \quad ...(i)$$

For m_2

$$2T' - m_2g = \frac{m_2}{2}(a_3 - a_2) \quad ...(ii)$$

For m_3

$$T - m_3 g = m_3 (a_1 + a_2) \quad ...(iii)$$

For m_4

$$T - m_4 g = m_4 (a_2 - a_1)$$

For m_5

$$m_5g - T' = m_5 a_3$$

Example 2

Solution

Assume,

a_A = Acceleration of block A

a_B = Acceleration of block B

α = Angular acceleration of pulley

T_1 = Tension in the horizontal part of pulley

T_2 = Tension in the vertical part of pulley

F = Force in the string

FBD of block A

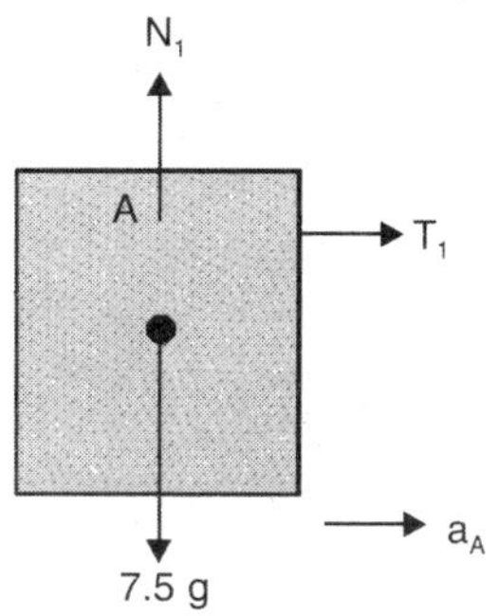

Fig. E2(a)

$$T_1 = 7.5\, a_A \quad ...(i)$$

FBD of pulley

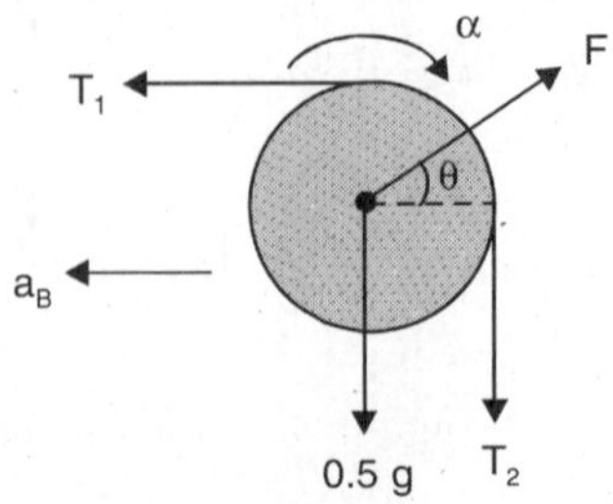

Fig. E2(b)

$$T_1 - F\cos\theta = 0.5\,a_B \qquad \ldots(ii)$$

FBD of block C

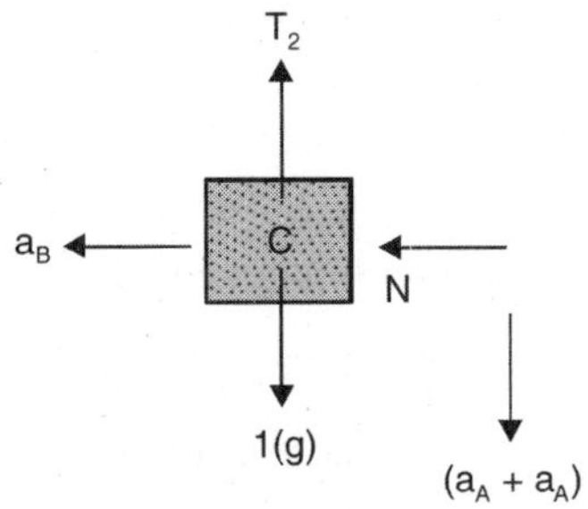

Fig. E2(c)

$$N = 1(a_B) \qquad \ldots(iii)$$

$$1(g) - T_2 = 1\,(a_A + a_B) \qquad \ldots(iv)$$

FBD of block B

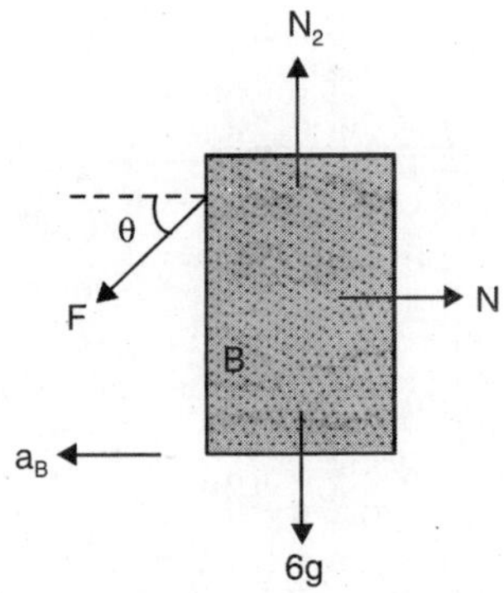

Fig. E2(d)

$$F\cos\theta - N = 6\,(a_B) \qquad \ldots(v)$$

Now for pulley

$$(T_2\,R - T_1\,R) = I_0\,\alpha$$

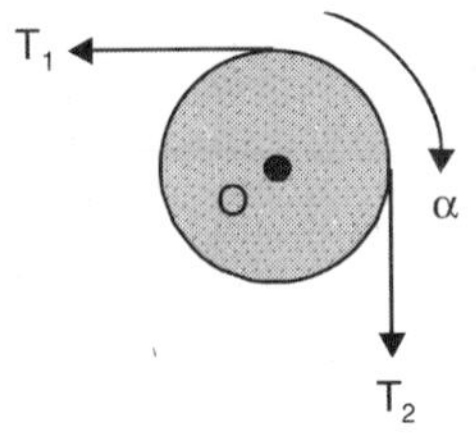

Fig. E2(e)

$$T_2 - T_1 = I_0\left(\frac{\alpha}{R}\right)$$

$$= \left(\frac{MR^2}{2}\right)\left(\frac{a_A + a_B}{R^2}\right)$$

$$T_2 - T_1 = \left(\frac{0.5}{2}\right)(a_A + a_B)$$

$$= (0.25)(a_A + a_B) \qquad \ldots(vi)$$

solving all above equations we get,

$$\mathbf{a_A = 1\ m/s^2}$$
$$\mathbf{a_B = 1\ m/s^2}$$

$$a_C = \sqrt{(a_B)^2 + (a_A + a_B)^2}$$

$$= \sqrt{(1)^2 + (1+1)^2}$$

$$= \sqrt{5}\ \mathbf{m/s^2}. \qquad \textbf{...Ans.}$$

Example 3

Solution

In this problem, only interaction force is demanding

N_0 =Interaction force between cylinder and wedge

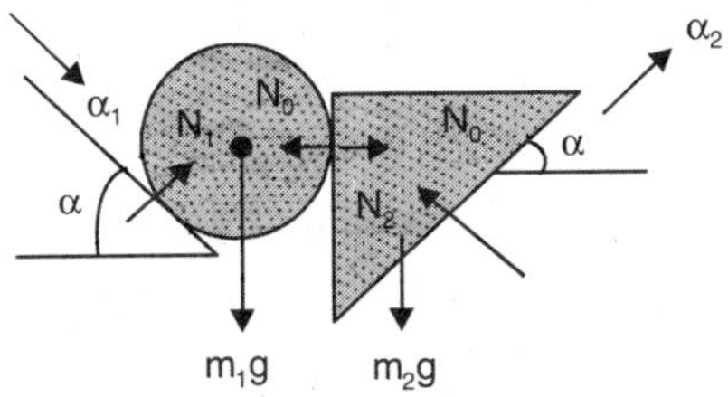

Fig. E3(a)

For cylinder along the inclined plane

$$m_1 g \sin \alpha - N_0 \cos \alpha = m_1 a_1 \quad \text{...(i)}$$

For wedge

$$N_0 \cos \alpha - m_2 g \sin \alpha = m_2 a_2 \quad \text{...(ii)}$$

$$\text{Now } a_1 = a_2 \text{ (think why)} \quad \text{...(iii)}$$

Using above all we get

$$\mathbf{N_0} = \left(\frac{\mathbf{2m_1m_2}}{\mathbf{m_1 + m_2}}\right)\mathbf{\tan\alpha} \quad \textbf{...Ans.}$$

Example 5

Solution

Total length of the chain = 2.5 m

$$\text{Length of } AC \text{ part of chain} = \left(\frac{\pi R}{2}\right) = \left(\frac{\pi}{2} \times \frac{4}{\pi}\right)$$

$$= 2$$

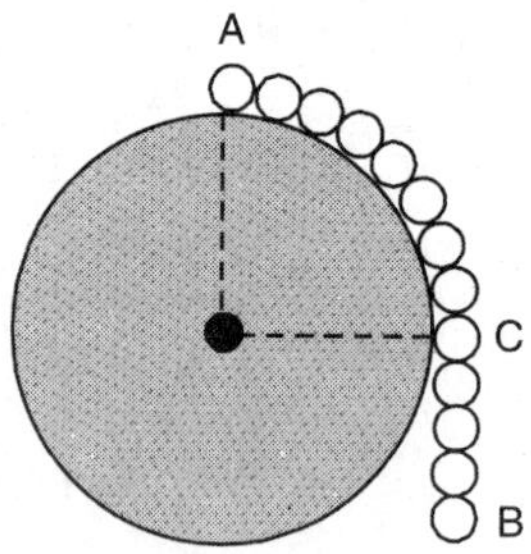

Fig.5(a)

Length of CB part of chain

$$= 2.5 \text{ m} - 2\text{m} = 0.5 \text{ m}$$

Now, concentrate on the part of AC chain

Consider a small element of chain of length $Rd\theta$.

$$\text{Mass of this element} = \lambda\,(Rd\theta)$$

where λ = Mass per unit length

Along tangent

$$(T_0 + dT) + \lambda\, Rg \sin\theta\, d\theta - T_0 = (\lambda R d\theta)\, a$$

a denote the acceleration of the chain

$$dT = \lambda R\,(a - g \sin\theta)\, d\theta$$

$$\int_0^T dT = \lambda R \int_0^{\pi/2} \left(a - g\sin\theta\right) d\theta$$

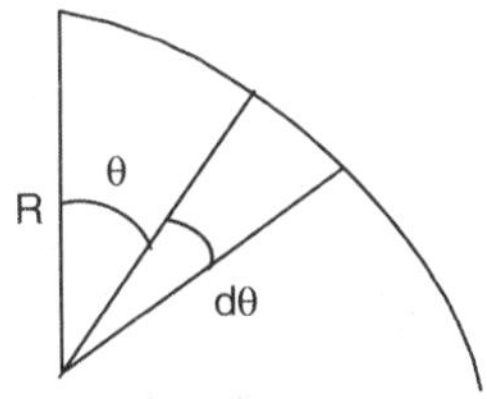

Fig. E5(b)

$$T = \lambda R\left(\frac{a\pi}{2} - g\right) \qquad \ldots(i)$$

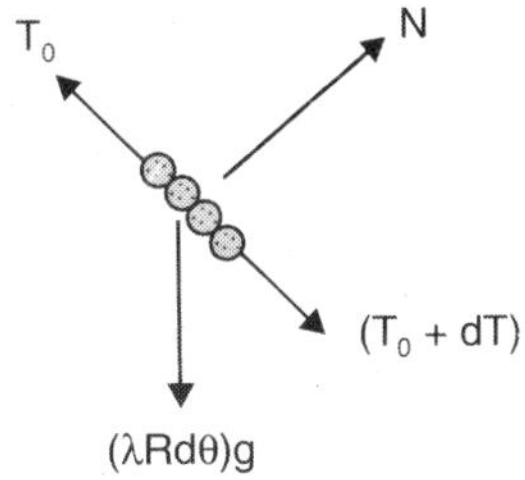

Fig. E5(c)

For part *CB*

$$(0.5\ \lambda)\ g - T = (0.5\ \lambda)\ a \qquad \ldots(ii)$$

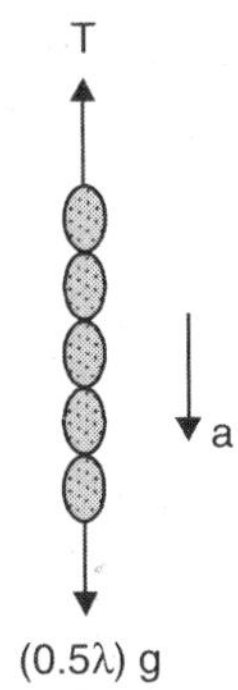

Fig. E5(d)

Using (*i*) and (*ii*) we get

Sought acceleration = **7.58 m/s²** **...Ans.**

Example 6

Solution

Suppose a_0 is the acceleration of the block relative to the accelerated board. Now,

$$V^2 = u^2 + 2\,a_0\,S$$

$$0 = \left(4\sqrt{2}\right)^2 + 2a_0(1.6)$$

$$a_0 = -\,10\text{ m/s}^2$$

$$\Rightarrow \text{Retardation.}$$

Now concentrate on the FBD of the block

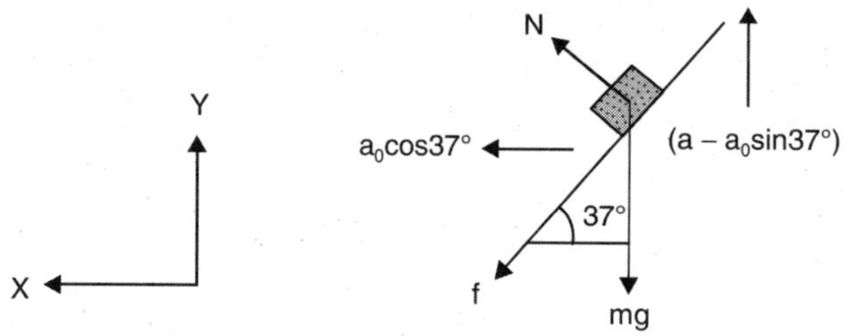

Fig. E6(a)

$$\Sigma F_y = m\,a_y$$

$$(N\cos 37° - mg - f\sin 37°) = m\,(a - a_0\sin 37°)$$

$$N\cos 37° - mg - \mu N\sin 37° = m\,(a - 6)$$

$$N = \frac{mg + ma - 6m}{(0.8 - 0.6\mu)} \qquad \ldots(i)$$

$$\Sigma F_x = m\,a_x$$

$$N_2\sin 37° + \mu N\cos 37° = ma_0\cos 37°$$

$$N\sin 37° + \mu N\cos 37° = m\,(8)$$

$$N = \frac{8m}{(0.6 + 0.8\mu)} \qquad \ldots(iii)$$

From equation (*i*) and (*ii*)

$$40\,\mu + 4\,\mu a + 3a = 20 \qquad \ldots(iii)$$

Similarly for down the plane.

$$\text{Assume } a_1 = \text{relative acceleration down the plane}$$

$$V^2 = u^2 + 2a_1S$$

$$(4 \times 4) = 2 \times a_1 \times (1.6)$$

$$a_1 = 5\text{ m/s}^2$$

Now after doing same thing for a_1 and drawing FBD, writing equations for block

We get

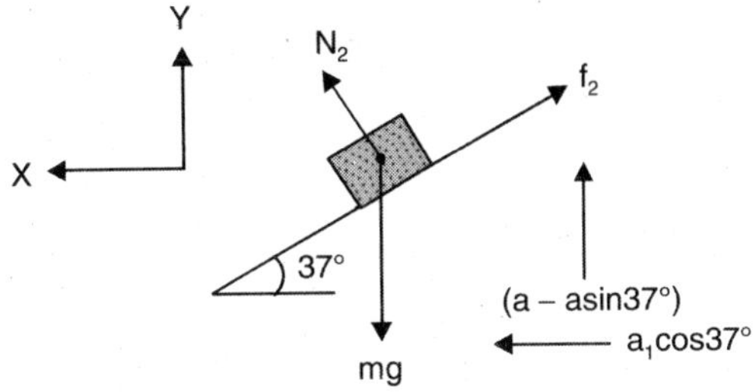

Fig. E6(b)

$$\Sigma F_y = m\, a_y$$

$$N_2 = \frac{mg + ma - 3m}{(0.8 + 0.6\mu)} \qquad \ldots(iv)$$

$$\text{And } \Sigma F_x = m\, a_x$$

$$N_2 = \frac{4m}{(0.6 - 0.8\mu)} \qquad \ldots(v)$$

By (*iv*) and (*v*)
We get

$$40\,\mu + 4\,\mu a - 3a = 5 \qquad \ldots(vi)$$

Now use (*vi*) and (*iii*)

$$\mathbf{a = 2.5\ m/s^2} \qquad \textbf{...Ans.}$$

$$\mu = \mathbf{0.25} \qquad \textbf{...Ans.}$$

Work, Energy and Momentum

Level 1

Example 4

Solution

Apply work-energy principle from A to B.

$$W_{ext} = \Delta T$$

$$mgR + \frac{1}{2}k\left[\left(\sqrt{2}R - R\right)^2\right] = \frac{1}{2}mV_B^2$$

We get

$$\mathbf{V_B} = \sqrt{2gR + \frac{kR^2}{m}(3 - 2 - \sqrt{2})} \qquad \text{...Ans.}$$

Similarly

$$\mathbf{V_C} = \sqrt{4gR + \frac{kR^2}{m}(3 - 2\sqrt{2})} \qquad \text{...Ans.}$$

Level 2

Example 5

Solution

Clearly no external force is acting on the system

$$\Sigma m_i x_i = 0$$

so, center of mass remains constant.

$$X_{\text{initial}}(C\cdot m) = \frac{m_2(-L)+m_1(-L)+X_0 M}{m_1+m_2+M}. \quad \ldots(i)$$

$$X_{\text{final}}(C\cdot m) = \frac{(l+X_0)M+lm_1+\left(l\frac{L}{2}\right)m_2}{m_1+m_2+M} \quad \ldots(ii)$$

Equate (*i*) and (*ii*) and get result as

$$l = \frac{-\left(\mathbf{m_1}+\frac{\mathbf{m_2}}{2}\right)\mathbf{L}}{\mathbf{m_1}+\mathbf{m_2}+\mathbf{M}} \quad \textbf{...Ans.}$$

Example 6

Solution

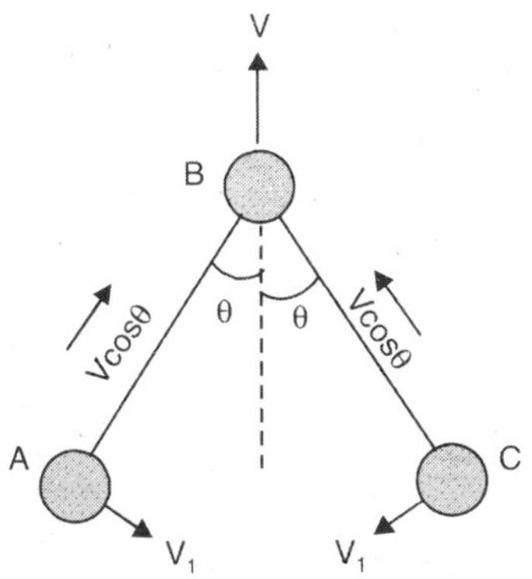

Fig. E6(a)

Let at a certain instant velocity of ball *B* be *V* and inclination of each string with the line of motion of *B* be θ.

V_1 represents the normal velocity of *A* and *C* because of the curve linear motion.

As string is inextensible and it remains taut at every moment so along the length *AB* velocity of each ball *A* and *B* or *A* and *C* remains same.

Now applying conservation of momentum when $\theta = 0$ and initial position that shown in problem

$$mv + mv + mv = mV_0$$

$$v = v_0/3 = 3 \text{ m/s}$$

Applying law of conservation of energy

$$\left[\frac{1}{2}m\left(\frac{v_0}{3}\right)^2\right] + \frac{1}{2}mv_1^2 + \frac{1}{2}m\left(\frac{v_0}{3}\right)^2 = \frac{1}{2}mv_0^2$$

$$v_1 = 3\sqrt{3} \text{ m/s}$$

$$V_A = \text{Resultant velocity of ball } A$$

$$V_A = \sqrt{v^2 + v_1^2}$$

$$= \sqrt{(3)^2 + (3\sqrt{3})^2}$$

$$= \mathbf{6 \text{ m/s}}$$ **...Ans.**

Example 11

Solution

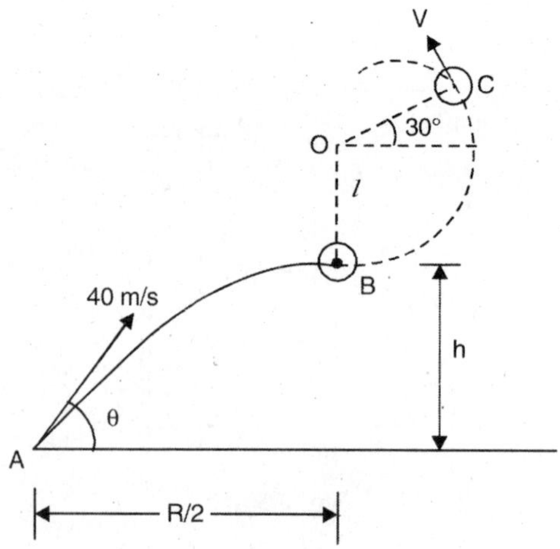

Fig. E11(a)

B is the highest point of trajectory so, at B bullet has velocity 40 cosθ in horizontal direction

Applying law of conservation of momentum

$$(1.3 \times 0) + (0.7 \times 40 \cos \theta) = (1.3 + 0.7)\, v_0$$

$$v_0 = 14 \cos \theta \qquad \ldots(i)$$

Because deviation of thread is 120° assume at point C tension of the thread disappears velocity of the combined body. Let it be v.

We can write at point C

$$(M + m)\, g \sin 30° = (M + m)\frac{v^2}{l}$$

$$v = \sqrt{7} \text{ m/s}$$

Applying work-energy theorem

$$w_{ext} = \Delta T$$

$$-(M + m)\, g\, l\, (1 + \sin 30°) = \frac{1}{2}(M + m)\, v^2 - \frac{1}{2}(M + m)\, v_0^2 \qquad \ldots(ii)$$

By using equation (i) and value at v in equation (ii) we get

$$\theta = 60°$$

Now,

Maximum height = raised by the body =

$$H_{max} = H + l\,(1 + \sin 30°) + y \qquad \ldots(iii)$$

Here y represents further height raised by the combined body

$$y = \frac{(v\cos 30°)^2}{2g} = 0.26 \text{ m}$$

$$H = \frac{(40 \sin 60°)^2}{2g} = 60 \text{ m}$$

$$l = 1.4 \text{ m}$$

Putting y, H and l in (iii)

We get

$$\mathbf{H_{max} = 62.36\ m} \qquad \textbf{...Ans.}$$

Now,

Distance between A and O is = AO

$$AO = \sqrt{(R/2)^2 + (H + l)^2}$$

$$\frac{R}{2} = \frac{(40)^2 \sin 2\theta}{2g} = 40\sqrt{3} \text{ m}$$

$$H = 60 \text{ m}$$
$$l = 1.4 \text{ m}$$

We get

$$\textbf{AO} = \textbf{92.57 m} \qquad \textbf{...Ans.}$$

Example 13

Solution

Clearly,

Velocity of m_1 when it has fallen by the distance h will be

$$u = \sqrt{2gh}$$

Let T be the tension in the string, v is the velocity of m_2 after jerk

By linear impulse momentum for m_2

Along Y-axis

$$T\Delta t = m_2 V \qquad ...(i)$$

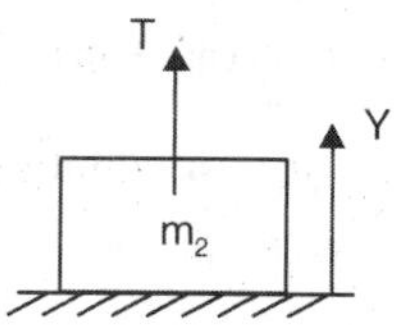

Fig. E13(a)

For m_1

Along Y-axis

$$T\Delta t = m_1 V + m_1 u \qquad ...(ii)$$

$$V = \frac{m_1 u}{m_1 + m_2} \qquad ...(iii)$$

Now,

Newton's law gives

$$a = \left(\frac{m_2 - m_1}{m_1 + m_2}\right) g$$

Applying kinematical relationship

$$V_0 = u_0 + a_0 t$$

$$0 = V - \left(\frac{m_2 - m_1}{m_1 + m_2}\right) gt$$

$$t = \left(\frac{m_1 + m_2}{m_2 - m_1}\right) \frac{v}{g}$$

$$\text{Total time} = 2t = \left(\frac{m_1 + m_2}{m_2 - m_1}\right)\frac{2v}{g}$$

$$= \frac{2}{g}\left(\frac{m_1 + m_2}{m_2 - m_1}\right)\left(\frac{m_1 u}{m_1 + m_2}\right) \qquad \text{...Ans.}$$

Example 14

Solution

Suppose u and v denote the initial velocities of A along horizontal direction after the impulsive reaction I.

Velocity of B must be in the horizontal direction with magnitude u because the string is inextensible.

Velocity of C must be along the line CA for a similar reason.

$$V_c = u \cos 60° - v \cos 30°$$

$$\text{Or } V_C = \left(\frac{u}{2} - \frac{\sqrt{3}V}{2}\right)$$

Now,

Impulse momentum theorem for every particle

For A

Along X-axis

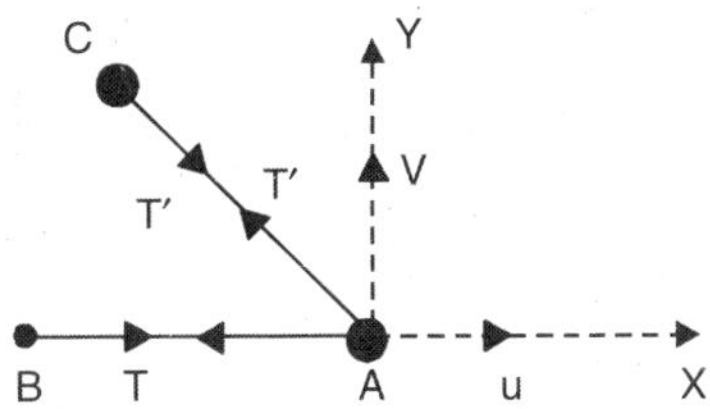

Fig. E14(a)

$$F_x \Delta t = \Delta P_x$$

$$I - T\Delta t - \frac{1}{2} T' \Delta t = mu \qquad \text{...(i)}$$

Along Y-axis

$$F_y \Delta t = \Delta P_y$$

$$\frac{\sqrt{3}}{2} T' \Delta t = mv \qquad \text{...(ii)}$$

For B

Along X-axis

$$F_x \Delta t = \Delta P_x$$

$$T\Delta t = mu \qquad \text{...(iii)}$$

For C

Along CA

$$F_{CA} \Delta t = \Delta P_{CA}$$

$$T'\Delta t = \frac{m}{2}\left(u - \sqrt{3}\, v\right) \qquad \text{...(iv)}$$

Using above all equations
We get,

$$\mathbf{u} = \frac{\mathbf{7I}}{\mathbf{15\ m}}$$

$$\mathbf{V} = \frac{\sqrt{\mathbf{3}}\mathbf{I}}{\mathbf{15\ m}} \qquad \textbf{...Ans.}$$

Direction of motion of *A* makes with *BA* an angle

$$= \tan^{-1}\left(\frac{\mathbf{v}}{\mathbf{u}}\right) = \tan^{-1}\left(\frac{\sqrt{3}}{7}\right) \qquad \textbf{...Ans.}$$

Example 15

Solution

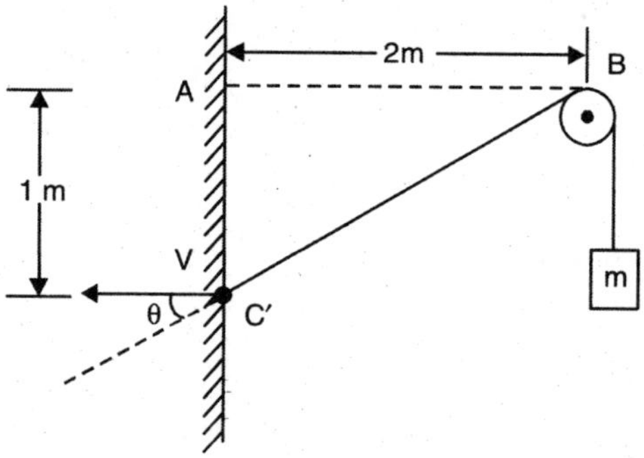

Fig. E15(a)

$C'B = \sqrt{(2)^2 + (1)^2} = \sqrt{5}$ vertical displacement by mass m is $\left(\sqrt{5} - 1\right)$ m.
String between two masses always remains taut so along the length (CB) velocity of both masses will be same.

Here $\cos\theta = \dfrac{2}{\sqrt{5}}$

Now,
Applying law of conservation of energy

$$Mg \times (1) = mg\left(\sqrt{5} - 1\right) + \frac{1}{2}Mv^2 + \frac{1}{2}m(v\cos\theta)^2$$

$$\mathbf{V} = 5\sqrt{\frac{5 - \sqrt{5}}{6}} \qquad \textbf{...Ans.}$$

Level 3

Example 2

Solution

Applying conservation of energy

$$mgl\sin\theta = \frac{MV^2}{2} + \frac{m}{2}\left[\left(V_r \sin\theta - V\right)^2 + V_r^2 \cos^2\theta\right] \qquad ...(i)$$

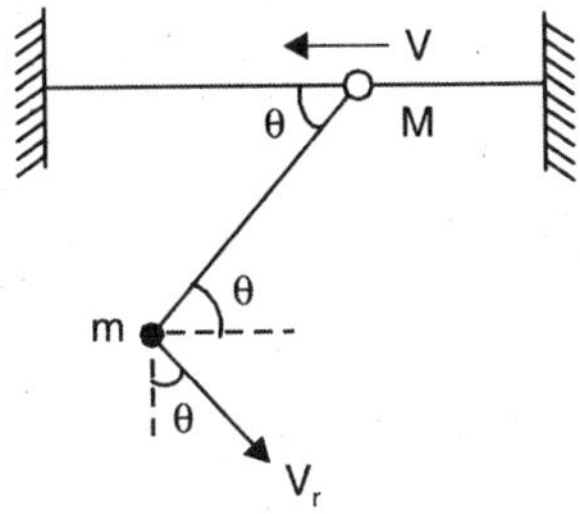

Fig. E2(a)

Conservation of linear momentum

$$m\,(V_r \sin\theta - V) = MV$$

$$V = \frac{mV_r \sin\theta}{M+m}$$

Applying (*i*) and (*ii*) we can get the result

$$\frac{V_r}{l} = \omega = \sqrt{\frac{2g(M+m)\sin\theta}{(M+m\cos^2\theta)l}} \qquad \textbf{...Ans.}$$

Example 3

Solution

V_{mM} = Velocity of ball of mass *m* with respect to hemisphere of mass *M*.

V_M = Velocity of hemisphere of mass *M*.

(*i*) Clearly by applying linear momentum conservation horizontally

$$\text{We get } V_M = \frac{mV_{mM}\cos\beta}{(m+M)} \qquad ...(i)$$

Energy conservation $\frac{1}{2}mV_m^2 + \frac{1}{2}mV_M^2 = mgR(\cos\alpha - \cos\beta)$ ...(*ii*)

We can also get a relation between V_m, V_{mM} and V_M as vector diagram of kinematics

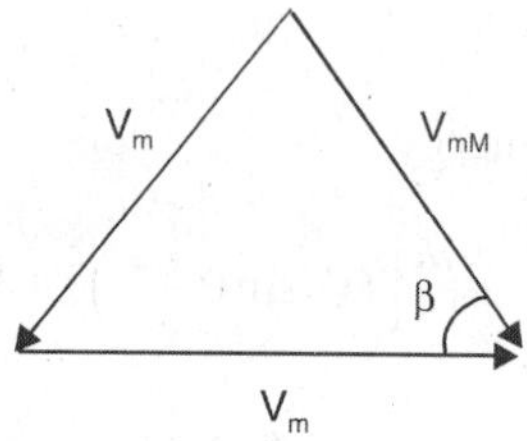

Fig. E3(a)

$$V_m^2 = V_{mM}^2 + V_m^2 - 2V_{mM}V_M\cos\beta \quad ...(iii)$$

Using above all equation
We get,

$$V_{mM} = \sqrt{\frac{2(m+M)(\cos\alpha\cos\beta)}{(M+m\sin^2\beta)}}$$

$$\omega = \frac{V_{mM}}{R} = \sqrt{\frac{2(m+M)(\cos\alpha\cos\beta)}{(M+m\sin^2\beta)R^2}} \quad \textbf{...Ans.}$$

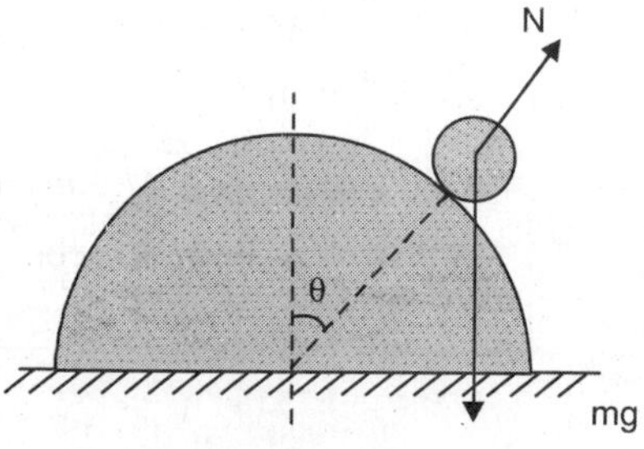

Fig. E3(b)

for ball

$$\Sigma F_n = m\,a_n$$

$$\text{mg}\cos\theta - N = m\,a_{mM} \quad ...(i)$$

for hemisphere

$$\Sigma F_x = m\,a_x$$

$$N\sin\theta = M\,a_M$$

$$a_M = \frac{N \sin\theta}{M} \qquad ...(ii)$$

$$\text{Suppose } \theta = \theta_0 \text{ and } N = 0$$

At the moment of break off.
We get by (*i*)

$$mg \cos \theta_0 = m\left(\frac{V_{mM}^2}{R}\right)$$

$$\cos \theta_0 = \frac{V_{mM}^2}{gR} \qquad ...(iii)$$

Now put V_{mM} value from above in (*iii*) we get sought result.

$$\theta_0 = \cos^{-1}\left(\frac{2(m+M)\cos\alpha}{[3M + m(2+\sin^2\beta)]}\right) \qquad \text{...Ans}$$

Example 4

Solution

Applying momentum conservation at bottom level

$$mu = Mv \qquad ...(i)$$

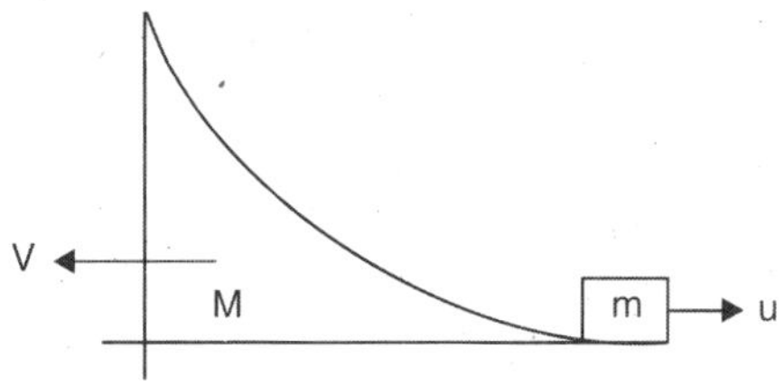

Fig. E4(a)

Energy conservation

$$mgh = \frac{Mv^2}{2} + \frac{mu^2}{2} \qquad ...(ii)$$

Now for the moment when block of mass *m* climbs on the curved surface of the wedge of mass *M*, applying momentum conservation for the system (wedge + block) in horizontal direction.

$$V_0 = \text{Total velocity of wedge and block}$$

$$mu = (M + m)\, V_0 \qquad ...(iii)$$

Energy conservation law

$$\frac{mv^2}{2} = \left(\frac{M+m}{2}\right)V_0^2 + mgh_{\max} \qquad \ldots(iv)$$

Using above all equations we get

$$h_{max} = \mathbf{h}\frac{\mathbf{M}^2}{(\mathbf{M}+\mathbf{m})^2} = \mathbf{h}\left(\frac{\mathbf{M}}{\mathbf{M}+\mathbf{m}}\right)^2 \qquad \textbf{...Ans.}$$

Example 5

Solution

Let the time of flight of first ball up to the instant of collision be n second for second ball
It is $(n - 1)$ second.

Suppose h is the height from ground. Then vertical displacement for first ball is $(65 - h)$ *m* and for second ball is $(30 - h)$ *m*.

For first ball

$$(65 - h) = \left(10\sqrt{2}\sin 45°\right)n + 5n^2 \qquad \ldots(i)$$

for second ball

$$(30 - h) = 5(n - 1)^2 \qquad \ldots(ii)$$

By (*i*) and (*ii*) we get

$$n = 2 \text{ sec}$$
$$h = 25 \text{ m}$$

Now for *AC*
Clearly

AC = Distance moved by first ball upto instant of collision + distance moved by second ball up to collision

$$AC = \left(10\sqrt{2}\cos 45°\right)(2) + 20(2-1)$$
$$= \mathbf{40\ m} \qquad \textbf{...Ans.}$$

Now, apply law of conservation of momentum in vertical direction

$$mv_y^1 + mv_y^2 = (m+m)v_y \qquad \ldots(iii)$$

Before collision

$$v_y^1 = \text{Velocity of first ball after 2 sec in vertical direction}$$
$$= (10\sqrt{2}\ \sin 45°) + 2g$$
$$v_y^2 = \text{Velocity of second ball after 2 sec in vertical direction}$$
$$= g\,(2 - 1) = g$$

After collision

v_y = Velocity of combined body of mass $(m + m)$ in vertical direction

Using in (*iii*)

We get $v_y = 20$ m/s

Similarly in horizontal direction

We get

$$v_x = 5\text{m/s}$$

For $h = 25$ m

$$25 = 20 \times t + \frac{1}{2} \times 10 \times t^2$$

$$t = 1 \text{ sec. For } S$$

$$S = 5 \times 1 = 5 \text{ m}$$

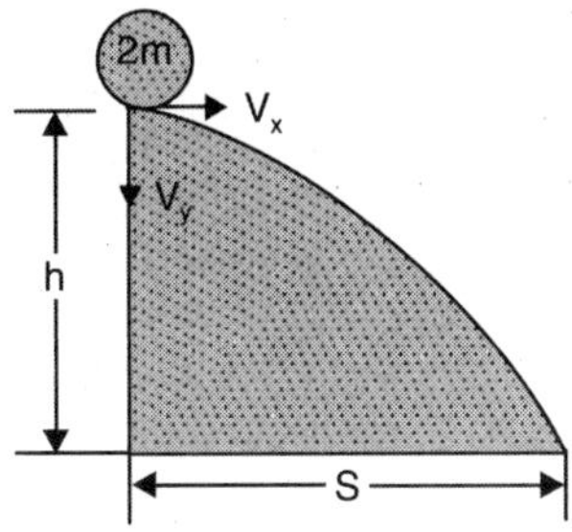

Fig. E5(a)

Distance from A to the point at which combined body strikes the ground = $(10\sqrt{2}\cos 45°) \times 2 - 5$

$$= \mathbf{15\ m}$$ **...Ans.**

Example 8

Solution

Let V = Velocity of either ball of mass m'

u' = velocity of ball of mass m. After collision

$$e = \frac{v\cos 60° - u'\cos 30°}{u\cos 30° - 0} \qquad ...(i)$$

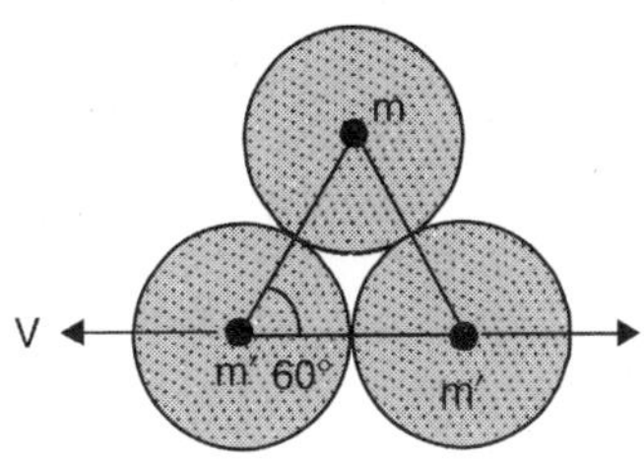

Fig. E8(a)

For m

$$2N\Delta t \cos 30° = m(u - u') \qquad ...(ii)$$

$$N\Delta t \cos 60° = m' v \qquad ...(iii)$$

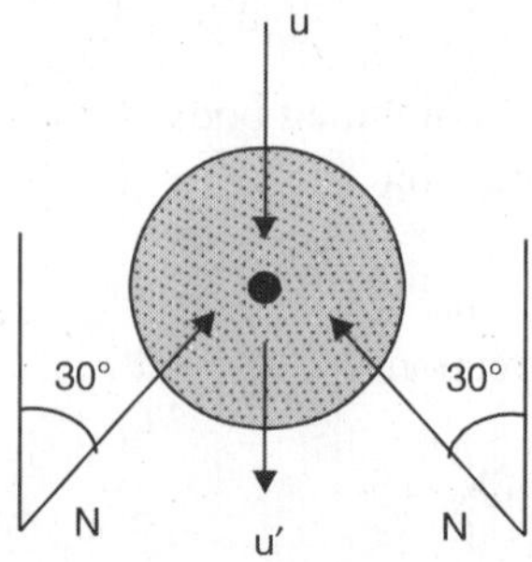

Fig. E8(b)

Using all above equations we get,

$$u' = u\,(m - 6em')\,/\,(m + 6m')$$

$$v = \sqrt{3}\,um\,(1 + e)\,/\,(m + 6m')$$

Now for impulsive tension T

$$T = N\Delta t \cos 30°$$

$$= \sqrt{3}\,m'\,v$$

By using equation (*iii*)
We get

$$\mathbf{T = 3\,mm'\,u\,(1 + e)/\,(m + 6m')} \qquad \text{...Ans.}$$

Example 10

Solution

Just concentrate on point A as

$$\Sigma F_n = ma_n$$

$$mg \sin\theta - N = m\frac{V_A^2}{R}$$

($N = 0$, break off condition)

$$mg \sin\theta = m\frac{V_A^2}{R} \qquad ...(i)$$

Applying conservation of energy

$$\frac{mV_A^2}{2} = mg\,R\,(1 - \sin\theta) + \frac{mV^2}{2} \qquad ...(ii)$$

Applying (*i*) and (*ii*)

$$\text{We get } V_A = \sqrt{\frac{V^2 + 2gR}{3}}$$

$$\sin\theta = \frac{2}{3} + \frac{V^2}{3gR}$$

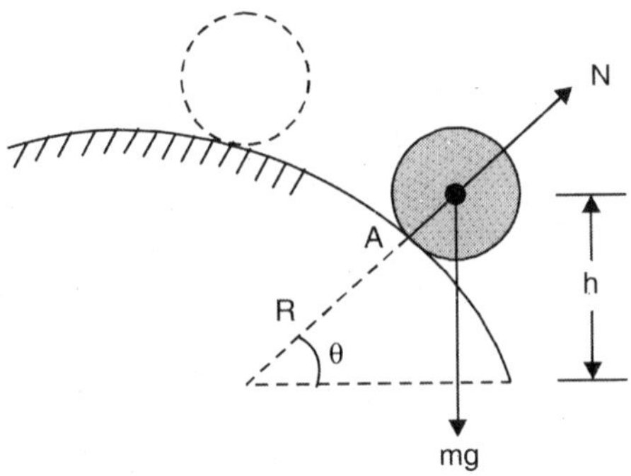

Fig. E10(a)

Equation of trajectory when a particle is thrown from a certain height y_0.

$$y = y_0 + x\tan\theta_0 - \frac{gx^2}{2u^2\cos^2\theta_0} \quad \text{...}(iii)$$

$$y_0 = R\sin\theta = \frac{2R}{3} + \frac{V^2}{3g} \quad \text{...}(iv)$$

θ_0 = Angle of projection by simple trigonometry
We get,

$$\cos\theta_0 = \frac{2}{3} + \frac{v^2}{2gR}$$

$$\text{So } \tan\theta_0 = -\sqrt{\frac{1\frac{V^2}{gR}}{2+\frac{V^2}{gR}}} \quad \text{...}(v)$$

$$u = V_A = \sqrt{\frac{V^2}{3} + \frac{2gR}{3}} \quad \text{...}(vi)$$

Applying (*iv*), (*v*) and (*vi*) in (*iii*) we get

$$y = \frac{2R}{3} + \frac{V^2}{3g} - \sqrt{\frac{gRV^2}{2gR+V^2}}\cdot x - \frac{27g^3R^2}{2(V^2+2gR)^3}\cdot x^2$$

Now when ball touches the ground $y = 0$

$$0 = \frac{2R}{3} + \frac{V^2}{3g} - \sqrt{\frac{gRV^2}{2gR + V^2}} \cdot x - \frac{27g^3R^2}{2(V^2 + 2gR)^3} \cdot x^2$$

By solving the quadratic equation in x

$$x = \frac{(V^2 + 2gR)^2}{(27g^3R^2)} \left(\sqrt{20g^2R^2gRV^2V^4} - \sqrt{2g^2R^2gRV^2V^4}\right)$$

Now distance d can be calculated as

$$d = x - R(1 - \cos\theta)$$

After putting the value of x and $\cos\theta$ we get the sought distance d as

$$\frac{(\mathbf{V}^2 + 2\mathbf{gR})^2}{(27\mathbf{g}^3\mathbf{R}^2)} \left(\sqrt{20\mathbf{g}^2\mathbf{R}^2\mathbf{gRV}^2\mathbf{V}^4} - \sqrt{2\mathbf{g}^2\mathbf{R}^2\mathbf{gRV}^2\mathbf{V}^4}\right) - \mathbf{R}\left(1\sqrt{\frac{1}{3}\frac{\mathbf{V}^2}{3\mathbf{gR}}}\right)$$

...Ans.

Example 11

Solution

V_0 = velocity of combined body just after collision.

Applying law of conservation of momentum

$$1 \times 9 = (2 + 1)\, v_0$$

$$v_0 = 3 \text{ m/s}$$

Here A is at rest position so string becomes slack and bodies move under gravity combined body upwards and A downwards till string again becomes taut.

Let after time t_1

Downward displacement of A (S_A) = upward displacement of combined body ($S_{B,C}$)

$$S_A = \frac{gt_1^2}{2} \text{ and}$$

$$S_{BC} = 3t_1 - \frac{gt_1^2}{2}$$

$$S_A = S_{B,C}$$

We get $t_1 = 0.3$ sec

Now after time t_1 each block $S_A = 0.45$ m distance. Velocity of $A = g \times t_1 = 10 \times 0.3 = 3$ m/s

Velocity of combined body =

$$3 - gt_1 = 3 - 10 \times 0.3 = 0$$

Velocity of block A is different than velocity of combined body B and C than a impulse is developed in the string due to developed tensions. After that velocities of each block will be equal.

For A

$$T\Delta t = 2 \times 3 - 2 \times v' \quad \ldots(i)$$

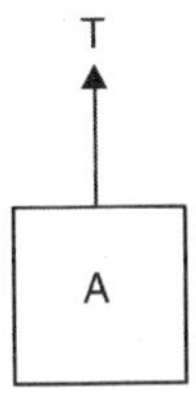

Fig. E11(a)

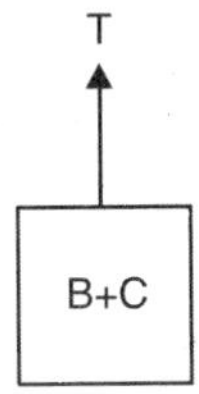

Fig. E11(b)

For combined body B and C

$$T\Delta t = 3v' \quad \ldots(ii)$$

Here v' is the velocity of each block after impulse is developed

Using these two equations

We get

$$v' = 1.2 \text{ m/s}$$

Now according to Newton's law

$$\vec{a} = \left(\frac{32}{3+2}\right)g(-\hat{j})$$

$$= -2 \text{ m/s}^2 \downarrow$$

Combined body comes to an instantaneous rest after time $t_2 = \dfrac{v'}{|a|}$

$$= 0.6 \text{ sec}$$

Height ascended by the combined body during time 0.6 sec

$$h = v't_2 - \frac{1}{2}t_2^2$$

$$= (1.2)(0.6) - \frac{1}{2}(2) \times (0.6)^2$$

$$= 0.36 \text{ m}$$

H_{max} (height raised by combined body)

$$= S_A + h = S_{(B,C)} + h$$

$$= 0.45 + 0.36$$

$$= \mathbf{0.81 \text{ m}} \quad \textbf{...Ans.}$$

Rotation and Simple Harmonic Motion

Level 1

Example 2

Solution

Let m be the mass of the square lamina for a certain instant.

According to the problem point A is suddenly jerked with velocity u. So angular momentum about 'A' remains conserved.

$$L_i = L_f$$

$$0 = ml^2\omega - ml\,(u - l\omega) + \frac{2}{3}ml^2\omega$$

$$\omega = \frac{\mathbf{3u}}{\mathbf{8}l} \qquad \text{...Ans.}$$

Example 3

Solution

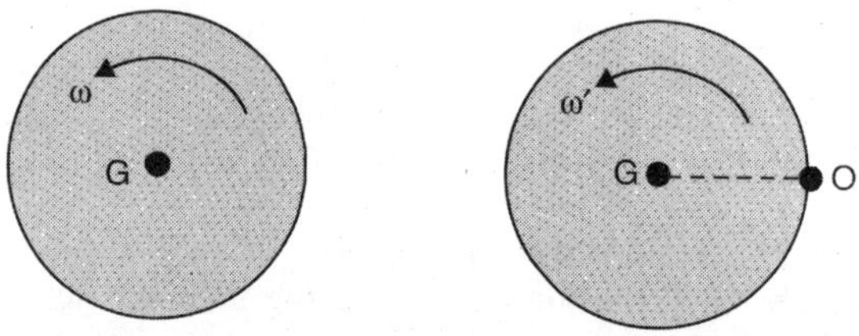

Fig. E3

Suppose at any instant point O is fixed.

Apply conservation of angular momentum about point 'O'.

$$L_i = L_f$$

$$\frac{ma^2\omega}{2} = \frac{ma^2}{2}\omega^1 + ma^2\omega^1$$

Sought result $\Rightarrow \omega^1 = \omega/3$ **...Ans.**

Example 5

Solution

Applying conservation of angular momentum about axis passing through its center.

About oo[1]

$$L_i = L_f$$

$$mv_0 r_0 = mvr \cos\theta \qquad \ldots(i)$$

Now, energy conservation at A and B

$$\frac{mv_0^2}{2} + mgh = \frac{mv^2}{2}$$

$$v = \sqrt{v_0^2 + 2gh} \qquad \ldots(ii)$$

$$r = \sqrt{r_0^2 - h^2} \qquad \ldots(iii)$$

Use (ii) and (iii) in (i) sought angle $\theta = \mathbf{\cos^{-1} \frac{1}{\sqrt{1+\frac{2gh}{v_0^2}}\sqrt{1-\frac{h^2}{r_0^2}}}}$ **...Ans.**

Example 8

Solution

Free body diagrams of the sphere and plank

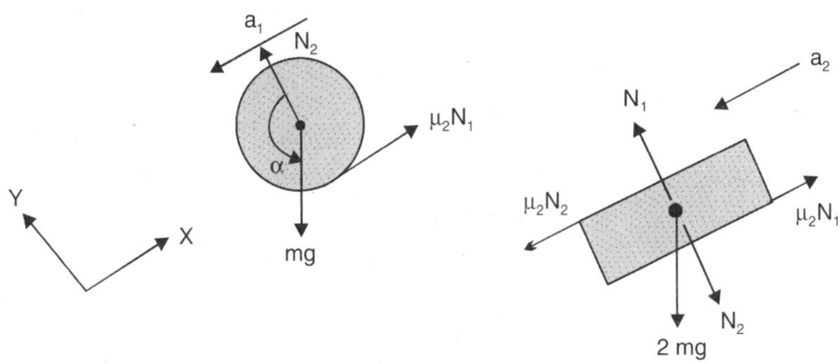

Fig. E8(a)

For sphere

Along y-axis

$N_2 = mg \cos\theta$

Along x-axis

$$ma_1 = mg \sin\theta - \mu_2 \, mg\cos\theta$$

$$a_1 = g\sin\theta - \mu_2 g\cos\theta$$

For plank

Along Y-axis

$$N_1 = N_2 + 2\,mg\cos\theta$$

Along X-axis

$$2ma_2 = 2mg\sin + \mu_2 N_2 - \mu_1 N_1$$

$$a_2 = \frac{2g\sin\theta + \mu_2 g\cos\theta 3 - \mu_1 g\cos\theta}{2}$$

For angular acceleration α of the sphere

$$\mu_2 N_2 \times R = \frac{2}{5} MR^2\alpha$$

$$\alpha = \frac{5\mu_2 g\cos\theta}{2R}$$

For no slipping condition

$$a_2 = a_1 - R\,\alpha$$

We get the desired result by using all above

$$\frac{\mu_1}{\mu_2} = \frac{\mathbf{8}}{\mathbf{3}}$$

Example 10

Solution

According to the figure as shown

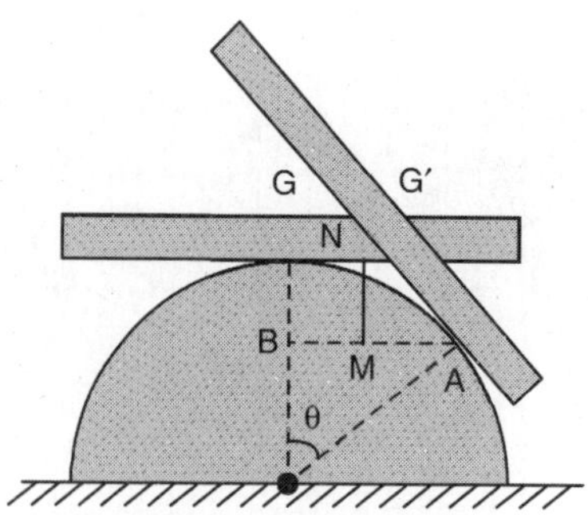

Fig. E10(a)

$$NG' = MG' - MN$$

$$= R(\theta\sin\theta + \cos\theta - 1)$$

$$W_{ex} = \Delta T$$

$$mgh = \frac{1}{2} Iw^2$$

$$mgR(\theta \sin\theta + \cos\theta - 1) = \frac{1}{2}\left(\frac{mL^2}{12}\right)\omega^2$$

$$\omega = \left(\frac{\mathbf{2}}{\mathbf{L}}\right)\sqrt{\mathbf{6gR(\theta sin\theta + cos\theta - 1)}} \quad \text{...Ans.}$$

Example 11

Solution

In the stretched position elongation in each spring is $x = R\theta$.
Spring in lower side is stretched by $2x$ and on upper side compressed by equal amount.
Restoring torque = τ
$\tau = I\alpha$

$$-(4kx)R = (mR^2)\alpha$$
$$-(4kR^2\theta) = (mR^2)\alpha$$
$$-(4k)\cdot\theta = m\alpha$$
$$\alpha = -\left(\frac{4k}{m}\right)\theta$$
$$\mathbf{f} = \frac{\mathbf{1}}{\mathbf{2\pi}}\sqrt{\frac{\mathbf{4k}}{\mathbf{m}}} \quad \text{...Ans.}$$

Example 13

Solution

This is a type of kater pendulum.

$$T = 2\pi\sqrt{\frac{I}{mgr}}$$

Here $r = L/2$
$I = mL^2/12 + m(L/2)^2 = mL^2/3$

$$T = 2\pi\sqrt{\frac{mL^2/3}{mg(L/2)}} = \mathbf{2\pi}\sqrt{\frac{\mathbf{2L}}{\mathbf{3g}}} \quad \text{...Ans.}$$

Level 2

Example 1

Solution

There are no external torques acting on the system as board and man. According to the angular momentum conservation

$$L_i = L_f$$

$$0 = mV_0 R - I\omega$$

$$mV_0 R = \frac{1}{2}MR^2\dot{\theta}\left(\dot{\theta} = \frac{\Delta\theta}{\Delta t}\right) s$$

$$\Delta\theta = 2\left(\frac{m}{M}\right)\left(\frac{V_0}{R}\right)\Delta t \qquad \ldots(i)$$

$$\Delta t = \frac{(2\pi RR\Delta\theta)}{v_0} \qquad \ldots(ii)$$

after putting the value of Δt in equation (i) we can get result as

$$\Delta\theta = \frac{4\pi m}{M + 2m} \qquad \textbf{...Ans.}$$

Example 4

Solution

Let the sphere is deflected by an angle θ_0 as shown in figure.

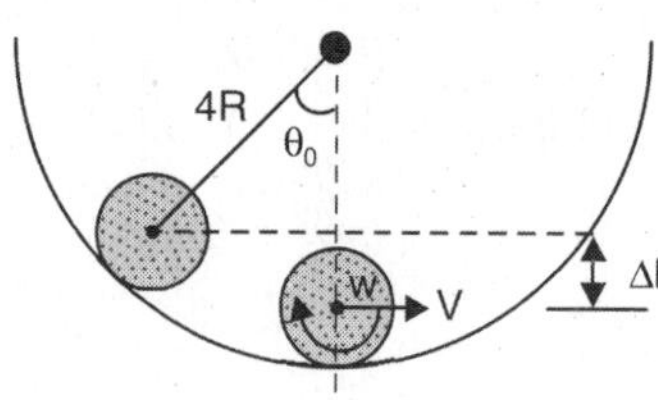

Fig. E3(a)

$$\Delta h = 4R\,(1 - \cos\theta_0) = 4R\left(2\sin^2\frac{\theta_0}{2}\right) \text{for small angle } \theta_0$$

$$\Delta h = 4R \times 2\frac{\theta_0^2}{4}$$

$$\Delta h = 2R\theta_0^2$$

let ω' is the angular velocity of center of mass of the sphere about O.

Then $$\omega' = \frac{V}{4R} = \frac{\omega}{4}$$

$$\left(\omega = \frac{V}{R}\right)$$

now, $$\omega_{ext} = \Delta T$$

$$mg\Delta h = \frac{1}{2}m(4R\omega')^2 + \frac{1}{2}\left(\frac{2}{5}mR^2\right)(4\omega')^2$$

we get by using above all

$$\omega' = \sqrt{\frac{5g}{28R}}\theta_0$$

$$\omega' = \left(\frac{2\pi}{T}\right)\theta_0$$

$$\mathbf{T} = 2\pi\sqrt{\frac{\mathbf{28R}}{\mathbf{5g}}} \qquad \text{...Ans.}$$

Example 5

Solution

Force method

Clearly $$\Delta l = \frac{mg}{k} \qquad \text{...}(i)$$

Now equilibrium is disturbed by small displacement y of the block of mass m

For block $$F_y = ma_y$$

$$Mg - T = m\ddot{y} \qquad \text{...}(ii)$$

For pulley $$\bar{z} = L\bar{\alpha}$$

$$TR - K(\Delta l + y)R = I\ddot{\theta} \qquad \text{...}(iii)$$

$$\ddot{y} = R\ddot{\theta} \qquad \text{...}(iv)$$

By using all above equations we get

$$TR - K(\Delta l + x)R = \frac{I}{R}\ddot{y} \quad \ldots(v)$$

Now use equations (*ii*) and (*v*)

We get
$$\ddot{y} = -\left(\frac{k}{m + \dfrac{I}{R^2}}\right)y$$

$$\omega^2 = \frac{k}{m + I/R^2}$$

$$\mathbf{T = 2\pi\sqrt{\frac{m + I/R^2}{K}}} \quad \textbf{...Ans.}$$

Example 6

Solution

Suppose mass *m* is displaced a small distance *y* along a line normal to plane as shown in the figure.

Since springs are identical, therefore increase in tension of each spring will be the same.

Restoring force produced by these two springs.

$$R.F. = (F + dF)\,2\sin\theta$$

$\because$ *y* is very small so,

$$\sin\theta = \frac{y}{l}$$

$$\text{Now } R.F. = 2F\left(\frac{y}{l}\right)$$

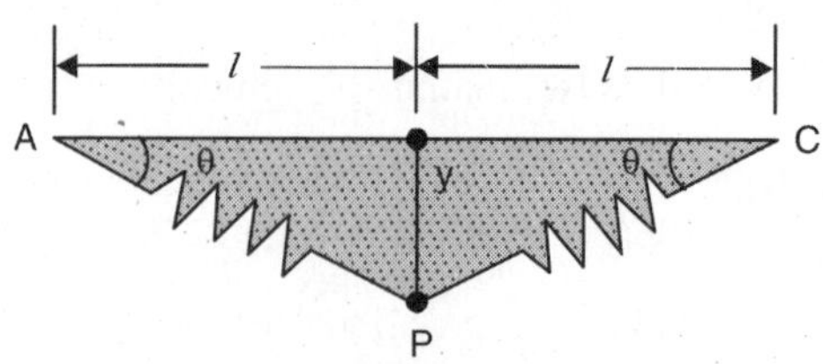

Fig. E5(a)

Similarly, restoring force produced by remaining springs will be same

$$\text{Resultant R.F.} = \frac{4F_0}{l} y$$

$$T = 2\pi\sqrt{\frac{ml}{4F}} = \pi\sqrt{\frac{\mathbf{ml}}{\mathbf{F}}} \quad \textbf{...Ans.}$$

Example 7

Solution

There is no horizontal force acting on the rod, so its center of gravity G does not acquire any horizontal velocity. Suppose after certain instant rod makes an angle θ with the vertical.

According to work energy theorem

$$W_{ext} = \Delta T$$

$$mga\,(\cos\alpha - \cos\theta) = \frac{1}{2}m\dot{y}^2 + \frac{1}{2}I\dot{\theta}^2$$

$$y = a\cos\theta \Rightarrow \dot{y}^2 = -(a\sin\theta)$$

$$I = \frac{ma^2}{3}$$

Sought results at $\theta = \dfrac{\pi}{2}$

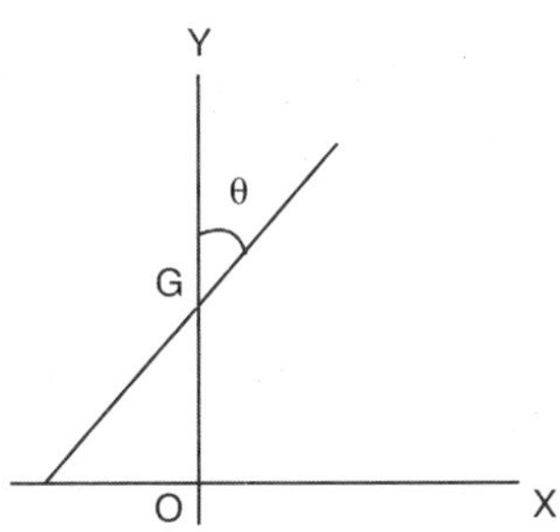

Fig. E7(a)

$$\dot{\theta} = \sqrt{(\mathbf{3g\cos\alpha})/\mathbf{2a}} \quad \textbf{...Ans.}$$

Example 8

Solution

Package suddenly feel the impact at point B and as the impact between the package and the support is perfectly plastic, it rotates about B.

Angular momentum conservation about point *B*

$$L_i = L_f$$

$$Mv_0 \frac{a}{2} = I_B \omega \qquad \ldots(i)$$

$$I_B = I_C + m\left(\frac{a}{\sqrt{2}}\right)^2 = \frac{ma^2}{6} + \frac{ma^2}{2}$$

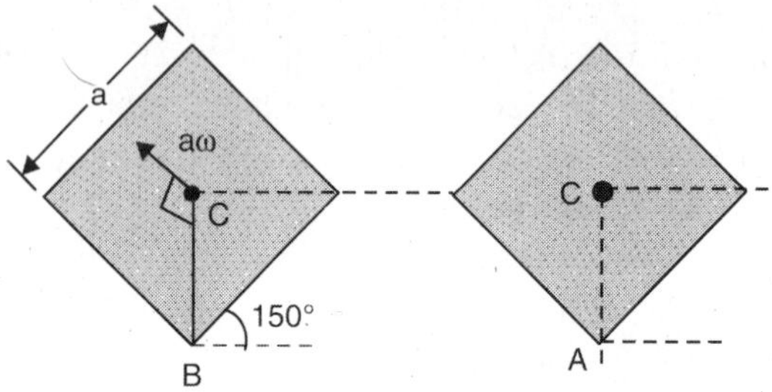

Fig. E8(a)

$$I_B = (2/3)\ ma^2$$

$$V_o = (4/3)\ a\omega \qquad \ldots(ii)$$

For the final destination at point *D*, package should attain zero linear and angular velocity.

$$\omega_{ext} = \Delta T$$

$$-mg\left[\frac{a}{\sqrt{2}} a \sin 60°\right] = 0 - \left[\frac{1}{2} m\left(\frac{a\omega}{\sqrt{2}}\right)^2 + \frac{1}{2}\left(\frac{ma^2}{6}\right)\omega^2\right]$$

$$\omega = \sqrt{0.2859/a}$$

Now, By using *(iii)*

$$\text{Sought velocity} = \mathbf{0.712\sqrt{ag}} \qquad \textbf{...Ans.}$$

Example 9

Solution

Let ω is the angular velocity after striking the block. According to conservation of angular momentum

$$L_i = L_f$$

$$mMv\ (a - h) + mk^2\ (v/a) = m\ (a^2 + k^2)\ \omega \qquad \ldots(i)$$

By this equation we get the value of ω with which the sphere begins to rise.

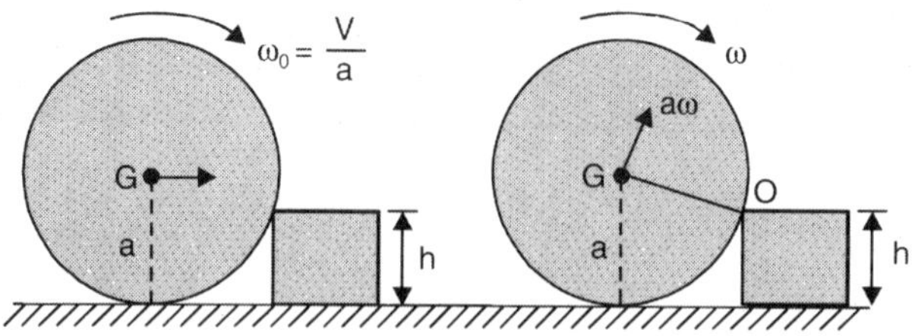

Fig. E9(a)

But to surmount the block, the kinetic energy of the sphere should be more than work that would have to be done for lifting the sphere through a height h.

So,

$$\frac{1}{2} m\,(a^2 + k^2)\,\omega^2 > mgh \qquad \ldots(ii)$$

By using (i) and (ii)

We can get the condition

Example 10

Solution

$$CG = \frac{3}{8}(R)$$

R denote radius of the hemisphere. Hemisphere begin to turn about 'O'. So using the equation

$$\Sigma\, M_0 = I_0\, \alpha$$

$$\alpha = \text{angular acceleration of hemisphere}$$

$$mg \times \frac{3}{8} R = m\,((OG)^2 + K^2)\,\alpha$$

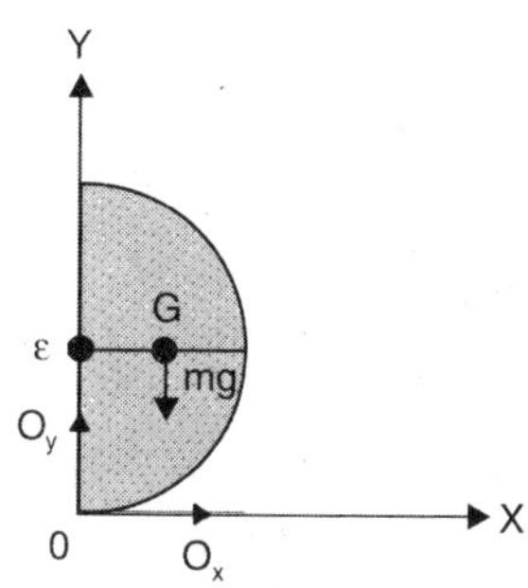

Fig. E10(a)

But
$$I_C = \frac{2}{5} mR^2$$

$$I_O = I_C + mR^2 = \frac{2}{5} mR^2 + mR^2 = \frac{7}{5} mR^2$$

$$\frac{3}{8} mgR = \frac{7}{5} mR^2\alpha$$

$$R\alpha = \frac{15}{16} g$$

Suppose O_x and O_y is the forces acting at point O in the horizontal and vertical directions respectively

$$\Sigma F_x = ma_x$$

$$O_x = m\,(a_x) \qquad \ldots(1)$$

$$\Sigma F_y = ma_y$$

$$mg - O_y = m\left(\frac{3}{8} l\right)\alpha \qquad \ldots(ii)$$

Using (*i*) and (*ii*)

We get
$$\mathbf{O_x = \frac{15}{56}\ mg}$$

$$\mathbf{O_y = \frac{403}{448}\ mg} \qquad \textbf{...Ans.}$$

Example12

Solution

Suppose α = angular acceleration of cylinder

a_p = acceleration of prism

a_{cp} = acceleration of cylinder axis with respect to prism $= R\,\alpha$

$$\vec{a}_{cp} = \vec{a}_c - \vec{a}_p$$

$$a_{cp}\lambda = R\,\alpha \cos 37°$$

$$a_{cpy} = R\,\alpha \sin 37°$$

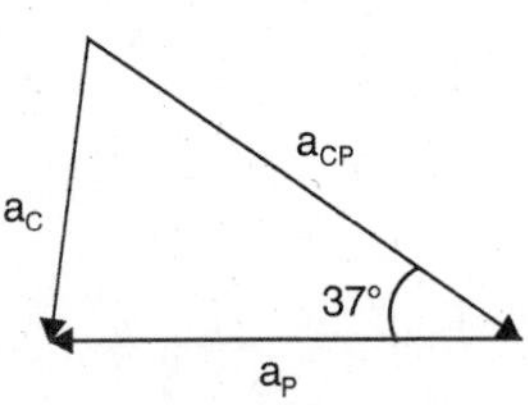

Fig. E12(a)

FBD of prism

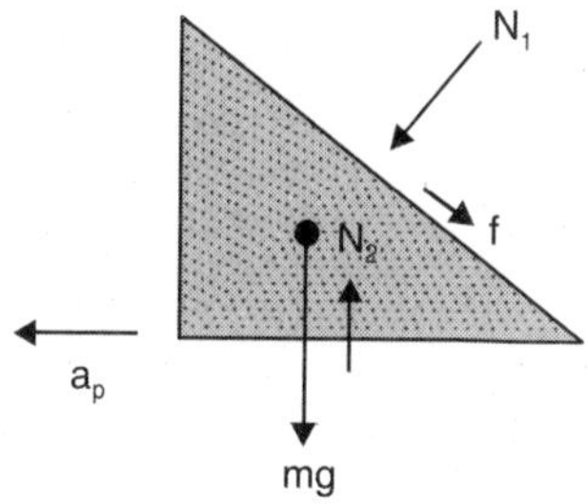

Fig. E12(b)

For horizontal forces

$$N_1 \sin 37° - f\cos 37° = Ma_p \quad \ldots(i)$$

FBD of cylinder

For horizontal forces

$$N_1 \sin 37° - f\phi \cos 37° = m\,(R\,\alpha \cos 37° - a_p) \quad \ldots(ii)$$

For vertical forces

$$mg - N_1 \cos 37° - f\sin 37° = m\,(R\,\alpha \sin 37°) \quad (iii)$$

Now, $\tau_C = I_C\alpha$

$$f_R = \frac{MR^2}{2}\alpha \quad \ldots(iv)$$

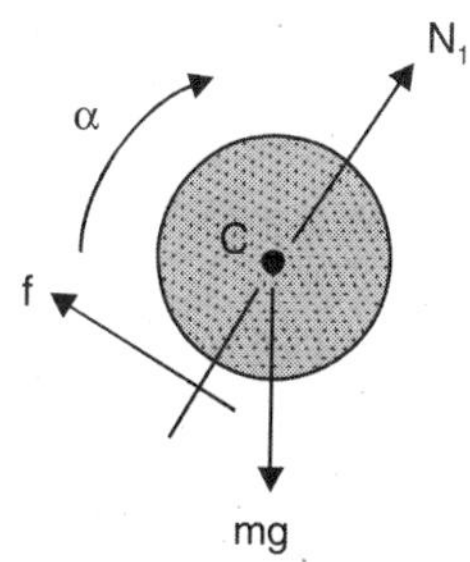

Fig. E12(c)

using all above equations we get

α = **30 rad/s²**

a = 3.5 m/s²

F = 12 N

...Ans.

Example 13

Solution

Clearly angular momentum of the system (disc + insect) is conserved at any instant.

ω = angular velocity of the disc.

θ = angle moved by insect relative to disc.

$$OB = r = 2\,R\cos\frac{\theta}{2} \quad \ldots(i)$$

Concentrating for angular momentum conservation about point O.

$$L_i = L_f$$

$$mv\sin(90° - \theta/2)r - mr^2\,\omega = I_o\,\omega$$

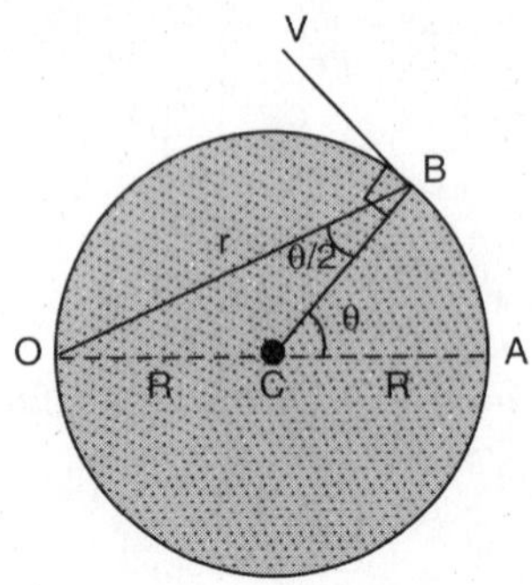

Fig. E13(a)

$$I_o = \frac{3}{2}MR^2$$

$$mv\,(\cos\theta/2)\cdot r - mr^2\,\omega = \frac{3}{2}MR^2\omega$$

ϕ = angle moved by disc when insect move by θ.

$$\left(\frac{3}{2}MR^2 + mr^2\right)\omega = mV_r\cos\theta/2 \qquad \ldots(ii)$$

$$\omega = \frac{d\phi}{dt} \qquad ..(iii)$$

$$V = R\frac{d\theta}{dt} \qquad \ldots(iv)$$

Putting the values of r, ω and v in (ii)
We get

$$d\phi = \frac{4m\cos^2\theta/2\cdot d\theta}{\left(3M + 8m\cos 2\frac{\theta}{2}\right)}$$

$$\int_0^\phi d\phi = \int_0^{2\pi}\frac{4m\cos^2\theta/2}{\left(3M + 8m\cos^2\frac{\theta}{2}\right)}d\theta,\ \phi = \pi\left[\mathbf{1} - \sqrt{\frac{\mathbf{3M}}{\mathbf{3M + 8m}}}\right] \qquad \textbf{...Ans.}$$

Example 14

Solution

There is no slipping at the bottom most point of the cylinder, so that is the instantaneous center of rotation

v^1 = velocity of rod

v = velocity of center of cylinder

According to work energy theorem

$$W_{ext} = \Delta T$$

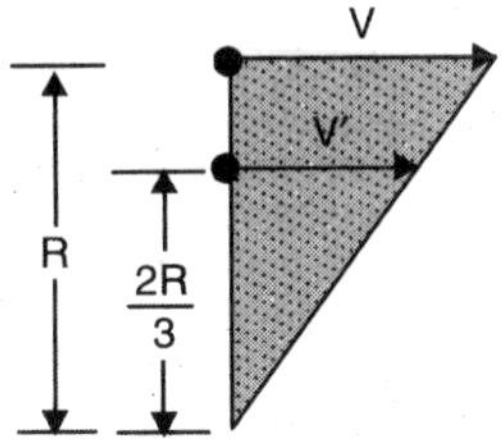

Fig. E14(b)

$$\frac{R}{3}(1-\cos\theta_0)\mathrm{Mg} = \left(\frac{1}{2}Mv^2 + \frac{1}{2}I_0\omega_0^2\right) + \frac{1}{2}Mv^2$$

$$I_0 = \frac{MR^2}{2},\ \omega_0 = \frac{v}{R}$$

We can get result as

$$\mathbf{V} = \sqrt{\frac{\mathbf{4R(1-\cos\theta_0)g}}{\mathbf{13}}} \quad \textbf{...Ans.}$$

Example 15

Solution

$\omega = 4\sqrt{\frac{g}{5a}}$ by applying conservation of angular momentum about point O.

$$L_i = L_f$$

$$Mua - \left(\frac{5}{4}Ma^2\right)\omega = \left(\frac{5}{4}Ma^2 + Ma^2\right)\omega^1$$

$$u - \frac{5}{4}\omega a = \frac{9}{4}\omega^1 a$$

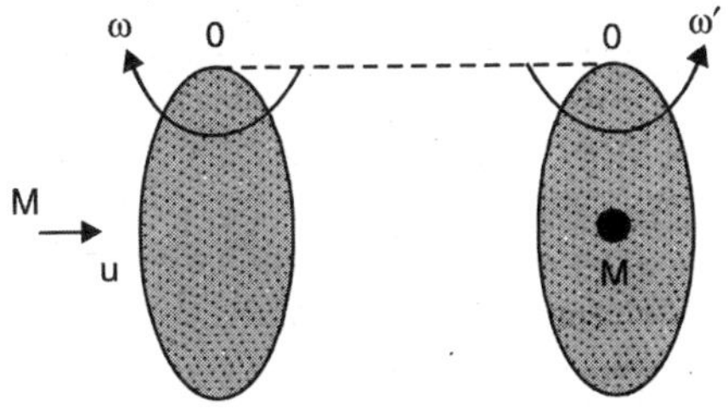

Fig. E15

$$\omega^1 = \frac{4u - 5\omega a}{9a}$$

Substitute the value of ω
we get

$$\omega^1 = \frac{4}{9a}\left(u - \sqrt{5ag}\right) \quad ...(i) \qquad \textbf{...Ans.}$$

Now

$$\omega_{ext} = \Delta T$$

$$2Mg\ (2a) = \frac{1}{2}\left(\frac{5}{4}Ma^2 + Ma^2\right)(\omega^1)^2 \quad ...\ (ii)$$

Now equate the value of ω^1 from both equations we can get

$$u = \left(3\sqrt{2} + \sqrt{5}\right)\sqrt{ag}$$

Example 17

Solution

For the small duration of impact, there is no force in the horizontal direction so, according to conservation of linear momentum

$$mV_1 = MV_2$$

Here

V_1 = velocity of ball in horizontal direction

V_2 = velocity of plank in horizontal direction opposite to V_1.

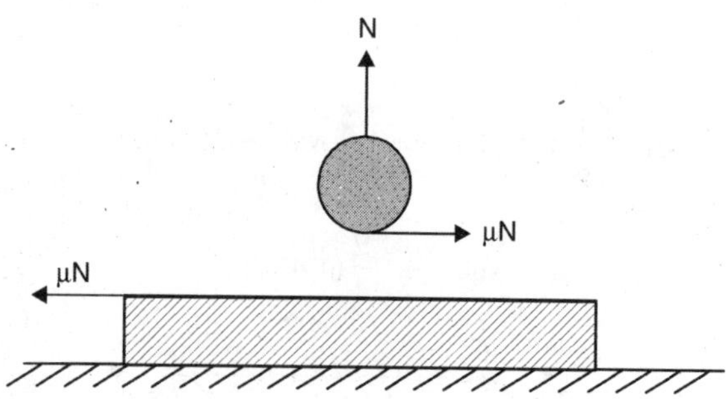

Fig. E17(a)

Linear impulse in vertical direction

$$Ndt = 2mV_0 \quad ...(ii)$$

Now, linear impulse in horizontal direction

$$\mu Ndt = mv_1 \quad ...(iii)$$

For angular impulse

$$(\mu N dt)(r) = \frac{2}{5} m r^2 \omega_0 \qquad \ldots(iv)$$

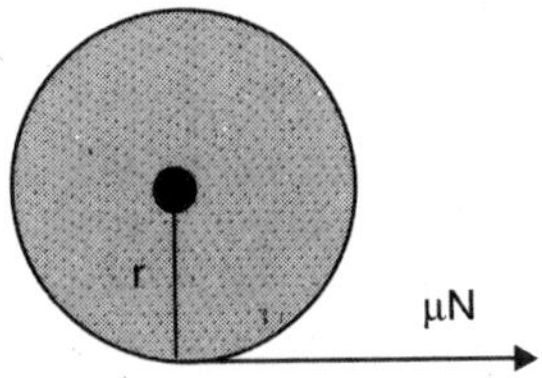

Fig. E17(b)

using all above equations we get

$$y = \frac{2}{5} r\omega_0 \qquad v_2 = \frac{m}{M} v_1 \qquad v_2 = \frac{m}{M}\left(\frac{2}{5} r\omega_0\right)$$

now the path of the ball is a projectile having time of flight

$$= T = \frac{2V_0}{g} \qquad \ldots(v)$$

Distance covered on the plank $= S$

$$S = (V_{b,P})\, T \qquad \ldots(vi)$$

$$\bar{V}_{bP} = \bar{V}_b - \bar{V}_p$$

$$= \left(\frac{2}{5} r w_0\right)\hat{i} - \frac{m}{M}\left(\frac{2}{5} r w_0\right)\cdot(\hat{i})$$

$$\left|\bar{V}_{b,P}\right| = \frac{2}{5}\left(\frac{M+m}{M}\right) r w_0 \qquad \ldots(vii)$$

putting the value of $V_{b,p}$ and T from equations(v) and (vii) in (vi)

$$\text{Sought distance} = \frac{4}{5}\left(\frac{\mathbf{m}+\mathbf{M}}{\mathbf{M}}\right)\frac{\mathbf{v}_0 \mathbf{r} \omega_0}{\mathbf{g}} \qquad \textbf{...Ans.}$$

Example 18

Solution

t_1 = time of ascent

t_2 = time of descent

Δt_1 = Time gained in time t_1

Δt_2 = Time lost in time t_2

The clock will show the right time when

$$\Delta t_1 = \Delta t_2 \quad \ldots(i)$$

Now,

$$\Delta t_1 = \left(\frac{\Delta T}{T}\right)t_1 = \left(\frac{TT'}{T}\right)t_1$$

$$= \left(1\frac{T'}{T}\right)t_1$$

$$\frac{T'}{T} = \frac{g}{g+a} \quad \because \left(T\alpha\frac{1}{\sqrt{g}}\right)$$

$$\text{So, } \Delta t_1 = \frac{2h}{a}\left(\sqrt{\frac{g+a}{g}}-1\right)$$

Similarly

$$\Delta t_2 = t_2\left(1-\sqrt{\frac{g-a}{g}}\right)$$

now use the equation (*ii*) and (*iii*) in (*i*)
we get,

$$t_2 = \frac{2h}{a}\left(\frac{\sqrt{g+a}-\sqrt{g}}{\sqrt{g}-\sqrt{g-a}}\right)$$

sought time $= t = t_1 + t_2$

$$\mathbf{t = \sqrt{\frac{2h}{a}}\left(\frac{\sqrt{g+a}-\sqrt{ga}}{\sqrt{g}-\sqrt{g-a}}\right)} \quad \textbf{...Ans.}$$

Example 20

Solution

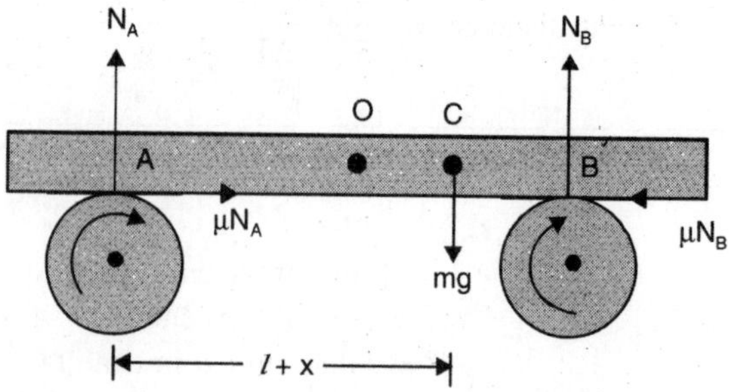

Fig. E20(a)

For equilibrium conditions

$$\Sigma F_y = 0$$

$$N_A + N_B - mg = 0 \quad ...(i)$$

$$\Sigma F_x = ma$$

$$\mu N_A - \mu N_B = ma \quad ...(ii)$$

$$\Sigma M_A = 0$$

$$2l\,(N_B) - (l + x)\,mg = 0 \quad ...(iii)$$

using (*i*), (*ii*) and (*iii*) we can get

$$a = -\frac{\mu g}{l}x$$

$$w = \sqrt{\frac{\mu g}{l}}$$

$$\mathbf{f} = \frac{1}{2\pi}\sqrt{\frac{\mu g}{l}} \quad \textbf{...Ans.}$$

Example 21

Solution

x = displacement of block from equilibrium position

x_1 = deformation of spring having constant k_1

x_2 = deformation of spring having constant k_2

Constraint relation

$$2x - x_1 - x_2 = 0 \quad ...(i)$$

$$k_1 x_1 = k_2 x_2 \quad ...(ii)$$

$$x_1 = \left(\frac{2k_2}{k_1 + k_2}\right)x \quad ...(iii)$$

$$x_2 = \left(\frac{2k_1}{k_1 + k_2}\right)x \quad ...(iv)$$

$$E = \frac{1}{2}k_1x_1^2 + \frac{1}{2}k_1x_2^2 + \frac{1}{2}mv^2$$

Putting the value of x_1 and x_2 from (*iii*) and (*iv*) in equation (*v*) we get

$$E = \left(\frac{2k_1k_2}{k_1 + k_2}\right)x^2 + \frac{1}{2}mv^2$$

$$\frac{dE}{dt} = 0$$

$$a = -\left(\frac{4k_1k_2}{m(k_1+k_2)}\right)x$$

$$\mathbf{T} = 2\pi\sqrt{\frac{\mathbf{m(k_1+k_2)}}{\mathbf{4k_1k_2}}}$$ **...Ans.**

Example 25

Solution

Torque produced by block B

$$\tau_B = m_2g \times R_2$$

Torque produced by block A

$$\tau_A = m_1g\ \sin 30° \times R_1$$

After calculation we get

$$\tau_B > \tau_A$$

So, pulley will rotate in clockwise direction. Now, notations used in the figure has its usual meanings FBD for m_1.

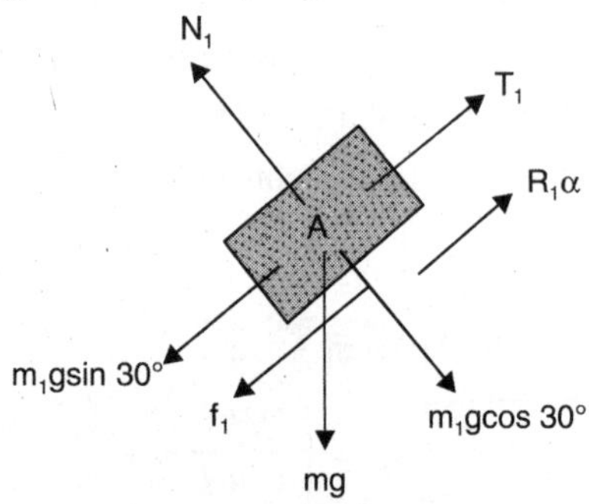

Fig. E25(a)

FBD for m_2

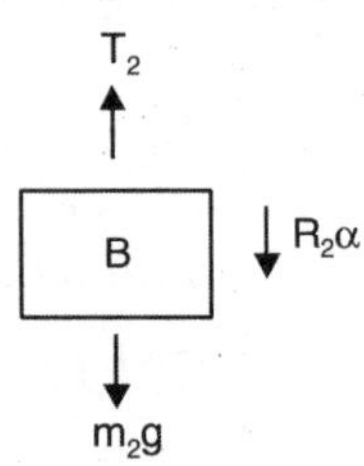

FBD for pulley

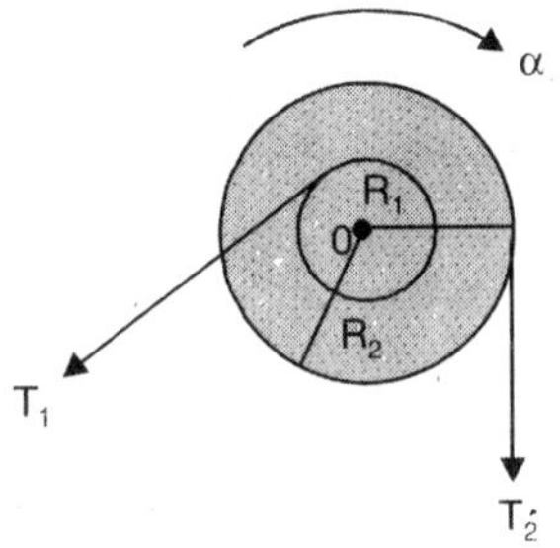

Fig. E25(b)

For block *A*

$$N_1 = m_1 g \cos 30° \qquad ...(i)$$

$$T_1 - m_1 g \sin 30° - f_1 = m_1 R_1 \alpha \qquad ...(ii)$$

For block *B*

$$m_2 g - T_2 = m_2 (R_2 \alpha) \qquad ...(iii)$$

For pulley

$$(T_2 R_2 - T_1 R_1) = I_0 \alpha \qquad ...(iv)$$

Use above all equations we get

$$\mathbf{T_1 = 17\ N}$$

$$\mathbf{T_2 = 26\ N}$$

$$\mathbf{a_A = R_1\ \alpha = 2\ m/s^2}$$

$$\mathbf{a_B = R_2\ \alpha = 3\ m/s^2} \qquad \textbf{...Ans.}$$

Example 27

Solution

Let $2l$ is the length of rod and α, β denote the initial accelerations of the string and rod respectively.

Acceleration of point *G* is the vector sum of acceleration of point *G* with respect to point *A* and acceleration of point *A*.

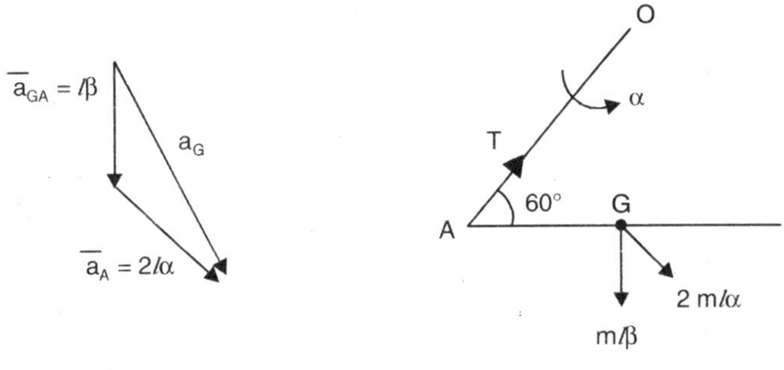

Fig. E27(a)

According to the figure

$$\Sigma F_x = ma_x$$

$$T \cos 60° = 2ml\alpha \cos 30° \qquad ...(i)$$

Moment about G

$$\Sigma M_G = I_G\alpha$$

$$Tl \sin 60° = \frac{ml^2}{3}\beta \qquad ...(ii)$$

$$\beta = 9\alpha \text{ (proved)}$$

Now for T.

$$\Sigma F_y = ma_y$$

$$mg - T \sin 60° = ml\beta + 2\,ml\alpha \cos 60°$$

we get

$$\mathbf{T = \frac{2\sqrt{3}}{13} mg} \qquad \textbf{...Ans.}$$

Level 3

Example 1

Solution

Suppose

$$\lambda = \text{modulus of elasticity}$$
$$m = \text{mass of the particle}$$
$$l_0 = \text{natural length of either string}$$
$$AP = BP = y$$
$$PC = x$$

For the equilibrium position

$$mg = 2\lambda\left(\frac{\Delta l}{\lambda_0}\right)\sin 30°$$

Here; $(\Delta l = l - l_0)$

$$mg = \lambda\frac{l - l_0}{l_0}$$

$$\because \lambda = mg$$
$$l_0 = l/2$$

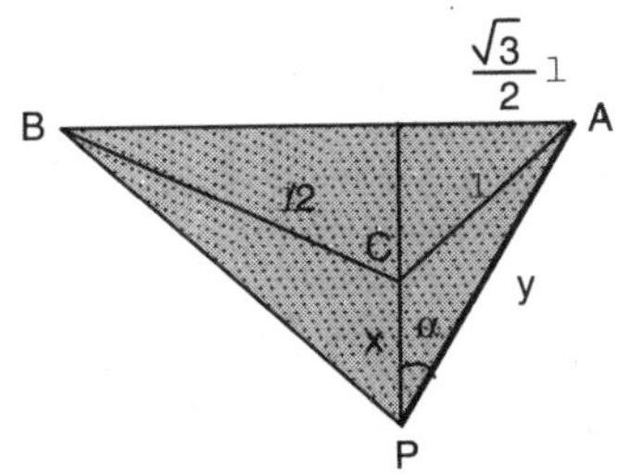

Fig. E1

now, we should write the equation of motion for particle of mass m when it is displaced by x at any time t, y = length of either string at that instant

$$ma = mg - 2\lambda\left(\frac{y - l_0}{l_0}\right)\cos\alpha \qquad \ldots(i)$$

$$y^2 = \left(x + \frac{1}{2}l\right)^2 + \frac{3}{4}l^2$$

$$y^2 = (l^2 + lx + x^2)$$

$$y = 1\left(1 + \frac{x}{l}\right)^{\frac{1}{2}}$$

$$= l + \frac{1}{2}x$$

$$\cos\alpha = \frac{l/2+x}{y} = \frac{l/2+x}{l+x/2}$$

$$= \frac{1}{2}\left(1+\frac{2x}{l}\right)\left(1-\frac{x}{2l}\right)$$

$$= \frac{1}{2}\left(1+\frac{3x}{l}\right) \text{(for first power of } x\text{)}$$

putting the y and $\cos\alpha$ values in the equation (i)

$$ma = mg - \frac{2\lambda + (x/2 + l/2)}{l/2}\frac{1}{2}\left(1+\frac{3x}{l}\right)$$

$$a = -\frac{5}{2}\left(\frac{g}{l}\right)x$$

$$\mathbf{T} = 2\pi\sqrt{2l/5g}$$ **...Ans.**

Example 4

Solution

(a) Angle moved by the ring for the distance s_0.

$$\theta = \frac{S_0}{R} \quad \text{...}(i)$$

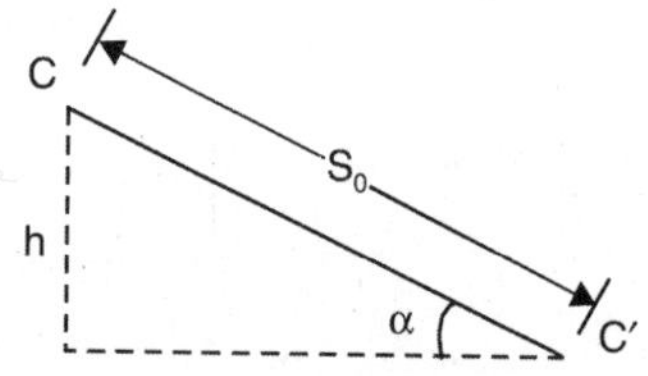

Fig. E4(a)

$$h_{ring} = S_0 \sin\alpha$$

$$h_{body} = S_0 \sin\alpha - R\sin\theta\sin\alpha$$

$$= (S_0 - R\sin\theta)\sin\alpha$$

Applying conservation of energy

$$w_{ext} = \Delta T$$

$$mg\,(h_{ring} + h_{body}) = \frac{mv^2}{2} + \frac{Iw^2}{2} + \frac{m}{2}V_A^2 \quad \text{...}(ii)$$

$$V_A = 2v \sin\frac{\theta}{2} \qquad \text{...(iii)}$$

By using above equations we can get

$$\mathbf{v} = \sqrt{\frac{2gs_0\sin\alpha - gR\sin\alpha\cdot\sin\left(\frac{S_0}{R}\right)}{1+2\sin^2\left(\frac{S_0}{2R}\right)}} \qquad \text{...Ans.}$$

(*b*) At $\theta = \pi$ ring will loose contact

$$S_0 = \pi R$$

$$V = \sqrt{\frac{2}{3}\pi g R\sin\alpha}$$

$$V_{cm,ring} = \frac{3v}{2} - v = \frac{v}{2}$$

$$\Sigma F_n = ma_n$$

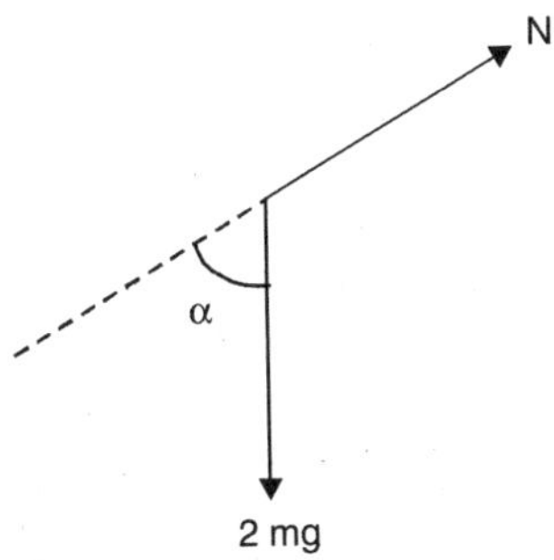

Fig. E4(b)

$$2mg\cos\alpha - N = \frac{2m}{R/2}\left(V_{cm,\ ring}\right)^2 mg$$

putting $N = 0$ and value of $V_{cm,ring}$
we get

$$\alpha = \tan^{-1}\left(\frac{6}{\pi g}\right) \qquad \text{...Ans.}$$

Example 5

Solution

Concept behind the maximal velocity of pole's end denoted by A is that the velocity of A in the center of mass frame will be in the direction of $\vec{v}_{cm}$.

$$\vec{v}_{cm} = \frac{\Sigma m_i v_i}{\Sigma m_i} = \frac{v_0}{2}\hat{j}$$

$$\vec{R}_{cm} = \frac{\Sigma m_i v_i}{\Sigma m_i} = \frac{1}{4}(\hat{i} - \hat{j})$$

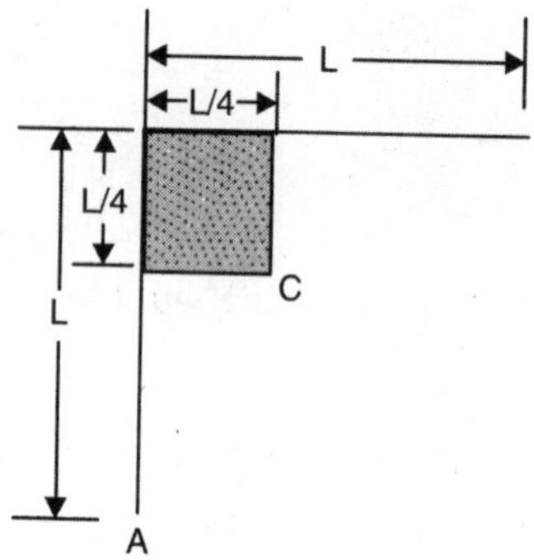

Fig. E5(a)

I_c = moment of inertia about cm

$$I_c = 2\left(\frac{1}{3}mL^2\right) - 2m\left[\left(\frac{L}{4}\right)^2 + \left(\frac{L}{4}\right)^2\right] = \frac{5}{12}mL^2$$

$$\vec{L}_{cm} = \vec{L}_0 + \vec{r} + \vec{v}$$

$$\bar{L}_{cm} = \left(\frac{L}{2}(-\hat{j}) - \vec{R}_{cm}\right) \times m_1(\vec{v}_1 - \vec{v}_{cm}) + \left(\frac{L}{2}(-\hat{i}) - \vec{R}_{cm}\right) \times m_2(\vec{v}_2 - \vec{v}_{cm})$$

After putting the values of $\vec{R}_{cm}, \vec{v}_1, \vec{v}_2, \vec{v}_{cm}$

We get $$\vec{L}_{cm} = -\frac{L}{4}mV_0\hat{k}$$

$$\vec{\omega}_{cm} = \frac{\bar{L}_{cm}}{I_c} = \frac{-\frac{1}{4}(mV_0)L}{\frac{5}{12}mL^2}\hat{k}$$

$$= -\frac{3}{5}\left(\frac{\mathbf{v_0}}{\mathbf{L}}\right)\hat{\mathbf{k}}$$ ...Ans.

Now we are on the way for finding maximal velocity of point A.

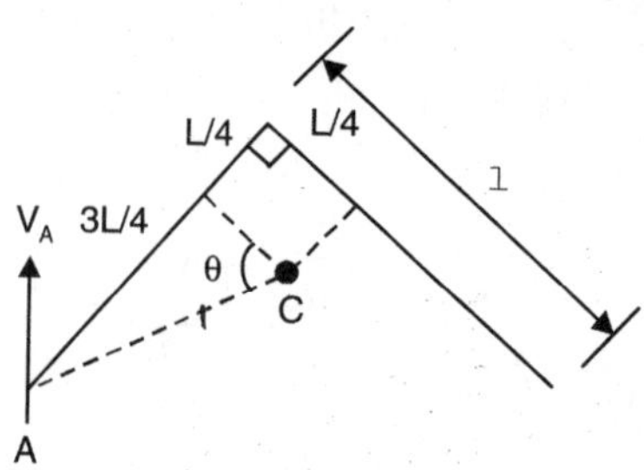

Fig. E5(b)

$$AC = \sqrt{\left(\frac{3L}{4}\right)^2 + \left(\frac{L}{4}\right)^2} = \frac{\sqrt{10}}{4}L$$

$$V_A\Big|_{max} = V_{cm} + \omega R_{AC}$$

$$V_A\Big|_{max} = \frac{v_0}{2} + \frac{3}{5}\frac{v_0}{L}\frac{\sqrt{10}}{4}L$$

$$= \frac{\mathbf{v_0}}{\mathbf{2}}\left(\mathbf{1} + \frac{\mathbf{3}}{\sqrt{\mathbf{10}}}\right) \qquad \textbf{...Ans.}$$

By figure

$$\tan\theta = \frac{3L/4}{L/4} = 3$$

$$\theta = \tan^{-1}(3)$$

$$\text{sought time} = \frac{\theta}{\omega} = \frac{\tan^1 3}{\frac{3}{5}\frac{v_0}{L}}$$

$$= \left(\frac{\mathbf{5}}{\mathbf{3}}\right)\left(\frac{\mathbf{L}}{\mathbf{V_0}}\right)\left(\mathbf{tan^{-1} 3}\right) \qquad \textbf{...Ans.}$$

Example 7

Solution

There are no torques acting at any instant so angular momentum is conserved. Suppose after some instant body of mass m reaches a new radius r.

$$L_i = L_f$$

$$mV_0 R = m\omega r^2$$

$$\omega = \frac{V_0 R}{r^2}$$

Table is smooth, there is no friction at any instant during motion and all forces are conservative. So we can apply conservation of mechanical energy

$$\frac{1}{2}mv_0^2 = \frac{1}{2}(M+m)v_r^2 + \frac{1}{2}m\omega^2 r^2 - Mg(R-r)$$

Here V_r denotes the velocity of M relative to m. Substituting the value of ω from equation (i) and put it into (ii)

We get

$$V_r = \frac{dr}{dt} = \pm\sqrt{\frac{mv_0^2}{(m+M)}\left(1-\frac{R^2}{r^2}\right)+\frac{2Mg(R-r)}{M+m}} \qquad \ldots(iii)$$

According to the question as M stops, we can relate $V_r = 0$
Now b put the $V_r = 0$ in equation (*iii*) we get

$$Mv_0^2\left(1-\frac{R^2}{r_1^2}\right)+2Mg(R-r_1)=0 \qquad \ldots(iv)$$

Here $r = r_1$ at the time when $V_r = 0$, i.e. Mstops. Using some algebraic treatment in equation (*iv*) we get a quadratic equation as

$$r_1^2 - \frac{V_0^2 - mr_1}{2Mg} - \frac{V_0^2 mR}{2Mg} = 0$$

sought distance $\mathbf{r_1} = \dfrac{\mathbf{mv_0^2}}{\mathbf{4Mg}} + \sqrt{\left(\dfrac{\mathbf{mV_0}}{\mathbf{4Mg}}\right)^2 + \dfrac{\mathbf{mv_0^2R}}{\mathbf{4Mg}}}$

(negative value of r_1 can be neglected)

Example 9

Solution

$$x_0 = \text{string extension}$$

$$mg \sin \alpha = kx_0 \qquad \ldots(i)$$

linear momentum conservation in the horizontal direction

$$MV_M = m\,(V_{mM} \cos \alpha - V_M)$$

$$V_M = \frac{mV_{mM} \cos\alpha}{M+m} \qquad \ldots(ii)$$

Now spring is extended by small amount x.

$$E = \frac{1}{2}\left(Mv_M^2\right)+\frac{1}{2}m\left[\left(V_{mM}\cos\alpha - V_M\right)^2 + (VmM\sin\alpha)^2\right]$$

$$-mgx\sin\alpha + \frac{1}{2}k\left(x+x_0\right)^2 \qquad \ldots(iii)$$

Use equation (*i*) and (*ii*) in (*iii*)
We get

$$E = \frac{1}{2}k\left(x+x_0\right)^2 - mg\times\sin\alpha + \frac{1}{2}m\left[\frac{M+m\sin^2\alpha}{M+m}\right]v_{mM}^2$$

Now,

$$\frac{dE}{dt} = 0$$

$$0 = K\left(x + x_0\right)\frac{dx}{dt} + mV_{mM}\left(\frac{dV_{mM}}{dt}\right)\left(\frac{M + m\sin^2\alpha}{M + m}\right) - mg\sin\alpha\left(\frac{dx}{dt}\right)$$

$$mg\sin\alpha = kx_0$$

$$\frac{dx}{dt} = V_{mM} \text{ and } \frac{dV_{mM}}{dt} = a_{mM}$$

we get

$$a_{mM} = -\frac{K(M+m)}{m(M + m\sin^2\alpha)}x$$

$$a_{mM} = -\omega^2 x$$

$$\omega = \sqrt{\frac{\mathbf{K(M+m)}}{\mathbf{m(M + m\sin^2\alpha)}}} \qquad \textbf{...Ans.}$$